Nutraceutics from Agri-Food By-Products

Scrivener Publishing
100 Cummings Center, Suite 541J
Beverly, MA 01915-6106

Publishers at Scrivener
Martin Scrivener (martin@scrivenerpublishing.com)
Phillip Carmical (pcarmical@scrivenerpublishing.com)

Nutraceutics from Agri-Food By-Products

Edited by

Umile Gianfranco Spizzirri

Department of Pharmacy, Health and Nutritional Sciences, University of Calabria, Rende, Italy

Scrivener Publishing

WILEY

This edition first published 2023 by John Wiley & Sons, Inc., 111 River Street, Hoboken, NJ 07030, USA and Scrivener Publishing LLC, 100 Cummings Center, Suite 541J, Beverly, MA 01915, USA

For more information about Scrivener publications please visit www.scrivenerpublishing.com.

Wiley Global Headquarters
111 River Street, Hoboken, NJ 07030, USA

For details of our global editorial offices, customer services, and more information about Wiley products visit us at www.wiley.com.

Library of Congress Cataloging-in-Publication Data

ISBN 978-1-394-17444-7

Cover image: Pixabay.Com
Cover design by Russell Richardson

Set in size of 11pt and Minion Pro by Manila Typesetting Company, Makati, Philippines

Printed in the USA

10 9 8 7 6 5 4 3 2 1

Contents

Preface

Global warming, lack of natural resources, and increasing world population over the next decades should motivate the implementation of innovative and sustainable food system approaches that involve the entire agri-food chain. Scientific communities and governments need to address these challenges with solutions that are sustainable, resilient, competitive, diverse, responsible, and inclusive. To accomplish this, different strategies should be engaged. These include improving access to food, increasing total food production, and reducing food waste by facilitating the circular economy through the valorisation of agri-food waste and by-products.

The Food and Agriculture Organization of the United Nations (FAO) estimates that approximately 1.3 billion tons of food produced globally is wasted across the food supply chain on an annual basis. In general, food waste includes materials lost during primary production and secondary processing, packaging, storage, and distribution, as well as those discarded by consumers. Food waste and by-products are mostly derived from animal and plant sources. Specifically, plant wastes, such as vegetable and fruit residues, remain after starch, oil, sugar, and juice extraction. Animal wastes are mainly the bones, tissues, and blood of meats and fish, as well as liquid and solid wastes usually related to dairy manufacturing.

In this regard, waste and by-products of the food industry represent a rich source of bioactive molecules, largely employed in the preparation of functional foods and manufacturing for the pharmaceutical and cosmetic industries. This book represents a complete overview covering different aspects (raw materials, technological innovations, and potential applications) concerning these topics.

Each chapter deeply analyses a specific agri-food chain, highlighting the main components recovered in the processing of wastes and by-products and their reuse in the production of high-value and innovative products. Specifically, a green approach to the extraction of active molecules is described, as well as the industrial application of agri-food wastes and by-products, and their chemical, physical, and biological properties. Such compounds are suitable for use in the food and pharmaceutical fields for the preparation of nutraceutical products with improved health properties. Indeed, this circular approach could be usefully employed in the industry to develop and commercialize new nutraceuticals and/or functional food that guarantee a considerable increase in the economic worth of the wastes, also producing beneficial effects on human health.

I would like to thank Wiley and Scrivener Publishing for their continuous support and guidance in the production of this volume.

Umile Gianfranco Spizzirri
University of Calabria, Rende, Italy
July 2023

1

Valorization of Industrial Coproducts From Tropical Fruit

Neith Pacheco[1], Daniel Castañeda-Valbuena[1], Iván Emanuel Herrera-Pool[1], Sara Herrera-Rodríguez[1], Tania González-Flores[1], Soledad Pech-Cohuo[1], Juan C. Cuevas-Bernardino[2] and Teresa Ayora-Talavera[1]*

[1]*Centro de Investigación y Asistencia en Tecnología y Diseño del Estado de Jalisco A.C. Unidad Sureste, Mérida, Yucatán, México*
[2]*CONACYT, Centro de Investigación y Asistencia en Tecnología y Diseño del Estado de Jalisco, A.C., Unidad Sureste, Mérida, Yucatán, México*

Abstract
This chapter provides and overview of the state-of-the-art valorization of several common and rare tropical fruits harvested in Mexico as a source of bioactive compounds with nutraceutical properties. The text includes an analysis and discussion focused on the content of the nutraceutical molecules present mainly in the coproducts (peels, seeds, pulp remains, internal tissues, and other unused parts), as well as the advantages and disadvantages of the methods used for their extraction. Identification techniques, *in vitro* and *in vivo* methods to evaluate the different biological properties with specific detail on its effect as a nutraceutical on the human microbiome is also discussed. Finally, a prospective vision about the production, new applications in medicine of these compounds is also presented.

Keywords: Nutraceuticals, tropical fruit, coproduct, valorization

1.1 Production of Tropical Fruit and Coproducts: Common and Uncommon Raw Materials

In recent years, the production of tropical fruits has increased considerably becoming the third most valuable group of fruits worldwide [1]. These

**Corresponding author*: tayora@ciatej.mx

Umile Gianfranco Spizzirri (ed.) Nutraceutics from Agri-Food By-Products, (1–44) © 2023 Scrivener Publishing LLC

fruits are potential sources of nutraceutical components that could be used to combat and reduce several health problems including metabolic syndrome. The health benefits and molecular mechanisms of nutraceuticals could provide commercial opportunities to the growers to allow them to expand their business opportunities by promoting fresh healthy fruits in the market and increase sustainability of food production [2].

Tropical fruits constitute a comparatively new group in global commodity trade that emerged in the international marketplace. Approximately, 99% of the tropical fruits production comes not only from low-income developing countries belonging mainly to Latin America and Caribbean regions but also from Asiatic and African areas [3]. Commercially available tropical fruits, such as banana, papaya, and mango, are important and widely cultivated fruits in tropical regions, however, there also are some others that have been little known and could also be considered potential sources of nutraceutical compounds, which can be capitalized and valorized in the international market of nutraceutical products (Table 1.1).

Every year, the processing of the most common tropical fruits generates large amounts of coproducts, which are composed of peels, endocarps, seeds, internal parts of the fruits and even pulp [22]. Most of them can have great potential to be used for extraction and isolation of value-added ingredients such as phenolic compounds, polyphenols, natural pigments, and pectins, that can be used as nutraceuticals for application in food fortification or in chemical, pharmaceutical, cosmetic and biotechnological industries [23]. For example, banana (*Musa paradisiaca*) peel represents approximately one third of the fruit (35%), and some studies have found higher amounts of nutraceutical compounds in the banana peel than in the edible part of the fruit [24]. Some less industrialized tropical fruits such as *Cordia dodecandra* A. DC. known as ciricote, are underutilized native tropical fruits that could be considered a source of different bioactive compounds with biological activities little studied [25]. Even though, some of these underutilized fruits have excellent sensory attributes, their use is still limited and artisanal, with little production, and marketing tends to be restricted to the local areas, in this sense *Byrsonima crassifolia, Annona squamosa, Bromelia pingui,* and *Cordia dodecandra* A. DC. are just commercialized in tropical areas as fresh or little processed products [14].

The high amount of coproducts generated during processing of common tropical fruits, as well as the importance of increasing the production of native species, and the increment on health problems related to a deficiencies in the quality of highly processed foods, have led to the development of new food ingredients or food products for human consumption, considering the tropical fruits coproducts and the underutilized products

Table 1.1 Common and uncommon tropical fruits produced in México.

Tropical fruit	Fruit production	By-products	References
Mangifera indica	54.83 million tons	35-60% (peel, seed)	[4]
Musa spp	116-117 million tons	18-33% peel	[5, 6]
Carica papaya	13.6 million tons	Peel and seeds	[1]
Psidium guajava L.	6.5 million tons	20% peel and 30% seed.	
Annona cherimola Mill	NR	Seed and peel.	[7]
Annona muricata	NR	8% seeds and 20% peel.	[8]
Hylocereus spp	NR	33-37.60% peels and 2.70-14-67% seeds.	[9, 10]
Pouteria sapota	14 024 tons*.	Seed (15-25%) and peel.	[11–13]
Byrsonima crassifolia	8470 tons *	Seed and husks (3-14%).	[14–16]
Annona squamosa	NR	Seed kernel (67.7%) and seed coat (32.4%).	[17, 18]
Bromelia pinguin	NR	20% peel and seeds.	[19, 20]
Muntingia calabura	NR	Peels.	[21]

NR = Data not reported. * In México.

raw materials for nutraceutical compounds that promotes their integral valorization and reincorporation to the food supply chain within the circular bioeconomy concept, creating revenue streams, business and job opportunities [26]. The high volume of discarded tropical fruit coproducts after the processing of edible fractions represents an opportunity for the development of biorefinery-based platforms that allow the use of the different high-value biomolecules present in this biomass [27]. Among the biomolecules that can be recovered from the coproducts of the fruits, those molecules that possess biological properties and whose use is of interest for the pharmaceutical, food and cosmetic industries stand out [28]. The biological properties reported for these molecules are varied and range from the ability to capture free radicals, to anti-inflammatory capacity, through the reduction of neuronal pain, as well as antimicrobial activity [3, 29, 30]. Nutraceutical compounds possess bioactivities that provides additional health benefits in addition to the basic nutritional value found in the food that contain them [31]. Within the group of molecules with nutraceutical properties, different compounds have been described, such as phenolic compounds, carotenoids, proteins (whole, peptides and amino acids), carbohydrates and fatty acids, among others [32], which are usually found distributed in the different tissues (peels, seeds and pulp) of the fruits [33], varying their concentrations according to the variety, tissue and fruit studied [34] (Figure 1.1).

1.2 Nutraceutical Compounds Recovery From Tropical Fruits Coproducts

1.2.1 Phenolic Compounds

Phenolic compounds are a group of secondary plant metabolites, synthesized by the metabolic pathways of phosphoenolpyruvate from glycolysis and erythrose-4-phosphate from the pentose phosphate cycle and shikimic acid [35]. These compounds are found in plants in the form of aglycones or linked by glycosidic bonds or ester bonds to macromolecules of the cell wall [36]. Its structure comprises an aromatic ring, which contains one or several hydroxyl groups and depending on its structure these compounds can be classified as phenolic acids, phenolic alcohols, flavonoids, lignans, stilbenes and tannins [37]. The biological properties and bioavailability of phenolic compounds largely depend on the food matrix from which they are derived, in this sense these phenolic compounds have been recovered and reported from different tropical fruits such as soursop (*Annona muricata*) [38], custard apple (*Annona cherimola*) [39], ciricote (*Cordia dodecandra*) [25] and the bananas [40] among others.

1.2.2 Carotenoids

Carotenoids are pigments found in almost all fruits that cause red, yellow and intermediate colorations, due to their chromophores. Its structure is

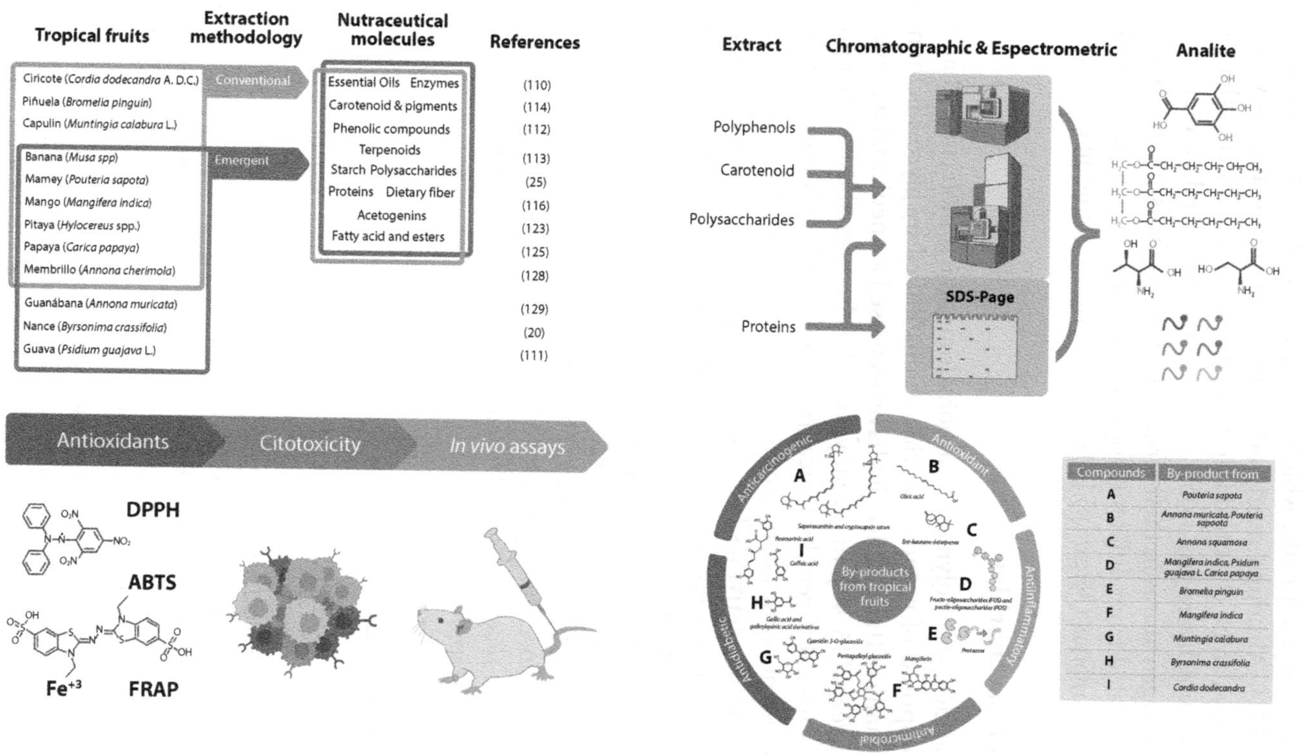

Figure 1.1 Schematic diagram of the identification and characterization of the nutraceutical compounds present in tropical fruits and coproducts produced in Mexico.

composed of the covalent bond of eight isoprene units, with the adhesion of rings at one or both ends, which can be replaced by hydroxyl groups or ketones [41]. Although nearly 600 carotenoid compounds have been identified, they can be classified into two large groups: carotenes that only have carbon and hydrogen atoms in their structure, and carotenes that have some oxygen atoms in their structure (Xanthophylls) [42]. Carotenes have been isolated from different tropical fruits such as mango (*Mangifera indica*) [43], bananas (*Musa* sp) [44] and papayas [45] among others.

1.2.3 Polysaccharides

Polysaccharides are polymeric carbohydrate macromolecules that are usually composed of monosaccharide chains linked by glycosidic bonds. Although many of these molecules have been used as medicinal treatments for several years [46, 47], in recent years it has been established that polysaccharides have different biological properties, such as immunomodulatory, antioxidant, antimutant, antidiabetic, antibiotic, anticoagulant and anti-inflammatory activities [48]. The use of polysaccharides has focused mainly on their use as dietary fiber that thanks to the enzymatic machinery of the intestinal microbiota they can be fermented to produce beneficial metabolites for human health [49] acting as prebiotic compounds [50]. In this sense, the most reported polysaccharides studied are oligosaccharides, including fructo-oligosaccharides (FOS), inulin, galacto-oligosaccharides (GOS) and lactulose, among others [51]. Some of tropical fruits studied for polysaccharides recovery for nutraceutical purposes are mango, guayaba, papaya and pineapple [43, 52].

1.2.4 Fatty Acids

Fatty acids are considered nutraceutical compounds as they presented different biological functions such as cardioprotective, hypolipidemic, antiatherogenic and anti-inflammatory that have been demonstrated [53]. Fatty acids can be classified based on their chemical structure, those that do not have double bonds (saturated), acids with one double bond (monounsaturated fatty acids) and those that have between 2 and 6 bonds (polyunsaturated fatty acids); They can also be classified according to the location of the first n-3 or n-6 double bond at the methyl end of the fatty acid or by the cis or trans configuration of the double bonds [54]. These compounds are usually found in the seeds of some fruits such as mango (*Mangifera indica*) (linolenic and linoleic) [55], Soursop (*Annona muricata*) (oleic acid,

10,13-octadecadienoic acid, linoleic acid, palmitoleic acid, and elaidic acid [56] and Mamey (*Pouteria sapota*) (oleic acid) [57].

1.2.5 Proteins (Peptide Isolate, Amino Acids)

Proteins represent an important interest as nutraceutical molecules, both in their native form and in the form of peptides or free amino acids [58]. However, its biological activity is mainly expressed in peptides, since they can be easily absorbed through the epithelium of the small intestine by passive means or transcellular mechanisms [59]. Bioactive peptides are characterized by having molecular masses less than 6000 Da and a size between 2 and 20 amino acids [60]. Bioactive peptides could prevent and/or treat some diseases such as cancer, some cardiovascular diseases and diabetes [61]. Although the fruits are not usually considered a source of protein, different authors have shown that some non-edible parts contain protein fractions that can be used for nutraceutical purposes, examples of which are mango peels [62], papaya seeds [63] and Saramuyo seeds [64].

For the isolation and recovery of the nutraceutical compounds, different techniques have been proposed based on characteristics of the by-products, the type of biomolecules, among others. The different approaches that have been used to the recovery of nutraceuticals in tropical fruits and by-products are presented below.

1.3 Nutraceutical Extraction Methods: Advantages and Drawbacks

Although the objective of the extraction process seems clear and is based on the ability to recover a certain target substance, intrinsically it must be borne in mind that the method used must be profitable, sustainable, highly selective, reducing impurities or other molecules that may interfere. in the identification and use of the extracted compounds [28]. These characteristics must be fulfilled while always maintaining the reproducibility and robustness of the method so that it can be used in different raw materials [43, 65]. Due to the above, in the extraction of nutraceutical compounds from fruits, multiple methodologies and techniques are used to assist the process, which allow not only the targeted obtaining of each of the compounds of interest, but also helps to reduce significantly energy consumption and the use of polluting solvents [57, 66, 67].

1.3.1 Traditional Methods

The basis of conventional extraction methods is based on the extractive capacity of solvents, the way they are heated and/or stirred [68]. These extraction procedures can be carried out in two ways, either liquid-liquid or liquid-solid [1] and their main disadvantages are long operating times, high energy consumption and large volumes of extraction. solvent used [69].

1.3.1.1 Maceration

Maceration is an extractive technique that is carried out by mixing the whole raw material, fractioned or ground, with the solvent at room temperature, using a closed container. After a long period of mixing, the mixture is filtered, recovering the liquid in which the compounds of interest are dissolved. Sometimes frequent agitation is used during the extraction process, which improves the diffusion of the compounds towards the solvent [70]. Although maceration is a simple method, of great importance for the recovery of thermolabile compounds [71], it has the disadvantages of long periods of extraction time and low extraction efficiency [72]. When the maceration process involves temperatures above >25°C, it is known as cooking. The final product resulting from the maceration, either at room temperature or hot, is known as an infusion [70].

1.3.1.2 Extraction Soxhlet

Soxhlet extraction, is a method that was originally used to determine the fat content of milk [73]. The procedure is carried out by placing the sample inside a thimble located in the main chamber of the extractor and the solvent in the boiling flask. When the solvent vaporizes, it enters the extraction chamber, which has a condenser in its upper part that produces the condensation of the solvent in the sample inside the thimble. When the solvent reaches the overflow level, it returns to the boiling flask through a siphon, where it restarts the cycle [74]. Soxhlet extraction has several advantages, among which are simplicity, high kinetics of the process due to the effect of temperature, low operating costs, and no need for filtering [72]. In contrast, its major disadvantages are long operating times, high energy consumption, low selectivity, low recovery yields, and large volumes of solvents [75].

1.3.1.3 Hydrodestilation

In the extraction of nutraceuticals from fruit coproducts, hydrodistillation is one of the oldest techniques. The hydrodistillation process is carried out by placing the vegetable raw material in a flask, then adding water and boiling the mixture. The nutraceuticals, water and oils, form a mixture that is vaporized and recovered by decantation in a flask [76]. Due to its high effectiveness extracting fat-soluble compounds, this technique is highly used to obtain essential oils [77]. The advantages of this extraction method are mainly the easy implementation of the method, the non-use of organic solvents, high efficiency in the separation of the essential oil, avoiding compaction and the formation of lumps, and low equipment cost [78]. While among its disadvantages are the long operating times and the high processing temperature [69, 79].

1.3.2 Innovative Methods

Problems related to traditional extraction methods such as safety risks, the toxicity of solvents, the presence of solvent residues in the extracts and the low yield in obtaining nutraceuticals, have been the justification for the development of extraction methodologies. known as clean or green technologies, which have as objectives: to reduce extraction times, optimize energy consumption, increase mass and heat transfer, the development of small-sized equipment and a reduction in unit operations of the entire system process [71].

Due to the above, several alternatives have emerged to replace the traditional extraction techniques of nutraceuticals.

1.3.2.1 Extraction With Pressurized Liquid

In this extraction methodology, liquid solvents are used whose surface tension has been reduced, density, diffusivity and viscosity have been increased by subjecting them to high temperature and pressure, to improve the ability of the solvent to penetrate the cell surface. raw material and cause the fracture of the cell matrix, improving the transfer of mass from the interior of the cell to the solvent [80]. The advantages of this method are mainly the rapid extraction and low solvent consumption, while its great disadvantage is focused on the sophisticated instrumentation required for temperature and pressure control [71].

1.3.2.2 Supercritical Fluids

This technique is based on the principle of replacing organic solvents with fluids in supercritical condition, that is, fluids that are above their critical temperature (Tc) and critical pressure (Pc) [69]. In the supercritical stage, the density, viscosity and diffusivity of the fluid are altered, exhibiting an improvement in its solvent capacity, with a notable improvement in transport properties, which increases the extraction yields of nutraceuticals [81]. The main advantages of this technology are the use of moderate extraction temperatures, no use of organic solvents and high extraction efficiency [82], while its disadvantages may focus more on the energy used during the process and the equipment installation cost [83].

1.3.2.3 *Extrusion Technology*

Although it is a technology that is already widely used in different industries of the agri-food sector, particularly in the production of food products such as wine. However, this technology has aroused interest in the use of agro-industrial biomass, because it is a technique that allows integrating different unit operations in the same process line, which facilitates the fractionation of the raw material [84]. In the extrusion process, the biomass is injected under pressure into a cylindrical chamber, where it is subjected to compression and shear forces using one or two screws [69], allowing liquid-solid extraction and separation [85], process that facilitates the subsequent recovery of biomolecules, both the dry matter and the liquid fraction.

1.3.2.4 *Microwave-Assisted Extraction*

Microwave-assisted extraction (MAE) is based on the application of radiation to produce the rotation of water molecules present in plant matrices, through the migration of ions and the rotation of dipoles. This phenomenon causes the evaporation of intracellular water, which causes the generation of holes in the cell membranes, favoring their permeability [86]. Currently, there are two types of MAE systems: One of these consists of the use of closed extraction vessels under controlled pressure and temperature, while the second type of system is open microwave ovens that work at atmospheric pressure [31]. The advantages of this method over traditional techniques are the short extraction times, low solvent volumes and high recovery rates of compounds of nutraceutical interest [87].

1.3.2.5 *Ultrasound-Assisted Extraction*

Ultrasound-assisted extraction (UAE) is a technique that has shown high effectiveness in recovering compounds with bioactive properties from different raw materials of plant origin [88]. UAE is based on the cavitation phenomenon that produces the propagation of ultrasonic waves and variations in temperature and pressure, which causes the cell wall of the matrix to fail, allowing intracellular compounds to enter the medium [89]. This phenomenon is known as "detexturing" [90] and it develops in 6 different stages: erosion, shear forces, sonoporation, fragmentation, capillary effect and detexturing [91].

UAE can be carried out in two different ways, directly or indirectly. In the direct sonication process, cylindrical probes are used that transmit ultrasonic waves to the solid-liquid system [92]. While the indirect UAE process is carried out in ultrasonic water baths, in which the bath water conducts the ultrasonic waves towards the container that contains the solid-liquid system [93]. The advantage of this method is basically that it can be considered a green technology since it does not require the use of organic solvents, reduces extraction times and reduces energy consumption [94]. The extraction by UAE carried out directly (probes) has the disadvantage of not being able to be used in large quantities [95], while the indirect extraction (tank) has the disadvantage that the water contained in the The sonicator tank and the walls of the container containing the solid-liquid system attenuate the supplied energy, reducing its transmissibility to the solute-solvent mixture [96]. Additionally, this extraction method, used both directly and indirectly, has a chemical disadvantage, which lies in the care that must be taken when carrying out the process, since the use of high power and/or the exposure of the system at prolonged sonication times, they can cause the degradation, oxidation, volatilization or re-absorption of the compounds of interest [97, 98].

1.3.2.6 Pulsed Electric Field (PEF) and High Voltage Electrical Discharge (HVED)

PEF and HVED are technologies based on the electroporation of the cell membrane, causing the release of biocompounds present inside the cell [96]. Specifically, PEFs use electrical pulses that range between 140 and 25,000 V, which are discharged for fractions of microseconds inside a chamber which contains the sample between two electrodes [99]. For its part, in HVED technology, energy is provided directly in a liquid through a plasma channel formed by applying a high voltage electrical discharge current between two electrodes, this energy causes the phenomenon of cavitation in the liquid, which leads to the formation of pores in the cell wall and subsequently to the release of cellular material [100]. The biggest advantage of this method is the low energy consumption, which is even lower than that used in other emerging technologies such as ultrasonic-assisted extraction and microwave-assisted extraction [101]. The limitations of these extraction methods are mainly based on the current difficulty in scaling up to an industrial level, since it is necessary to develop reliable and affordable high-pulse generators, as well as the development of manuals and regulations that allow their extended use [102].

1.3.2.7 Deep Eutectic Solvents Hashemi

Deep eutectic solvents are a group of solvents produced by mixing two (a hydrogen donor and a halide ion) solid components that interact with each other through hydrogen bonds [103]. The preparation of these solvents is done by mixing the exact amounts of salt and hydrogen donor in the absence of moisture, which will form a homogeneous liquid when melted [104]. Multiple DES have been described, some formed from the mixture of a quaternary salt (ammonium, sulfonium and phosphonium) and a metal chloride, hydrated metal chloride or HBD and another group composed of mixtures of a metal chloride and hydrogen donor [105]. The advantage of using these solvents with respect to traditional solvents in the extraction of nutraceuticals from plant material lies in two main aspects, one is that they have a low viscosity, a characteristic that allows them to more efficiently remove the intracellular material; the second aspect is presented in a lower toxicity of DES associated with the application of natural and biorenewable components [75, 106].

1.3.3 Recovery Bioactive Natural Products by Using Enzymatic Process

Biocatalysis is like a green, sustainable and environmentally friendly technique, enzyme-assisted extraction [66], is based on the great hydrolytic capacity of the enzymes, which can be used in a directed manner against the different constituents of the plant cell wall, allowing an improved recovery of the compounds of interest even when some are linked by covalent bonds [107]. Biocatalysis can be subdivided into three types of processing: fermentation, microbial transformation, and enzyme-assisted extraction [108]. One of the most outstanding advantages of this method is that the hydrolysis process is most often carried out at neutral pH, room temperature and atmospheric pressure, avoiding the degradation of the compounds of interest and facilitating process control [82, 108]. While the disadvantages of biocatalysis can be concentrated in the high commercial cost of the enzymes, their low recovery rate and the interaction of the enzymes used with enzymes from the raw material [109, 110].

1.4 Identification Techniques of Nutraceuticals Present in Tropical Fruits By-Products

As it has been specified in the above topic, tropical fruits and its by-products contain significant amounts of biomolecules that have been studied by their biological activities [111]. Recently, the development of advanced scientific tools and the analytical methods used for biomolecules studies have allow the physical, molecular, morphological and structural identification of several compounds, as well as their quantification [111]. The principal biomolecules from tropical fruits that are generally studied with analytical tools are the phenolic compounds and the polysaccharides, as they have shown different biological activities associated to their structures and functional groups [111, 112]. Additionally, some compounds that are presented in minor quantities on tropical fruits, but also play an important role in cellular functions are the vitamins and carotenoids [113]. Likewise, the presence of some proteins that can be converted to peptides are also interesting due to their health benefits [114]. Finally, there are by-products of tropical fruits such as the seeds that possess certain quantity of oil that content monounsaturated fatty acids that can be used in food formulations due to their characteristics, and their biological function is now under study [115].

1.4.1 Phenolic Compound Quantification and Identification

Depending on the characteristics of the molecules that are present in tropical fruits or coproducts, there are several techniques that have been used to determine its presence and quantity. The phenolic compounds as they are present in peels, leaves, seeds, skin and pulp are the most commonly molecules studied, an increasing number of studies related to its identification and quantification has been observed in the past decades that focus also on the correlation of specific compounds to specific biological activities [24], among them the phenolic acids, flavonols, flavanols, tannins, anthocyanins and flavones have been the most detected compounds. Nevertheless, it exists an important fraction of this kind of compounds that has been underestimated as it is retained in the extraction residues that are generally high molecular weight polyphenols, or phenolic compounds bounded to macromolecules such as polysaccharides and proteins that are called non-extractable polyphenols (NEPs) [112]. For the determination of phenolic compounds and its quantification from tropical fruits, the extractable compounds with low molecular weight are principally evaluated by chromatographic techniques that are based on the separation on the compounds from a mixture of molecules extracted from the fruit of the by-products by the use of a stationary phase where the molecules can be separated from each other while the aid of a mobile phase, after a previous cleaning procedure. In this sense, high-performance liquid chromatography (HPLC) has been the most evaluated technique that enables the profiling of phenolic compounds present in the sample, different stationary of mobile phase combinations and detectors have been evaluated and have provided specific information on the composition of the different tropical fruits evaluated, among them, reverse-phase liquid chromatography (RP-HPLC) has been the most widely studied using UV-Vis and diode-array detectors (DAD) [24]. Another technique that has been very widespread due to its capacity to improve sensitivity is the High-performance liquid chromatography-tandem mass spectrometry (HPLC/MS) or some variation as HPLC coupled to an electrospray ionization (ESI) source operating in the negative mode and a time of flight (TOF) mass detector (HPLC-ESI-TOF-MS) that enables the possibility to determine a higher quantity of compounds. Electrochemical detection (ECD) has been also evaluated to improve specificity as well as pulsed amperometric detection (PAD) (García Alonso *et al.*). For phenolic compounds identification, electrospray ionization mass spectrometry (ESI-MS), ultra-high-performance liquid chromatography (UHPLC) coupled to ESI with different high-resolution mass analyzers have been used, some examples are: Orbitrap, quadrupole-time of flight (Q-TOF-MS), triple quadrupole-TOF (TQ-TOF-MS), Matrix-assisted laser desorption/ionization (MALDI) [24]. Gas chromatography (GC) couple to mass spectrometry (GC/MS) has been also used for phenolic compounds identification as it is a fast, versatile, and highly sensitive technique, nevertheless the strong polarity of these compounds makes necessary the inclusion of derivatization of the samples. Additionally some new techniques as direct analysis in real-time (DART) has been and attractive technique to analyze solid, liquid and gaseous samples, its combination with MS has been applied to obtain a big spectrum of compounds without the use of chromatography, when it is used with high-resolution mass spectra (HMRS) it could show instantly several low and high polarity compounds with a minimum o no fragmentation spectra [112]. There are also documents that report the application of near-infrared and mid-infrared spectroscopy for the determination of the total content of phenolic compounds, as it is a fast and non-destructive analytical method based on the absorption of the IR light at different wavelengths that helps to identify the functional groups of the phenolic compounds that exist in the extract [24]. On the other hand, size-exclusion chromatography (SEC), preparative-HPLC or high-speed countercurrent chromatography (HSCCC) have been used to evaluated non extractable phenolic compounds [112].

1.4.2 Polysaccharides Quantification and Identification

In the case of polysaccharides, the use of Fourier transform infrared spectroscopy (FT-IR) and nuclear magnetic resonance spectroscopy (NMR) have been used to identify the molecular and structural characteristics, additionally, X-ray diffraction (XRD) have been used for surface morphology and physical characterization and differential scanning calorimetry (DSC) and scanning electron microscopy (SEM) the physical and surface morphology characterization of this compounds extracted from tropical and sub-tropical fruits [111]. After extraction, the separation process is one of the most difficult stages prior identification or quantification techniques, but it is the step that will allow a better analysis of the samples. Related to polysaccharides, the functional and structural characterization that includes their monosaccharide composition, molecular weight, homogeneity, glycosidic linkages among other characteristics, have been studies by FT-IR, gas chromatography-mass spectroscopy (GC-MS), gas chromatography (GC), high-performance liquid chromatography (HPLC), gas-liquid chromatography (GLC), liquid chromatography (LC) and NMR, focusing on the chemical structure's determinations. High-performance anion-exchange chromatography (HPAEC), high-performance anion-exchange chromatography coupled with pulsed amperometric detection (HPAEC-PAD), ion exchange chromatography (IEC) Further specific analysis for MW homogeneity, such as high-performance gel permeation chromatography (HPGPC), high performance size exclusion chromatography (SEC), and high-performance gel filtration chromatography (HPGFC) have also been evaluated [111]. The functional groups linked to polysaccharides has been evaluated preferably by FT-IR, for the analysis of the backbone structures the GC-MS has been wily used, NMR is commonly used to analyze the numeric configurations of glycoside residues and confirmed by correlated spectroscopy (COSY), total correlated spectroscopy (TCOSY), double-quantum-filtered correlation spectroscopy (DQF-COSY), heteronuclear simple bond and multiple bond correlation spectroscopy (HSQC and HMBC), and rotating-frame over Hauser spectroscopy (ROESY). For the advanced study of physical characterization to indicate parameters as morphology, physical strength, rheology, crystallization and viscosity of polysaccharides, methods such as SEM, XRD, transmission electron microscopy (TEM), atomic force microscopy (AFM) ultraviolet-visible spectroscopy (UV), thermos-gravimetric analysis (TGA) and matrix-assisted laser desorption/ionization time of flight mass spectrometry (MALDI-TOF-MS) [111].

1.4.3 Vitamins, Peptides and Fatty Acids Quantification and Identification

For the vitamin identification and quantification in tropical fruits several techniques have been proposed, in this sense, the reversed phase HPLC (RP-HPLC) with non-polar stationary phase and a polar mobile phase has been the most popular methodology for the separation and further quantification of vitamins, with detector as DAD, FLD, ECD and MS, depending on the nature of the molecule the fluorescence detector has been also used. The hydrophilic interaction liquid chromatography (HILIC) has been used as the best alternative of RPLC for vitamins. Ultra-high-performance liquid chromatography (UHPLC) couple with mass detectors are techniques that have increased resolution and sensitivity of the chromatographic analysis as well as the reduction on time a detrimental effect on MS caused by organic solvents. Additionally, the capillary electrophoresis (CE) using detector like laser-induced fluorescence (LIF), capillary zone electrophoresis (CZE) and capillary electron-chromatography (CEC), has been used to improve detection limits of several vitamins [113]. For protein detection and specific peptides quantification techniques such as SDS-page, SEC, HILIC, RP-HPLC, HPLC-MS/MS, ESI-Q-TOF-MS/MS, RP-HPLC-ESI-MS/MS has been used particularly for the determination of molecular weight, but also for the identification of the compounds and amino acids present on the samples [114]. Land plants including tropical fruits, produce a tremendous diversity of fatty acids with several unusual functional groups, different chain length, position and number of C-C double bonds, the study on the elucidation of the structure of this compounds will help to the knowledge of specific biological functions when accumulated in the oils of seeds and fruits, in particular the monounsaturated acyl chains have demonstrated a considerable potential in the food and chemical industries due to their physicochemical properties related to their structures [115]. The quantification and identification of fatty acids have been generally obtained by the use of GC with detector such as flame ionization detector (FID) and GC couple to MS/MS, nevertheless there is few information related to the identification of this compounds from tropical fruits by-products.

1.5 *In vitro* and *In vivo* Methods to Evaluate Biological Activity of Nutraceuticals

The most studied activity is the antioxidant effect that is determined after the total phenolic content with the use of different techniques [116], DPPH, ABTS, FRAP, superoxide radical scavenging activity (SRSA), the metal chelating activity (MCA), the total reducing power (TRP) assays are the most common. Table 1.2 shows different and useful *in vitro* assays to determine *in vitro* antioxidant effect, directly related to polyphenolic content of diverse tropical fruits.

Another way to evaluate the nutraceutical potential of tropical fruits is considering the relevance on base of the evaluation on health with the focus in some relevant diseases as cancer which worldwide distributed. Annona Chelimola is a good example because it contains Annonaceous acetogenins, with potential antineoplastic activity. As well as the value of *in vitro* cells models important to the evaluation of nutraceutical potential in a controlled manner, the truth is, that the final effect must be validated in a physiological complete system, such as animal models. For example, some results had been shown in Mango peel by its antitumorigenic effect proved in xenograft animal's model. And other diseases with equal relevance are those related to non-infectious diseases such as obesity and cardiovascular diseases (Table 1.2).

1.6 Effect of Nutraceuticals on the Microbiome

A community of microorganisms localized in a specific area is called microbiome that plays an important role in health and diseases. The human intestinal microbiome represents the largest mass and concentration of microorganisms in the body, the microbiome composition varies among individuals based on diet, environment, life habits, age, gender, genetics, health status, drug exposure, pollutant exposure, physical activity, and lifestyle practices. It has been demonstrated that a perturbation of the microbiome is implicated in the development of multiple diseases that includes autoimmunity, allergic disorders, brain health, depression, Alzheimer`s, metabolic modulation, and some cancers [140, 141]. The human diet can modify the composition of the microbiome and therefore it can influence the biochemical processes in the body. A balance diet is essential for microbial homeostasis that promotes health, the nutritional composition of diet is one of the most critical factors for regulating intestinal microbiota, as

Table 1.2 *In vitro* and *in vivo* assays for characterization of bioactivities of nutraceutical compounds recovered from coproducts of Mexican tropical fruits.

Tropical fruit	Properties	Assay	Molecules correlated	Ref.
Cordia dodecandra.	Antioxidant activity	TPC (Total phenolic content), DPPH, ABTS	Phenolic compounds	[25]
Annona Cherimola.		ABTS, DPPH (radical scavenging activity), FRAP (ferric reducing antioxidant Power)		[117]
Byrsonima crassifolia.		TPC, TFC (Total flavonoid content), ORAC (Oxygen Radical Absorbance Capacity), DPPH		[118]
Annona Cherimola Mill.		ABTS, DPPH, FRAP		[119]
Psidium guajava L.		DPPH, CUPRAC (cupric ion reducing antioxidant capacity), FRAP, TPC, TFC		[120]
Carica papaya.		DPPH, ABTS, reducing power assays	Polyphenols & flavonoids	[121]

(Continued)

Table 1.2 *In vitro* and *in vivo* assays for characterization of bioactivities of nutraceutical compounds recovered from coproducts of Mexican tropical fruits. (*Continued*)

Tropical fruit	Properties	Assay	Molecules correlated	Ref.
Mangifera indica.	Cellular antioxidant activity (CAA)	permeability coefficient across the Caco2/HT29 monolayer	Phenolic compounds	[122]
Musa acuminata.	Antioxidant activity	TPC, TFC, DPPH, CUPRA,		[123]
		FRAP, TPC, DPPH, ABTS βcarotene bleaching (BCB)		[124]
Mangifera indica L.		TPC, DPPH		[125]
Musa acuminata.		TPC, TFC, DPPH, SRSA, (Superoxide radical scavenging activity), MCA (Metal chelating activity), TRP (Total reducing power)	Polyphenols & flavonoids	[126]
Mangifera indica L.	Rat erythrocytes protection	Protection against hydrogen peroxide induced oxidative damage	Phenolic compounds	[127]
Annona muricata.	Crown gall tumor inhibition, tumoral cytotoxicity	A549 (brine shrimp, lung), MCF7 (breast) and HT29 (colon)	Annomuricin A, B and C	[128]
Musa acuminata.	Cytotoxic effect. SulphoRhodamine-B method)	HepG2	Polyphenols & flavonoids	[129]

(*Continued*)

Table 1.2 *In vitro* and *in vivo* assays for characterization of bioactivities of nutraceutical compounds recovered from coproducts of Mexican tropical fruits. (*Continued*)

Tropical fruit	Properties	Assay	Molecules correlated	Ref.
Mangifera indica.	Cell viability	HT29, Caco2 and HCT116 (colon),	Gallic acid, mangiferin, citric acid, quinic acid, pentagalloylglucose and methyl gallate	[130]
Annona Cherimola Mill.	Antiproliferative activity	HeLa (cervical carcinoma), HepG2 (hepatocellular carcinoma)	Alkaloids	[119]
A. squamosa.	Cytotoxicity effect	Jurkat and MCF7 (breast), HL60 (Human promyelocytic leukemia), VERO (kidney monkey), HCT116	Anonaine, asimilobine, nornuciferine, and reticuline.	[131]
Muntingia calabura L.	Cytotoxicity effect	HL60	Flavones & chalcone	[132]
Carica Papaya.	Inhibition of tumor cell line proliferation	T cell lines: Jurkat, Molt4, CCRFCEM and HPBALL. Burkitt's lymphoma cell lines. A chronic myelogenous leukemia cell line (K562), (Hela), (HepG2 and Huh7), a lung adenocarcinoma cell line (PC14) a pancreatic epithelioid carcinoma cell line (Panc1), and mesothelioma cell lines (H2452, H226, and MESO4)	Tocopherol, lycopene, flavonoid and benzyl-isothiocyanate	[133]

(*Continued*)

Table 1.2 *In vitro* and *in vivo* assays for characterization of bioactivities of nutraceutical compounds recovered from coproducts of Mexican tropical fruits. (*Continued*)

Tropical fruit	Properties	Assay	Molecules correlated	Ref.
Psidium guajava L.	Anti-diabetic effects	Diabetic mice induced by streptozotocin combined with high fat diet.	Polysaccharides	[134]
Mangifera indica L.	Lipid-lowering effect	Hyperlipidemia model induced in Wistar rats	Phenolic compounds	[135]
Mangifera indica L.	Suppress breast cancer ductal carcinoma. Increased (ROS)	Mice, xenografted with MCF10DCIS	Pyrogallolyl	[136]
Mangifera indica L.	Anti-obesity effects	Obese rats fed a high fat diet (HFD)	Mangiferin	[137]
Mangifera indica L.).	Decreased lipid peroxidation in plasma, kidney and liver	Streptozotocin induced diabetic rats	Phenolic compounds	[125]
Annona muricata.	Anti-inflammatory activities reductions in the leukocyte migration and exudate volume	Carrageenan induced edema in rat paws		[138]
Annona muricata	Acute and chronic inflammation. By attenuation in the TNF-α and IL-1β protein expression	Complete Freund's adjuvant (CFA) induced arthritis in rats and xylene induced ear edema in mice.		[139]

an example changes in diet can alter the viability of intestinal population by consuming ultra-processed foods in large quantities, on the other hand a diversity of components presents in a diet with variety of good groups, may positively influence physiological parameters that help on the prevention of metabolic syndromes. In this sense, the alterations in the consumption pattern favoring the inclusion of foods with functional properties and nutraceuticals that are known to have therapeutic effects, could contribute positively to the changes in the population structure of microbiome [141].

Phenolic compounds are the nutraceuticals most associated with the development of a favorable gut microbiota that presents a significant impact to the development and maintenance of human health, thus the importance of the studies related to the extraction, identification and consumption of tropical fruits that present different kind of phenolic compounds are becoming higher [141]. In this sense, the hypothesis of the action of phenolic compound indicates the binding to the membranes of bacterial cells, and the effect on the metabolism is related on the strain of the microorganism, the research suggest that polyphenols are related in the production of substances that can interfere with pathogenic bacterial synthesis pathways and prevent the exponential growth of pathogenic bacteria [141].

Large polysaccharides, specifically the non-digestible such as pectin could not be directly transported to the microbial cell without prior breakdown to the respective sugar monomer or oligomers. Pectin with large molecular weight arrive to the intestine to be degraded by the 500-1000 distinct species that possess the enzymatic machinery for its degradation. Additionally, fructo-oligosaccharides, galacto-oligosacharides, as well as the pectin, are non-digestible carbohydrates, commonly known as fiber that can be metabolized by the microbiota to support their health and growth [140]. These polysaccharides known as prebiotics may be delivered to the human body by diet or through formulations of orally supplements promoting health care.

1.7 Future Trends of Nutraceuticals Obtained From Tropical Fruit By-Products

Local and exotic fruits from different regions can help meet the needs of health-conscious consumers and ensure regional food security and sustainable production [68, 142]. The consumption of some fruits is strongly associated with several health benefits due to their high nutritional values and properties for reducing the risks of certain diseases, such as different types of cancer and cardiovascular diseases. The above is attributed to nutraceuticals in the plant matrices [143, 144]. Nutraceuticals obtained from fruit coproducts, such as proteins, polysaccharides, fibers, flavor compounds, and phytochemicals, can be used as a source of nutraceuticals [145, 146]. Nutraceuticals are medicinal food additives that enhance immunity and treat specific diseases such as cancer [147]. Apple pulp and peel, grape seeds and skin, citrus and banana peel, and mango seed have been used to extract phenolic compounds as nutraceuticals [145, 148]. Therefore, some novel sources of tropical fruit coproducts to obtain nutraceutical compounds, as well as trends in their application and future aspects to be considered, will be described below.

1.8 Novel Nutraceutical Origins From Tropical Fruit Coproducts

The culture of health care through fruit consumption is an area in constant growth due to the investment of producers or consumers. Therefore, the consumption of known fruits has been added to those not so well known, seeing a significant increase in the demand for exotic fruits. Rambutan (*Nephelium lappaceum* L.) is one of the most admired exotic tropical fruits and has recently spread to other humid tropical regions worldwide. It belongs to the *Sapindaceae* family and is believed to be native to Southeast Asia, especially Indonesia and Malaysia [149]. Several studies have confirmed that the different parts of Rambutan plants contain a wide range of antioxidant compounds, such as polyphenols, xanthophylls, tocopherol, carotene, tannins, and ascorbic acid [150]. Rambutan has antibacterial properties for different types of gram-positive and gram-negative bacteria [151]. The effect of an ethanolic extract of Rambutan peel on antihyperglycemic activity in an *in vitro* model has also been reported [152]. The presence of flavonoids, polyphenols, and tannins has the possibility that the mechanism of blood glucose reduction in animals was evaluated mediated by inhibiting glucose absorption, which stimulates insulin release [153]. Also, due to the presence of antioxidant compounds, Rambutan can protect against oxidative damage in cells and tissues of the human body [154]. In addition, some of the primary chemicals in the Rambutan fruit peel extract include ellagic acid, flavonoids, and tannins. These compounds are effective in controlling obesity [155]. Another benefit discovered is that the antioxidant compounds in Rambutan peel improve total sperm count, with a dose of 15 mg/kg BW having the highest efficiency [156].

Another fruit cultivated in tropical countries and considered a valuable cash crop worldwide is the melon (*Cucumis melo* L.) from the Cucurbitaceae family, which is native to Iran and Pakistan [143, 157]. The edible part of the melon represents at least 65% of the total weight, while peels and seeds represent 25% and 7%, respectively [158]. *Melon coproducts* are peels, seeds, and bagasse, usually rejected in fruit processing. However, it has been reported that they maintain a high concentration of nutraceuticals and represent a significant percentage of the total weight used to extract nutraceuticals [143]. One of the options for the valorization of melon coproducts is the production of functional ingredients. The above due to these raw materials still retain a high content of relevant nutrients, namely fiber and protein, as well as several nutraceuticals (carotenoids and polyphenols) [159]. For example, these organic materials can be used directly as a source of dietary fiber (flour) and for the modulation of intestinal microbiota, such as bifidobacterium. These bacteria have been associated with colon, stomach, breast, and prostate cancer prevention [160]. Moreover, polyphenols inhibit digestive enzymes (lipase, amylase, and glucosidase) that minimize hyperglycemia associated with diabetic and obesity disorders. Under these functional properties and based on their richness in nutrients and bioactive components, melon coproducts could be used as active ingredients to develop novel foods, such as cookies, fiber bars, or cereals [161].

The area known as Mesoamerica, a historic region in North America, extends from central Mexico to Belize, Guatemala, El Salvador, Honduras, Nicaragua, and northern Costa Rica. From 2500 BC to AD 1521 Mesoamerica was a cultural area defined by great civilizations such as the Olmec, Maya, and Aztecs, among others [162]. This area is considered one of the world's leading centers of domestication and an essential center of genetic diversity, where species such as corn (*Zea mays* L.), beans (*Phaseolus* spp.), and squash (*Cucurbita* spp.) were domesticated and integrated into a multi-crop system [162]. The importance of fruits native to the Mesoamerican region lies in their contribution to today's agriculture and cuisine worldwide [162]. Therefore, there are several studies on the nutraceutical properties of the native fruits of the region, but many others are still analyzing their potential [162]. It has been reported that these fruits' detailed and expanded phenolic composition has not yet been compiled and analyzed. In addition, the health benefits related to decreased risk of metabolic syndrome associated with the phenolic compounds of these fruits require further study. Another understudied aspect of some crops is the effects of innovative technologies on phenolic compounds in Mesoamerican fruits, and the mechanisms of the observed changes in phenolic content are often unknown [162]. Even some other fruits, due to lack of knowledge or interest of the population, are disappearing.

An example of these fruits native to the region is the black sapote (*Diospyros digyna* Jacq.), also called "black persimmon," "chocolate fruit pudding," or in the Mayan language "*Tauch*" [163, 164]. It is a species that comes from the Ebenaceae family. *D. digyna* is a native tree of South America, also found on the coast of Mexico and Central America [165]. In this region's traditional medicine, the roots, wood, leaves, fruits, and seeds have been used to treat arteriosclerosis, asthma, diabetes, dermatitis, hypertension, and insomnia [166]. This fruit has been studied little, but it has been reported to have an interesting nutritional and nutraceutical profile. However, it may be underutilized due to its unique appearance. It is an oval or nearly round fruit with a diameter of 5 to 15 cm, having an olive epicarp, a dark pulp, and a considerable number of smooth brown seeds [166]. The pulp is consumed fresh and can be used to prepare ice cream, jams, and creams [166]. However, it is difficult to export as it is highly perishable after harvest [167]. A few studies have found that this fruit's pulp, seed, and peel are an essential source of polyphenols, particularly sinapic acid, fumaric acid, and proanthocyanidins [168]. The studies that have been carried out suggest that the coproducts of black sapote may have potential use as alternative raw materials for preparing food supplements or functional foods, but more research is needed [168]. Likewise, as it is considered an abandoned and underutilized fruit in the Yucatan Peninsula, its further study and application would allow its rescue [163]. Other fruits that are considered abandoned and underutilized in the Yucatan Peninsula and with potential use for nutraceutical extraction are: Pool box/Blackhead (*Annona purpurea* MOC. And Sessé ex Dunal); cocoyol (*Acrocomia mexicana* Karw. Ex Mart); Pepino kat (*Parmentiera aculeata* (H.b. & K.) Seeman), Tauch/Black sapote (*Diospyrus dygina* Jacq); Wuayum/Huaya país (*Meliccocus olivaeformis* (Kunth.) Radlk); and Cho' och/mountain sapote (*Pouteria hypoglauca* (Standl.) Baehni) [163].

1.9 Trends in the Application of Nutraceuticals Obtained From Tropical Fruit Coproducts

Several studies on nutraceuticals extracted from tropical fruits show an antioxidant capacity to reduce oxidative stress in chronic diseases, such as neurological disorders, diabetes, obesity, cardiovascular risk, and cancer [169]. Therefore, the interest in applying nutraceuticals obtained from tropical fruits has increased the attention to their effect on treating these diseases [3]. It has been reported that consuming nutraceutical drinks from fruits and vegetables can reduce the effects of neurodegenerative diseases [170]. Alzheimer's and Parkinson's diseases are the most common neurodegenerative disorders, which increase with age [170]. In this sense, the neuroprotective effects of tropical juices, such as a mixture of pomegranate, white guava, and fermented papaya, have been endorsed in treating Alzheimer's disease [171]. However, the number of studies on using tropical fruit coproducts against this type of disorder is still limited, or only preliminary studies have been conducted [169]. Concerning cardiovascular diseases, different types of tropical fruit coproducts have been tested worldwide to treat these diseases [169]. For example, camu-camu seeds (*Myrciaria dubia*), a fruit from the Amazon region rich in phenolic compounds, have been reported to inhibit lipid peroxidation in tests using male Wistar rat egg homogenates from 24% to 86% [172]. Another tropical fruit that has been reported to reverse the symptoms of metabolic syndrome, which increases the risk of cardiovascular disease, is *Garcinia dulcis* [173]. This fruit is native to Asia, consisting of morelloflavones and garcinol, which have antioxidant, anti-inflammatory, and cardioprotective properties [174]. Diabetes is considered a global pandemic, as it is estimated that by 2035 there will be 471 million people with this condition [175]. Type 2 diabetes is the most common type; it has increased due to rapid social and lifestyle changes, such as low physical activity, low fruit, and vegetable intake, and increased consumption of junk food [175]. Several tropical fruits have shown antidiabetic properties, such as the residue of the fruit *Ananas comosus* (L.) Merr. var. Mauritius. These were evaluated *in vitro* assays, which could be an individual or synergistic effect of the phenolic compounds present in this fruit [176]. Tropical fruit coproducts represent a rich source of natural compounds with preventive and therapeutic potential against cancer, intending to find new alternative therapies to prevent and treat this disease [177]. For example, some biomolecules in the Cucurbitaceae family, such as cucurbitacin, are natural anticancer agents [178]. However, information about its biological activity is limited, but some reports indicate the potential of cucurbitacins to treat pathologies, such as cancer [178]. Cucurbitacin B has been found to inhibit five different leukemia cells [179].

1.10 Future Aspects of Nutraceuticals

According to nutraceutical production trends, these will likely focus on products for healthy aging, healthy breakfasts, and snacks, improving cognitive function, muscle development, and personalized nutrition [180]. The growing demand for specific ingredients, such as whole grains and gluten-free cereals, is rising. So, to help improve waste disposal in the interest of corporate sustainability and increasing regulatory pressure around disposal, food manufacturers have used the valorization of tropical fruit coproducts as a source of ingredients in food processing [143, 162]. In this sense, Sachdeva *et al.* [142] describe the following aspects that will direct the demand for nutraceuticals (Figure 1.2):

A. Production of functional foods
Nowadays, people are more concerned about their health and prefer to consume foods that have specific benefits; therefore, the demand for nutritional foods will increase in the future. Therefore, it is necessary to find new sources and tropical fruit coproducts are a good option.

B. New entries
Further developments in food supplements enriched in microorganisms, including probiotics, prebiotics, and symbiotic bacteria, will be required.

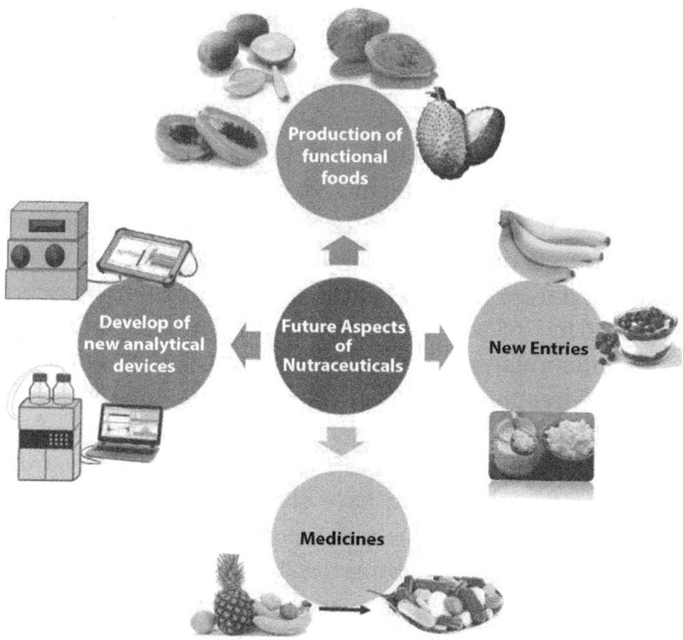

Figure 1.2 Critical aspect for future nutraceutical demand related to tropical fruit coproducts. Adapted from Sachdeva *et al.* [142].

C. Development of new analytical devices
Several devices are currently used for the analysis of the composition of plant extracts, such as chromatographic analysis. However, the future demand for nutraceuticals will require further advances in developing new devices.

D. Medicines
Similarly, the demand for nutraceuticals will encourage the development of drugs based on compounds obtained from tropical fruits.

Growing consumer interest in long-term nutritional and dietary benefits and increased valorization of tropical fruit coproducts have contributed to a favorable market for the nutraceutical industry. Therefore, nutraceuticals from these sources are expected to be our favorite recipe of the future for their multiple health benefits.

1.11 Conclusion

This chapter provides an overview of the nutraceutical compounds present in coproducts of some common and uncommon tropical fruits produced in Mexico, based on an approach to the methodology used for their recovery, identification, and biological characterization. Emerging extraction techniques have shown a significant advance in the study of biomolecules with nutraceutical properties, because they allow larger amounts to be recovered using G.R.A.S solvents and with lower energy consumption. Likewise, the different chromatographic techniques coupled to mass spectroscopy result in the set of techniques most used for the identification of the different biocompounds recovered from fruit coproducts. The bioactivity of mixtures extracted from tropical fruits has been shown to possess different biological properties of nutraceutical interest, such as antioxidant, antidiabetic, anti-inflammatory, and anticancer activity, as well as a detailed beneficial effect when interacting with the intestinal microbiota. However, clinical studies are needed to make way for the development of new products. The most studied biomolecules extracted from fruit coproducts are the phenolic compounds, which have shown promising results both *in vitro* and *in vivo* for their use as nutraceuticals. However, more detailed research into the nutraceutical properties of all the other molecules present in the different parts of the fruits, as well as the study of rare and unused fruits, is still needed.

All the work compiled in this chapter allows us to infer that the development of nutraceutical products is seen as an alternative to the use of coproducts generated during the processing of fruits, which today represent a risk for ecosystems and a problem for the big industrialists.

References

1. García-Villegas, A. *et al.*, Cosmeceutical potential of major tropical and subtropical fruit by-products for a sustainable revalorization. *Antioxidants*, 11, 2, 203, 2022.
2. Panchal, S.K. and Brown, L., Tropical fruits from Australia as potential treatments for metabolic syndrome. *Curr. Opin. Pharmacol.*, 63, 1–6, 2022.
3. de la L. Cádiz-Gurrea, M. *et al.*, Revalorization of bioactive compounds from tropical fruit by-products and industrial applications by means of sustainable approaches. *Food Res. Int.*, 138, 109786, 2020.
4. Ojeda, G.A., Sgroppo, S.C., Sánchez, M.C., de Ancos, S.B., Mango 'criollo' by-products as a source of polyphenols with antioxidant capacity. Ultrasound assisted extraction evaluated by response surface methodology and HPLC-ESI-QTOF-MS/MS characterization. *Food Chem.*, 396, 133738, 2022.
5. Eixenberger, D. *et al.*, Tropical agroindustrial biowaste revalorization through integrative biorefineries—Review part II: Pineapple, sugarcane and banana by-products in Costa Rica. *Biomass Convers. Biorefin.*, 2022.
6. Hikal, W.M. *et al.*, Banana peels: A waste treasure for human being. *Evid.-Based Complement. Alternat. Med.*, 2022, 7616452, 2022.
7. Perrone, A., Yousefi, S., Salami, A., Papini, A., Martinelli, F., Botanical, genetic, phytochemical and pharmaceutical aspects of Annona cherimola Mill. *Sci. Hortic.* Amsterdam, 296, 110896, 2022.
8. Abbas, S., Sharif, M.K., Sibt-e-Abbas, M., Fikre Teferra, T., Sultan, M.T., Anwar, M.J., Nutritional and therapeutic potential of sesame seeds. *J. Food Qual.*, 2022, 6163753, 2022.
9. Jalgaonkar, K., Mahawar, M.K., Bibwe, B., Kannaujia, P., Postharvest profile, processing and waste utilization of dragon fruit (*Hylocereus Spp.*): A review. *Food Rev. Int.*, 38, 4, 733–759, May 2022.
10. Jimenez-Garcia, S.N., Ramirez-Gomez, X.S., Aguirre-Becerra, H., Escobar-Ortiz, A., Contreras-Medina, L.M., Garcia-Trejo, J.F., Feregrino-Perez, A.A., Pitahaya peel: A by-product with great phytochemical potential, biological activity, and functional application. *Molecules*, 27, 16, 5339, 2022.
11. Bernardino-Nicanor, A. *et al.*, Preparation, composition, and functional properties of a protein isolate from a defatted mamey sapote (Pouteria sapota) seed meal. *CyTA - J. Food*, 12, 2, 176–182, Apr. 2014.
12. Espinosa-Zaragoza, S., Villegas-Monter, Á., Saucedo-Veloz, C., Corrales-García, J., Avendaño Arrazate, C.H., Alia-Tejacal, I., Quality features of zapote mamey (Pouteria sapota jacq.) selected in Alpoyeca, Guerrero, Mexico. *Rev. Mex. Cienc. Agríc.*, 7, 4, 953–959, 2016.

13. Garcia-Amezquita, L.E., Tejada-Ortigoza, V., Heredia-Olea, E., Serna-Saldívar, S.O., Welti-Chanes, J., Differences in the dietary fiber content of fruits and their by-products quantified by conventional and integrated AOAC official methodologies. *J. Food Compos. Anal.*, 67, 77–85, 2018.
14. Agredano-De la Garza, C.S. *et al.*, Physicochemical characterization and dietary fiber of 15 Nance (*Byrsonima crassifolia* L.) fruits selections from Nayarit. *Sci. Hortic.* (Amsterdam), 289, 110460, July, 2021.
15. Pérez Gutiérrez, R.M., Anti-inflammatory effect of birsonimadiol from seeds of Byrsonima crassifolia. *Food Sci. Biotechnol.*, 25, 2, 561–566, 2016.
16. Robles-Melchor, L. *et al.*, Removal of methylene blue from aqueous solutions by using nance (*Byrsonima crassifolia*) seeds and peels as natural biosorbents. *J. Chem.*, 2021, 5556940, 2021.
17. Adam Yahya Abdualrahman, M. *et al.*, Postharvest physicochemical properties of the pulp and seed oil from *Annona squamosa* L. (Gishta) fruit grown in Darfur region, Sudan. *Arab. J. Chem.*, 12, 8, 4514–4521, 2019.
18. Rana, V.S., Fatty oil and fatty acid composition of Annona squamosa Linn. Seed kernels. *Int. J. Fruit Sci.*, 15, 1, 79–84, Jan. 2015.
19. Pío-León, J.F., López-Angulo, G., Paredes-López, O., de J. Uribe-Beltrán, M., Díaz-Camacho, S.P., Delgado-Vargas, F., Physicochemical, nutritional and antibacterial characteristics of the fruit of Bromelia pinguin L. *Plant Foods Hum. Nutr.*, 64, 3, 181–187, 2009.
20. Ragazzo-Sánchez, J.A. *et al.*, Utilization of by-products of endemic fruits: Encapsulation of proteolytic extracts of guamara (Bromelia pinguin) and cocuixtle (Bromelia karatas) by electrospraying. *LWT - Food Sci. Technol.*, 149, 111670, 2021.
21. Rotta, E.M., Haminiuk, C.W.I., Maldaner, L., Visentainer, J.V., Determination of antioxidant activity and phenolic compounds of Muntingia calabura Linn. peel by HPLC-DAD and UPLC-ESI-MS/MS. *Int. J. Food Sci. Technol.*, 52, 4, 954–963, 2017.
22. de Sousa Sabino, B. *et al.*, Polysaccharides from acerola, cashew apple, pineapple, mango and passion fruit co-products: Structure, cytotoxicity and gastroprotective effects. *Bioact. Carbohydr. Diet. Fibre*, 24, 100228, 2020.
23. Dinani, S.T. and van der Goot, A.J., Challenges and solutions of extracting value-added ingredients from fruit and vegetable by-products: A review. *Crit. Rev. Food Sci. Nutr.*, 2022.
24. Garcia-Alonso, A., Sánchez-Paniagua López, M., Manzanares-Palenzuela, C.L., Redondo-Cuenca, A., López-Ruíz, B., Edible plant by-products as source of polyphenols: Prebiotic effect and analytical methods. *Crit. Rev. Food Sci. Nutr.*, 2022.
25. Pacheco, N. *et al.*, Physicochemical composition, phytochemical analysis and biological activity of ciricote (Cordia dodecandra A. D.C.) fruit from Yucatán. *Nat. Prod. Res.*, 36, 1, 440–444, 2022.

26. Gómez-García, R., Campos, D.A., Aguilar, C.N., Madureira, A.R., Pintado, M., Valorisation of food agro-industrial by-products: From the past to the present and perspectives. *J. Environ. Manage.*, 299, 113571, 2021.
27. Leong, Y.K. and Chang, J.-S., Valorization of fruit wastes for circular bioeconomy: Current advances, challenges, and opportunities. *Bioresour. Technol.*, 359, 127459, 2022.
28. Patra, A., Abdullah, S., Pradhan, R.C., Review on the extraction of bioactive compounds and characterization of fruit industry by-products. *Bioresour. Bioprocess.*, 9, 1, 14, 2022.
29. Padayachee, A., Day, L., Howell, K., Gidley, M.J., Complexity and health functionality of plant cell wall fibers from fruits and vegetables. *Crit. Rev. Food Sci. Nutr.*, 57, 1, 59–81, Jan. 2017.
30. González-Sarrías, A., Larrosa, M., García-Conesa, M.T., Tomás-Barberán, F.A., Espín, J.C., Nutraceuticals for older people: Facts, fictions and gaps in knowledge. *Maturitas*, 75, 4, 313–334, 2013.
31. Kaur, K., Functional nutraceuticals: Past, present, and future, in: *Nutraceuticals. Nanotechnology in the Agri-Food Industry*, A. M. Grumezescu (Ed.), vol. 4, pp. 41–78, Elsevier B.V., London, UK, 2016.
32. Sharma, M., Dwivedi, P., Singh Rawat, A.K., Dwivedi, A.K., Nutrition nutraceuticals: A proactive approach for healthcare, in: *Nutraceuticals. Nanotechnology in the Agri-Food Industry*, A. M. Grumezescu (Ed.), vol. 4, pp. 79–116, Elsevier B.V, London, UK, 2016.
33. Oladzad, S., Fallah, N., Mahboubi, A., Afsham, N., Taherzadeh, M.J., Date fruit processing waste and approaches to its valorization: A review. *Bioresour. Technol.*, 340, 125625, 2021.
34. Karasawa, M.M.G. and Mohan, C., Fruits as prospective reserves of bioactive compounds: A review. *Nat. Prod. Bioprospect.*, 8, 5, 335–346, 2018.
35. Kołton, A., Długosz-Grochowska, O., Wojciechowska, R., Czaja, M., Biosynthesis regulation of folates and phenols in plants. *Sci. Hortic.* (Amsterdam), 291, 110561, Jan. 2022.
36. Das, P.R., Islam, M.T., Lee, S.H., Lee, M.K., Kim, J.B., Eun, J.B., UPLC-DAD-QToF/MS analysis of green tea phenolic metabolites in their free, esterified, glycosylated, and cell wall-bound forms by ultra-sonication, agitation, and conventional extraction techniques. *LWT*, 127, October 2019, 109440, 2020.
37. Vuolo, M.M., Lima, V.S., Maróstica-Junior, M.R., *Phenolic compounds: Structure, classification, and antioxidant power*, in: *Bioactive Compounds: Health Benefits and Potential Applications*, M. R. Segura-Campos (Ed.), pp. 33–50, Elsevier Inc., Duxford, UK, 2018.
38. Edet, U.O. *et al.*, Evaluation of Annona muricata extract against Staphylococcus aureus isolate and in-silico activity of bioactive compounds against Capsular protein (Cap5O). *BMC Complement. Med. Ther.*, 22, 1, 192, 2022.

39. Fuel, M. *et al.*, Antioxidant and antiproliferative potential of ethanolic extracts from Moringa oleifera, Tropaeolum tuberosum and Annona cherimola in colorrectal cancer cells. *Biomed. Pharmacother.*, 143, 112248, 2021.
40. Chaudhry, F., Muhammad, L., Hayat, Z., Ranjha, M.M., Chaudhry, K., Elboughdiri, N., Asmari, M., Uddin, J., Extraction and evaluation of the antimicrobial activity of polyphenols from banana peels employing different extraction techniques. *Separations*, 9, 7, 165, 2022.
41. Meléndez-Martínez, A.J. *et al.*, A comprehensive review on carotenoids in foods and feeds: Status quo, applications, patents, and research needs. *Crit. Rev. Food Sci. Nutr.*, 62, 8, 1999–2049, Mar. 2022.
42. Rodríguez-Mena, A., Ochoa-Martínez, L.A., González-Herrera, S.M., Rutiaga-Quiñones, O.M., González-Laredo, R.F., Olmedilla-Alonso, B., Natural pigments of plant origin: Classification, extraction and application in foods. *Food Chem.*, 398, 133908, 2023.
43. Gupta, A.K. *et al.*, A review on valorization of different by-products of mango (Mangifera indica L.) for functional food and human health. *Food Biosci.*, 48, 101783, 2022.
44. Li, Z. *et al.*, The Musa troglodytarum L. genome provides insights into the mechanism of non-climacteric behaviour and enrichment of carotenoids. *BMC Biol.*, 20, 1, 186, 2022.
45. Lara-Abia Jorge, S.-W.-C. and Cano, M.P., Effect of ultrasound-assisted extraction of carotenoids from papaya (Carica papaya L. cv. Sweet Mary) using vegetable oils. *Molecules*, 27, 3, 638, 2022.
46. Wang, P.-C., Zhao, S., Yang, B.-Y., Wang, Q.-H., Kuang, H.-X., Anti-diabetic polysaccharides from natural sources: A review. *Carbohydr. Polym.*, 148, 86–97, 2016.
47. Xie, J.-H. *et al.*, Advances on bioactive polysaccharides from medicinal plants. *Crit. Rev. Food Sci. Nutr.*, 56, sup1, S60–S84, Jul. 2016.
48. Yu, Y., Shen, M., Song, Q., Xie, J., Biological activities and pharmaceutical applications of polysaccharide from natural resources: A review. *Carbohydr. Polym.*, 183, 91–101, 2018.
49. Yao, H., Flanagan, B.M., Williams, B.A., Mikkelsen, D., Gidley, M.J., Particle size of dietary fibre has diverse effects on *in vitro* gut fermentation rate and end-products depending on food source. *Food Hydrocolloids*, 134, 108096, 2023.
50. Gibson, G.R. *et al.*, Expert consensus document: The International Scientific Association for Probiotics and Prebiotics (ISAPP) consensus statement on the definition and scope of prebiotics. *Nat. Rev. Gastroenterol. Hepatol.*, 14, 8, 491–502, 2017.
51. Bhadra, S., Chettri, D., Verma, A.K., Microbes in fructooligosaccharides production. *Bioresour. Technol. Rep.*, 20, 101159, 2022.
52. Martínez, R., Torres, P., Meneses, M.A., Figueroa, J.G., Pérez-Álvarez, J.A., Viuda-Martos, M., Chemical, technological and *in vitro* antioxidant

properties of mango, guava, pineapple and passion fruit dietary fibre concentrate. *Food Chem.*, 135, 3, 1520–1526, 2012.

53. Langyan, S. *et al.*, Food and nutraceutical functions of sesame oil: An underutilized crop for nutritional and health benefits. *Food Chem.*, 389, 132990, 2022.
54. Orsavova, J., Misurcova, L., Ambrozova, J.V., Vicha, R., Mlcek, J., Fatty acids composition of vegetable oils and its contribution to dietary energy intake and dependence of cardiovascular mortality on dietary intake of fatty acids. *Int. J. Mol. Sci.*, 16, 6, 12871–12890, Jun. 2015.
55. Lenucci, M.S., Riccardo, T., Mita, G., Durante, M., Bioactive compounds and antioxidant activities in different fractions of mango fruits (Mangifera indica L., Cultivar Tommy Atkins and Keitt). *Antioxidants*, 11, 3, 484, 2022.
56. Mesquita, P.C., Rodrigues, L.G.G., Mazzutti, S., da Silva, M., Vitali, L., Lanza, M., Intensified green-based extraction process as a circular economy approach to recover bioactive compounds from soursop seeds (Annona muricata L.). *Food Chem. X*, 12, 100164, 2021.
57. Tacias-Pascacio, V.G. *et al.*, Aqueous extraction of seed oil from mamey sapote (Pouteria sapota) after viscozyme l treatment. *Catalysts*, 11, 6, 748, 2021.
58. Tacias-Pascacio, V.G., Morellon-Sterling, R., Siar, E.H., Tavano, O., Berenguer-Murcia, Á., Fernandez-Lafuente, R., Use of Alcalase in the production of bioactive peptides: A review. *Int. J. Biol. Macromol.*, 165, 2143–2196, 2020.
59. Sun, X., Acquah, C., Aluko, R.E., Udenigwe, C.C., Considering food matrix and gastrointestinal effects in enhancing bioactive peptide absorption and bioavailability. *J. Funct. Foods*, 64, 103680, 2020.
60. Tacias-Pascacio, V.G. *et al.*, Bioactive peptides from fisheries residues: A review of use of papain in proteolysis reactions. *Int. J. Biol. Macromol.*, 184, 415–428, 2021.
61. Castañeda-Valbuena, D., Berenguer-Murcia, Á., Fernandez-Lafuente, R., Morellon-Sterling, R., Tacias-Pascacio, V.G., Biological activities of peptides obtained by pepsin hydrolysis of fishery products. *Process Biochem.*, 120, 53–63, 2022.
62. Parniakov, O., Barba, F.J., Grimi, N., Lebovka, N., Vorobiev, E., Extraction assisted by pulsed electric energy as a potential tool for green and sustainable recovery of nutritionally valuable compounds from mango peels. *Food Chem.*, 192, 842–848, 2016.
63. Parniakov, O., Roselló-Soto, E., Barba, F.J., Grimi, N., Lebovka, N., Vorobiev, E., New approaches for the effective valorization of papaya seeds: Extraction of proteins, phenolic compounds, carbohydrates, and isothiocyanates assisted by pulsed electric energy. *Food Res. Int.*, 77, 711–717, 2015.

64. Vaidya, A.M., Saxena, D.C., Somani, V.B., Response surface methodology for optimization of protein isolate preparation from custard apple seed flour (Annona squmosa). *Int. J. Curr. Microbiol. Appl. Sci.*, 6, 9, 3079–3089, 2017.
65. Azmir, J. *et al.*, Techniques for extraction of bioactive compounds from plant materials: A review. *J. Food Eng.*, 117, 4, 426–436, 2013.
66. Díaz-Suárez, P. *et al.*, Aqueous enzymatic extraction of Ricinus communis seeds oil using Viscozyme L. *Ind. Crops Prod.*, 170, 113811, 2021.
67. Claux, O., Santerre, C., Abert-Vian, M., Touboul, D., Vallet, N., Chemat, F., Alternative and sustainable solvents for green analytical chemistry. *Curr. Opin. Green Sustain. Chem.*, 31, 100510, Oct. 2021.
68. Enríquez-Valencia, S.A., Ayala-Zavala, J.F., González-Aguilar, G.A., López-Martínez, L.X., Valorization of industrial by-products and waste from tropical fruits for the recovery of bioactive compounds, recent advances, and future perspectives, in: *Valorization of Agri-Food Wastes and By-Products: Recent Trends, Innovations and Sustainability Challenges*, R. Bhat (Ed.), pp. 29–46, Elsevier B.V, London, UK, 2021.
69. Chemat, F. *et al.*, A review of sustainable and intensified techniques for extraction of food and natural products. *Green Chem.*, 22, 8, 2325–2353, 2020.
70. Srivastava, N. *et al.*, Chapter 21 - Advances in extraction technologies: Isolation and purification of bioactive compounds from biological materials, in: *Natural Bioactive Compounds. Technological Advancements*, R. p. Sinha and D. Häder, (Eds.), pp. 409–433, Elsevier B.V, London, UK, 2021.
71. Soquetta, M.B., de M. Terra, L., Bastos, C.P., Green technologies for the extraction of bioactive compounds in fruits and vegetables. *CYTA - J. Food*, 16, 1, 400–412, 2018.
72. Jha, A.K. and Sit, N., Extraction of bioactive compounds from plant materials using combination of various novel methods: A review. *Trends Food Sci. Technol.*, 119, 579–591, 2022.
73. von Soxhlet, F., Die gewichtsanalytische bestimmung des milchfettes. *Dingler's Polytech. J.*, 232, 461–465, 1879.
74. Daud, N.M. *et al.*, Valorisation of plant seed as natural bioactive compounds by various extraction methods: A review. *Trends Food Sci. Technol.*, 119, 201–214, 2022.
75. Hashemi, B., Shiri, F., Švec, F., Nováková, L., Green solvents and approaches recently applied for extraction of natural bioactive compounds. *TrAC Trends Anal. Chem.*, 157, 116732, 2022.
76. Suresh, A., Velusamy, S., Ayyasamy, S., Rathinasamy, M., Techniques for essential oil extraction from kaffir lime and its application in health care products—A review. *Flavour Fragr. J.*, 36, 1, 5–21, 2021.

77. Rassem, H.H.A., Nour, A.H., Yunus, R.M., Techniques for extraction of essential oils from plants: A review. *Aust. J. Basic Appl. Sci.*, 10, 16, 117–127, 2016.
78. El Asbahani, A. *et al.*, Essential oils: From extraction to encapsulation. *Int. J. Pharm.*, 483, 1–2, 220–243, 2015.
79. Petigny, L., Périno, S., Minuti, M., Visinoni, F., Wajsman, J., Chemat, F., Simultaneous microwave extraction and separation of volatile and non-volatile organic compounds of boldo leaves. From lab to industrial scale. *Int. J. Mol. Sci.*, 15, 5, 7183–7198, Apr. 2014.
80. Garcia-Mendoza, M.P., Paula, J.T., Paviani, L.C., Cabral, F.A., Martinez-Correa, H.A., Extracts from mango peel by-product obtained by supercritical CO2 and pressurized solvent processes. *LWT - Food Sci. Technol.*, 62, 1, 131–137, 2015.
81. Granone, L.I., Hegel, P.E., Pereda, S., Citrus fruit processing by pressure intensified technologies: A review. *J. Supercrit. Fluids*, 188, 105646, 2022.
82. Shang, A., Luo, M., Gan, R.-Y., Li, B.-Y., Li, H.-Y., Li, H.-B., Extraction and assessment methods as well as resources of natural antioxidants in foods and herbs, in: *Plant Antioxidants and Health*, H.M. Ekiert, K.G. Ramawat, J. Arora (Eds.), pp. 679–707, Springer International Publishing, Cham, 2022.
83. Yılmaz, F.M., Görgüç, A., Gençdağ, E., Recovery and purification of antioxidant compounds from plant origin agro-industrial by-products, in: *Plant Antioxidants and Health*, H.M. Ekiert, K.G. Ramawat, J. Arora (Eds.), pp. 775–797, Springer International Publishing, Cham, 2022.
84. Dey, D., Richter, J.K., Ek, P., Gu, B.-J., Ganjyal, G.M., Utilization of food processing by-products in extrusion processing: A review. *Front. Sustain. Food Syst.*, 4, 2020, 2021.
85. Delvar, A., de Caro, P., Candy, L., Caro, Y., Cheong Sing, A.S., Raynaud, C., Integrated process for extraction and formulation in emulsions of active molecules from fresh passion fruits (Passiflora edulis Sims). *J. Food Eng.*, 263, 388–397, 2019.
86. Morya, S., Menaa, F., Jiménez-López, C., Lourenço-Lopes, C., BinMowyna, M.N., Alqahtani, A., Nutraceutical and pharmaceutical behavior of bioactive compounds of miracle oilseeds: An overview. *Foods*, 11, 13, 1824, 2022.
87. Sadeghi, A., Hakimzadeh, V., Karimifar, B., Microwave assisted extraction of bioactive compounds from food: A review. *Int. J. Food Sci. Nutr. Eng*, 7, 1, 19–27, 2017.
88. More, P.R. and Arya, S.S., Intensification of bio-actives extraction from pomegranate peel using pulsed ultrasound: Effect of factors, correlation, optimization and antioxidant bioactivities. *Ultrason. Sonochem.*, 72, 105423, Apr. 2021.
89. Tacias-Pascacio, V.G. *et al.*, Phenolic compounds in mango fruit: A review. *J. Food Meas. Charact.*, 16, 1, 619–636, 2022.

90. Khadhraoui, B. *et al.*, Histo-cytochemistry and scanning electron microscopy for studying spatial and temporal extraction of metabolites induced by ultrasound. Towards chain detexturation mechanism. *Ultrason. Sonochem.*, 42, November 2017, 482–492, 2018.
91. Castañeda-Valbuena, D., Ayora-Talavera, T., Luján-Hidalgo, C., Álvarez-Gutiérrez, P., Martínez-Galero, N., Meza-Gordillo, R., Ultrasound extraction conditions effect on antioxidant capacity of mango by-product extracts. *Food Bioprod. Process.*, 127, 212–224, 2021.
92. Kumar, K., Srivastav, S., Sharanagat, V.S., Ultrasound assisted extraction (UAE) of bioactive compounds from fruit and vegetable processing by-products: A review. *Ultrason. Sonochem.*, 70, August 2020, 105325, 2021.
93. Wen, C. *et al.*, Advances in ultrasound assisted extraction of bioactive compounds from cash crops – A review. *Ultrason. Sonochem.*, 48, 538–549, 2018.
94. Medina-Torres, N., Ayora-Talavera, T., Espinosa-Andrews, H., Sánchez-Contreras, A., Pacheco, N., Ultrasound assisted extraction for the recovery of phenolic compounds from vegetable sources. *Agronomy*, 7, 3, 47, 2017.
95. Chemat, F., Rombaut, N., Sicaire, A.G., Meullemiestre, A., Fabiano-Tixier, A.S., Abert-Vian, M., Ultrasound assisted extraction of food and natural products. Mechanisms, techniques, combinations, protocols and applications. A review. *Ultrason. Sonochem.*, 34, 540–560, 2017.
96. Rodríguez De Luna, S.L., Ramírez-Garza, R.E., Serna Saldívar, S.O., Environmentally friendly methods for flavonoid extraction from plant material: Impact of their operating conditions on yield and antioxidant properties. *Sci. World J.*, 2020, 6792069, 2020.
97. Monteiro, M. *et al.*, Effect of extraction method and solvent system on the phenolic content and antioxidant activity of selected macro- and microalgae extracts. *J. Appl. Phycol.*, 32, 1, 349–362, 2020.
98. Vinatoru, M., Mason, T.J., Calinescu, I., Ultrasonically assisted extraction (UAE) and microwave assisted extraction (MAE) of functional compounds from plant materials. *TrAC - Trends Anal. Chem.*, 97, 159–178, 2017.
99. Gil-Martín, E., Forbes-Hernández, T., Romero, A., Cianciosi, D., Giampieri, F., Battino, M., Influence of the extraction method on the recovery of bioactive phenolic compounds from food industry by-products. *Food Chem.*, 378, 131918, 2022.
100. Sarkis, J.R., Boussetta, N., Blouet, C., Tessaro, I.C., Marczak, L.D.F., Vorobiev, E., Effect of pulsed electric fields and high voltage electrical discharges on polyphenol and protein extraction from sesame cake. *Innov. Food Sci. Emerg. Technol.*, 29, 170–177, 2015.
101. Barba, F. J. *et al.*, Emerging extraction, in: *Food Waste Recovery. Processing Technologies and Industrial Techniques*, C. M. Galanakis, (Ed.), pp. 249–272, Elsevier B.V, London, UK, 2015.

102. Carpentieri, S., Soltanipour, F., Ferrari, G., Pataro, G., Donsì, F., Emerging green techniques for the extraction of antioxidants from agri-food by-products as promising ingredients for the food industry. *Antioxidants*, 10, 9, 1417, 2021.
103. Clarke, C.J., Tu, W.-C., Levers, O., Bröhl, A., Hallett, J.P., Green and sustainable solvents in chemical processes. *Chem. Rev.*, 118, 2, 747–800, Jan. 2018.
104. Smith, E.L., Abbott, A.P., Ryder, K.S., Deep eutectic solvents (DESs) and their applications. *Chem. Rev.*, 114, 21, 11060–11082, Nov. 2014.
105. Zhang, Q., De Oliveira Vigier, K., Royer, S., Jérôme, F., Deep eutectic solvents: Syntheses, properties and applications. *Chem. Soc. Rev.*, 41, 21, 7108–7146, 2012.
106. Hashemi, B., Zohrabi, P., Dehdashtian, S., Application of green solvents as sorbent modifiers in sorptive-based extraction techniques for extraction of environmental pollutants. *TrAC Trends Anal. Chem.*, 109, 50–61, 2018.
107. Qadir, R., Anwar, F., Batool, F., Mushtaq, M., Jabbar, A., Enzyme-assisted extraction of Momordica balsamina L. fruit phenolics: Process optimized by response surface methodology. *J. Food Meas. Charact.*, 13, 1, 697–706, 2019.
108. Larios-Cruz, R. *et al.*, Extraction of bioactive molecules through fermentation and enzymatic assisted technologies, in: *High Value Fermentation Products*, pp. 27–59, Mar. 2019.
109. Gligor, O., Mocan, A., Moldovan, C., Locatelli, M., Crişan, G., Ferreira, I.C.F.R., Enzyme-assisted extractions of polyphenols – A comprehensive review. *Trends Food Sci. Technol.*, 88, 302–315, 2019.
110. Puri, M., Sharma, D., Barrow, C.J., Enzyme-assisted extraction of bioactives from plants. *Trends Biotechnol.*, 30, 1, 37–44, 2012.
111. Mohan, K. *et al.*, Recent insights into the extraction, characterization, and bioactivities of chitin and chitosan from insects. *Trends Food Sci. Technol.*, 105, September, 17–42, 2020.
112. Domínguez-Rodríguez, G., Marina, M.L., Plaza, M., Rapid fingerprinting of extractable and non-extractable polyphenols from tropical fruit peels using direct analysis in real time coupled to orbitrap mass spectrometry. *Food Chem.*, 371, 131191, 2022.
113. Fatima, Z., Jin, X., Zou, Y., Kaw, H.Y., Quinto, M., Li, D., Recent trends in analytical methods for water-soluble vitamins. *J. Chromatogr. A*, 1606, 360245, 2019.
114. Sosalagere, C., Adesegun Kehinde, B., Sharma, P., Isolation and functionalities of bioactive peptides from fruits and vegetables: A reviews. *Food Chem.*, 366, 130494, 2022.
115. Kazaz, S., Miray, R., Lepiniec, L., Baud, S., Plant monounsaturated fatty acids: Diversity, biosynthesis, functions and uses. *Prog. Lipid Res.*, 85, 101138, November 2021, 2022.

116. Singleton, V.L., Orthofer, R., Lamuela-Raventós, R.M., Analysis of total phenols and other oxidation substrates and antioxidants by means of folin-ciocalteu reagent. *Methods Enzymol.*, 299, 1974, 152–178, 1999.
117. Gentile, C., Giuseppe, M., Palazzolo, E., Gianguzzi, G., Perrone, A., Serio, G., Farina, V., Pomological, sensorial, nutritional and nutraceutical profile of seven cultivars of cherimoya (Annona cherimola Mill). *Foods*, 10, 1, 35, 2021.
118. Seabra Pires, F. C. *et al.*, Bioactive compounds and evaluation of antioxidant, cytotoxic and cytoprotective effects of murici pulp extracts (Byrsonima crassifolia) obtained by supercritical extraction in HepG2 cells treated with H2O2. *Foods*, 10, 737, 2021.
119. Mannino, G., Carla, G., Porcu, A., Agliassa, C., Caradonna, F., Bertea, C.M., Chemical profile and biological activity of cherimoya (Annona cherimola Mill.) and atemoya (Annona atemoya) leaves. *Molecules*, 25, 11, 2612, 2020.
120. Hartati, R., Nadifan, H.I., Fidrianny, I., Crystal guava (*Psidium guajava* L. 'Crystal'): Evaluation of *in vitro* antioxidant capacities and phytochemical content. *Sci. World J.*, 2020, 9413727, 2020.
121. Khadam, S. *et al.*, Phytochemical screening and bioactivity assessment of leaves and fruits extract of Carica papaya. *Pak. J. Pharm. Sci.*, 32, 5, 1941–1948, Sep. 2019.
122. Pacheco-Ordaz, R., Wall-Medrano, A., Goñi, M.G., Ramos-Clamont-Montfort, G., Ayala-Zavala, J.F., González-Aguilar, G.A., Effect of phenolic compounds on the growth of selected probiotic and pathogenic bacteria. *Lett. Appl. Microbiol.*, 66, 1, 25–31, Jan. 2018.
123. Vu, H.T., Scarlett, C.J., Vuong, Q.V., Optimization of ultrasound-assisted extraction conditions for recovery of phenolic compounds and antioxidant capacity from banana (Musa cavendish) peel. *J. Food Process. Preserv.*, 41, 5, e13148, Oct. 2017.
124. Toh, P.Y., Leong, F.S., Chang, S.K., Khoo, H.E., Yim, H.S., Optimization of extraction parameters on the antioxidant properties of banana waste. *Acta Sci. Pol. Technol. Aliment.*, 15, 1, 65–78, 2016.
125. Gondi, M., Basha, S.A., Bhaskar, J.J., Salimath, P.V., Prasada Rao, U.J.S., Anti-diabetic effect of dietary mango (Mangifera indica L.) peel in streptozotocin-induced diabetic rats. *J. Sci. Food Agric.*, 95, 5, 991–999, 2015.
126. Saravanan, K. and Aradhya, S.M., Potential nutraceutical food beverage with antioxidant properties from banana plant bio-waste (pseudostem and rhizome). *Food Funct.*, 2, 10, 603–610, 2011.
127. Ajila, C.M. and Prasada Rao, U.J.S., Protection against hydrogen peroxide induced oxidative damage in rat erythrocytes by Mangifera indica L. peel extract. *Food Chem. Toxicol.*, 46, 1, 303–309, 2008.

128. Al Kazman, B.S.M., Joanna, E., Hanrahan, J.R., Traditional uses, phytochemistry and pharmacological activities of annonacae. *Molecules*, 27, 11, 3462, 2022.
129. El-Sawi, S.A., Ibrahim, M.E., Sleem, A.A., Farghaly, A.A., Awad, G.E.A., Merghany, R.M., Development of alternative medicinal sources from golden berry, bananas and carrot wastes as antioxidant, cytotoxic and antimicrobial agents. *Acta Ecol. Sin.*, 42, 3, 224–232, 2022.
130. Lo Galbo, V., Marianna, L., Giuliano, M., Emanuele, S., Carlisi, D., Calvaruso, G., De Blasio, A., Di Liberto, D., D'Anneo, A., Redox imbalance and mitochondrial release of apoptogenic factors at the forefront of the antitumor action of mango peel extract. *Molecules*, 26, 14, 4328, 2021.
131. Pinto, N.C.C. *et al.*, Cytotoxicity and bacterial membrane destabilization induced by Annona squamosa L. extracts. *An. Acad. Bras. Cienc.*, 89, 3 Suppl, 2053–2073, 2017.
132. Sufian, A.S., Ramasamy, K., Ahmat, N., Zakaria, Z.A., Yusof, M.I.M., Isolation and identification of antibacterial and cytotoxic compounds from the leaves of Muntingia calabura L. *J. Ethnopharmacol.*, 146, 1, 198–204, 2013.
133. Otsuki, N., Dang, N.H., Kumagai, E., Kondo, A., Iwata, S., Morimoto, C., Aqueous extract of Carica papaya leaves exhibits anti-tumor activity and immunomodulatory effects. *J. Ethnopharmacol.*, 127, 3, 760–767, 2010.
134. Luo, Y., Bin, P., Wei, W., Tian, X., Wu, Z., Antioxidant and anti-diabetic activities of polysaccharides from guava leaves. *Molecules*, 24, 7, 1343, 2019.
135. Sandoval-Gallegos, E.M. *et al.*, *In vitro* bioaccessibility and effect of Mangifera indica (Ataulfo) leaf extract on induced dyslipidemia. *J. Med. Food*, 21, 1, 47–56, Aug. 2017.
136. Nemec, M.J. *et al.*, Polyphenolics from mango (Mangifera indica L.) suppress breast cancer ductal carcinoma *in situ* proliferation through activation of AMPK pathway and suppression of mTOR in athymic nude mice. *J. Nutr. Biochem.*, 41, 12–19, 2017.
137. Ramírez, N.M. *et al.*, Anti-obesity effects of tea from Mangifera indica L. leaves of the Ubá variety in high-fat diet-induced obese rats. *Biomed. Pharmacother.*, 91, 938–945, 2017.
138. Roslida, A.H., Tay, C.E., Zuraini, A., Chan, P.F., Anti-inflammatory and anti-nociceptive activities of the ethanolic extract of Annona muricata leaf. *J. Nat. Remedies*, 10, 2, 97–104, 2010.
139. De Sousa, O. V., Vieira, G. D.-V., De Pinho, J. de J. R. G., Yamamoto, C. H., Alves, M. S., Antinociceptive and anti-inflammatory activities of the ethanol extract of Annona muricata L. leaves in animal models. *Int. J. Mol. Sci.*, 11, 5, 2067–2078, 2010.
140. Juarez, V.M., Montalbine, A.N., Singh, A., Microbiome as an immune regulator in health, disease, and therapeutics. *Adv. Drug Deliv. Rev.*, 188, 114400, 2022.

141. Santos, D., Frota, E.G., Vargas, B.K., Tonieto Gris, C.C., dos Santos, L.F., Bertolin, T.E., What is the role of phenolic compounds of yerba mate (Ilex paraguariensis) in gut microbiota? *Phytochemistry*, 203, February, 113341, 2022.
142. Sachdeva, H., Khaturia, S., Chahar, M., Nutraceuticals as therapeutic agents for prevention and treatment of diseases, in: *Prebiotics, Probiotics and Nutraceuticals*, K.K. Behera, R. Bist, S. Mohanty, M. Bhattacharya (Eds.), pp. 45–68, Springer Nature Singapore, Singapore, 2022.
143. Gómez-García, R., Campos, D.A., Aguilar, C.N., Madureira, A.R., Pintado, M., Valorization of melon fruit (Cucumis melo L.) by-products: Phytochemical and biofunctional properties with emphasis on recent trends and advances. *Trends Food Sci. Technol.*, 99, 507–519, 2020.
144. Shofian, N.M. *et al.*, Effect of freeze-drying on the antioxidant compounds and antioxidant activity of selected tropical fruits. *Int. J. Mol. Sci.*, 12, 7, 4678–4692, 2011.
145. Xiong, X. *et al.*, Value-added chemicals from food supply chain wastes: State-of-the-art review and future prospects. *Chem. Eng. J.*, 375, 121983, 2019.
146. Baiano, A.R., Recovery of biomolecules from food wastes — A review. *Molecules*, 19, 9, 14821–14842, 2014.
147. Kumar, K., Yadav, A.N., Kumar, V., Vyas, P., Dhaliwal, H.S., Food waste: A potential bioresource for extraction of nutraceuticals and bioactive compounds. *Bioresour. Bioprocess.*, 4, 1, 18, 2017.
148. Varzakas, T., George, Z., Verpoort, F., Plant food residues as a source of nutraceuticals and functional foods. *Foods*, 5, 4, 88, 2016.
149. Windarsih, G.U.T., Characterization of leaf morphology on several rambutan (Nephelium lappaceum) cultivars from Serang City, Banten, Indonesia, in: *Prosiding Seminar Nasional Masyarakat Biodiversitas Indonesia*, vol. 8, no. 1, pp. 18–23, 2022.
150. Bhattacharjee, P., Das, S., Das, S.K., Chander, S., Rambutan (Nephelium lappaceum L.): A potential fruit for industrial use, serving nutraceutical, livelihood interests and enhancing climate resilience. *S. Afr. J. Bot.*, 150, 26–33, 2022.
151. Phuong, N.N.M., Le, T.T., Van Camp, J., Raes, K., Evaluation of antimicrobial activity of rambutan (Nephelium lappaceum L.) peel extracts. *Int. J. Food Microbiol.*, 321, 108539, 2020.
152. Manaharan, T., Palanisamy, U.D., Ming, C.H., Tropical plant extracts as potential antihyperglycemic agents. *Molecules*, 17, 5, 5915–5923, May 2012.
153. Muhtadi, M., Haryoto, H., Sujono, T.A., Suhendi, A., Antidiabetic and antihypercholesterolemia activities of rambutan (Nephelium lappaceum L.) and durian (Durio zibethinus Murr.) fruit peel extracts. *J. Appl. Pharm. Sci.*, 6, 4, 190–194, 2016.

154. Thitilertdecha, N., Teerawutgulrag, A., Kilburn, J.D., Rakariyatham, N., Identification of major phenolic compounds from Nephelium lappaceum L. and their antioxidant activities. *Molecules*, 15, 3, 1453–1465, Mar. 2010.
155. Lestari, S.R., Djati, M.S., Rudijanto, A., Fatchiyah, F., The physiological response of obese rat model with rambutan peel extract treatment. *Asian Pac. J. Trop. Dis.*, 4, S780–S785, 2014.
156. Sipahutar, K.V., Nugraha, Y., Fauziah, C., Effect of rambutan fruit peel extract on total sperm counts of wistar rats with obesity. *Biota Biol. dan Pendidik. Biol.*, 13, 1, 21–29, 2020.
157. Rashid, U., Rehman, H.A., Hussain, I., Ibrahim, M., Haider, M.S., Muskmelon (Cucumis melo) seed oil: A potential non-food oil source for biodiesel production. *Energy*, 36, 9, 5632–5639, 2011.
158. Fundo, J.F., Miller, F.A., Garcia, E., Santos, J.R., Silva, C.L.M., Brandão, T.R.S., Physicochemical characteristics, bioactive compounds and antioxidant activity in juice, pulp, peel and seeds of Cantaloupe melon. *J. Food Meas. Charact.*, 12, 1, 292–300, 2018.
159. Silva, M.A., Albuquerque, T.G., Alves, R.C., Oliveira, M.B.P.P., Costa, H.S., Melon (Cucumis melo L.) by-products: Potential food ingredients for novel functional foods? *Trends Food Sci. Technol.*, 98, 181–189, 2020.
160. Veiga, M., Costa, E.M., Silva, S., Pintado, M., Impact of plant extracts upon human health: A review. *Crit. Rev. Food Sci. Nutr.*, 60, 5, 873–886, Mar. 2020.
161. Sulaiman, S.F. and Ooi, K.L., Antioxidant and α-glucosidase inhibitory activities of 40 tropical juices from Malaysia and identification of phenolics from the bioactive fruit juices of Barringtonia racemosa and Phyllanthus acidus. *J. Agric. Food Chem.*, 62, 39, 9576–9585, Oct. 2014.
162. Gómez-Maqueo, A., Zamantha, E.A., Welti-Chanes, J., phenolic compounds in Mesoamerican fruits-characterization, health potential and processing with innovative technologies. *Int. J. Mol. Sci.*, 21, 21, 8357, 2020.
163. Ancona, J. J., Montañez Escalante, P. I., Ek Rodríguez, I. L., Ruenes Morales, M. del R., *Los frutales abandonados y subutilizados en la Península de Yucatán*. México: Grafica Peninsular, 2015. [Online]. Available: https://patrimoniobiocultural.com/producto/los-frutales-abandonados-y-subutilizados-en-la-peninsula-de-yucatan/
164. Ramírez-Briones, E. *et al.*, Fruits of wild and semi-domesticated Diospyros tree species have contrasting phenological, metabolic, and antioxidant activity profiles. *J. Sci. Food Agric.*, 99, 13, 6020–6031, Oct. 2019.
165. Lim, T.K., Introduction, in: *Edible Medicinal and Non-Medicinal Plants*, vol. 1, L.T.K. Fruits, (Ed.), pp. 1–4, Springer Netherlands, Dordrecht, 2012.
166. Jiménez-González, O. and Guerrero-Beltrán, J. Á., Diospyros digyna (black sapote), an undervalued fruit: A review. *ACS Food Sci. Technol.*, 1, 1, 3–11, Feb. 2021.

167. Arellano Gómez, L. A., Saucedo Veloz, C., Arévalo Galarza, L., Cambios bioquímicos y fisiológicos durante la maduración de frutos de zapote negro (Diospyros digyna Jacq.). *Agrociencia*, 39, 2, 173–181, Sep. 2005.
168. Mannino, G., Serio, G., Bertea, C.M., Chiarelli, R., Lauria, A., Gentile, C., Phytochemical profile and antioxidant properties of the edible and non-edible portions of black sapote (Diospyros digyna Jacq.). *Food Chem.*, 380, 132137, 2022.
169. Correia, R.T.P., Borges, K.C., Medeiros, M.F., Genovese, M.I., Bioactive compounds and phenolic-linked functionality of powdered tropical fruit residues. *Food Sci. Technol. Int.*, 18, 6, 539–547, Sep. 2012.
170. Olaniran, A.F., Taiwo, A.E., Bamidele, O.P., Iranloye, Y.M., Malomo, A.A., Olaniran, O.D., The role of nutraceutical fruit drink on neurodegenerative diseases: A review. *Int. J. Food Sci. Technol.*, 57, 3, 1442–1450, Mar. 2022.
171. Ooi, T.C. *et al.*, Neuroprotection of tropical fruit juice mixture via the reduction of iNOS expression and CRH level in β-amyloid-induced rats model of Alzheimer's disease. *Evid.-Based Complement. Alternat. Med.*, 2020, 5126457, 2020.
172. Fidelis, M. *et al.*, Camu-camu seed (Myrciaria dubia) – from side stream to anantioxidant, antihyperglycemic, antiproliferative, antimicrobial, antihemolytic, anti-inflammatory, and antihypertensive ingredient. *Food Chem.*, 310, 125909, 2020.
173. John, O. D., Mouatt, P., Majzoub, M. E., Thomas, T., Panchal, S. K., Brown, L., Physiological and metabolic effects of yellow mangosteen (Garcinia dulcis) rind in rats with diet-induced metabolic syndrome. *Int. J. Mol. Sci.*, 21, 272, 2020.
174. Murthy, H.N., Dayanand, D., Dewir, Y.H., Ibrahim, A., Phytochemicals and biological activities of garcinia morella (gaertn.) desr.: A review. *Molecules*, 25, 23, 2020.
175. Adefegha, S.A., Oyeleye, S.I., Oboh, G., Distribution of phenolic contents, antidiabetic potentials, antihypertensive properties, and antioxidative effects of soursop (*Annona muricata* L.) fruit parts *in vitro*. *Biochem. Res. Int.*, 2015, 347673, 2015.
176. Riya, M.P., Antu, K.A., Vinu, T., Chandrakanth, K.C., Anilkumar, K.S., Raghu, K.G., An *in vitro* study reveals nutraceutical properties of Ananas comosus (L.) Merr. var. Mauritius fruit residue beneficial to diabetes. *J. Sci. Food Agric.*, 94, 5, 943–950, Mar. 2014.
177. Mirza, B. *et al.*, Mango (Mangifera indica L.): A magnificent plant with cancer preventive and anticancer therapeutic potential. *Crit. Rev. Food Sci. Nutr.*, 61, 13, 2125–2151, Jul. 2021.

178. Amaro, A.L., Oliveira, A., Almeida, D.P.F., Biologically active compounds in melon: Modulation by preharvest, post-harvest, and processing factors, in: *Processing and Impact on Active Components in Food*, V. Preedy (Ed.), pp. 165–171, Elsevier B.V, London, UK, 2015.
179. Chan, K.T., Li, K., Liu, S.L., Chu, K.H., Toh, M., Xie, W.D., Cucurbitacin B inhibits STAT3 and the Raf/MEK/ERK pathway in leukemia cell line K562. *Cancer Lett.*, 289, 1, 46–52, 2010.
180. Aryee, A.N.A. and Boye, J.I., Current and emerging trends in the formulation and manufacture of nutraceuticals and functional food products, in: *Nutraceutical and Functional Food Processing Technology*, pp. 1–63, 2014.

2

Prebiotic Properties of Fruit By-Products and Their Metabolic Benefits

Maria Margareth Veloso Naves[1]*, Aline Medeiros Alves-Santos[1], Clara Sandra Araújo Sugizaki[1] and Rafaela G. Feresin[2]

[1]*School of Nutrition, Federal University of Goiás (UFG), Goiânia, Brazil*
[2]*Department of Nutrition, Georgia State University (GSU), Atlanta, USA*

Abstract
Prebiotics, such as soluble fibers, and prebiotic candidates, such as polyphenols, may prevent or control metabolic disorders by promoting gut health. Prebiotics are fermented by beneficial bacteria in the colon, producing metabolites that can act locally to maintain and improve gut health and immune function, and elicit systemic benefits to the host health, including antioxidant and anti-inflammatory effects. Agro-food by-products or residues from fruit processing are rich sources of prebiotic compounds and can stimulate growth and metabolite production of beneficial bacteria in human gut microbiota. *In vitro* and *in vivo* prebiotic activity of fruit by-products like acerola, apple, cashew apple, guava, grape, jaboticaba, mango, melon, orange, passion fruit, and pineapple has been demonstrated. Although animal and clinical studies on the prebiotic effects of fruit processing wastes are warranted, their use by agro-industry as potential prebiotic ingredients for novel functional foods, nutraceuticals and cosmeceuticals may be recommended.

Keywords: Fruit residues, dietary fibers, phenolic compounds, hindgut microbiota, gut health, prebiotic activity

2.1 Introduction

The fruit by-products or residues from agro-industry are being explored as potential ingredients for formulating new functional food products and nutraceuticals [1]. In recent years, the prebiotic activity of by-products from

**Corresponding author*: maria_margareth_veloso@ufg.br

Umile Gianfranco Spizzirri (ed.) Nutraceutics from Agri-Food By-Products, (45–68) © 2023 Scrivener Publishing LLC

fruits like acerola, apple, cashew apple, grape, guava, jaboticaba, mango, melon, orange, passion fruit, and pineapple have been investigated. The most evaluated fruit by-products are peel, shell (external mesocarp plus peel), pomace, seed, and skin. These fractions are rich sources of nutrients and bioactive compounds, which confer to them high antioxidant capacity and prebiotic potential [2, 3].

Prebiotics are substrates selectively used by hindgut microorganisms generating health benefits to the host, as described in the expert consensus document published by the International Scientific Association for Probiotics and Prebiotics (ISAPP) [4]. Some dietary fibers fit the prebiotic concept, and, thus, may be classified as prebiotics. Currently, only the resistant oligosaccharides—fructans, such as fructooligosaccharides (FOS), oligofructose, and inulin; and galactans—are well documented in literature as prebiotics. Other fibers are candidates to prebiotics or have prebiotic potential [5], as well as some phenolic compounds, such as catechins, anthocyanins, and proanthocyanidins [6]. Prebiotics are fermented by beneficial bacteria in the colon, producing metabolites like short-chain fatty acids (SCFA), which can act locally to maintain and improve the gut epithelium homeostasis, intestinal barrier, and immune function, besides generating systemic benefits to the host health, including antioxidant and anti-inflammatory effects [4, 7].

The fruit by-products from agro-industry, such as peel, shell, pomace, and skin have high content of dietary fibers and phenolic compounds. Further, the concentration of dietary fiber and phenolics is usually higher in fruit by-products than in the edible fraction of the fruit [2]. The main fiber types found in fruit by-products are non-starch polysaccharides (cellulose, hemicellulose, pectin, and gums), and lignin [3, 5]. Flavonoids, tannins, and phenolic acids are the major classes of phenolic compounds found in fruit by-products [8].

The high density of bioactive compounds, the increasing demand for natural products with high nutritional and therapeutic potential, and the economic, social, and environmental impact generated by losses in the fruit processing industries, have stimulated research on by-products [9]. In this context, the prebiotic potential of fruit by-products has been investigated, mainly in *in vitro* and animal studies. In *in vitro* studies in monoculture assays with probiotics, fruit by-products promoted the growth of *Lactobacillus* and *Bifidobacterium* strains and suppressed the growth of pathogenic bacteria. In addition to favorable modulation of gut microorganisms, some *in vitro* studies reported increases in SCFA production, and a prebiotic potential of fruit by-products similar to that of FOS, a standard prebiotic, in jaboticaba pomace [10], and even superior to FOS, in orange albedo [11]. In rat and mouse studies, the prebiotic potential of some fruit by-products like acerola, cashew apple, and peach have been shown to increase abundance and diversity of gut beneficial bacteria [12, 13]; however, human clinical trials on the prebiotic effects of fruit by-products are lacking.

Even though there are prebiotics available in the market, as FOS and inulin, the search for novel food matrices abundant in beneficial microbiota modulator compounds is of great interest. Agro-food by-products present economic, social, and environmental advantages [9], and have broad application as potential prebiotic ingredients in the development of novel food products, nutraceuticals, and cosmeceuticals.

2.2 Prebiotic Compounds

2.2.1 Dietary Fiber

Dietary fibers are complex and heterogeneous substances, which resist hydrolysis by digestive enzymes and absorption in the mammalian small intestine, despite some of them being fermented by hindgut bacteria [5]. They are grouped into subgroups according to their chemical structure, physicochemical properties, and degree of polymerization. The usual classification includes four subgroups: resistant oligosaccharides, non-starch polysaccharides, resistant starches, and associated substances or non-carbohydrates [14].

The subgroup of resistant oligosaccharides (with 3–9 monomeric units) includes galactosides (raffinose, stachyose, verbascose); FOS; galactans like galactooligossacharides (GOS), transgalactooligosaccharides (TOS), and mannanoligosaccharides (MOS); and xylooligosaccharides (XOS). The main non-starch polysaccharides (with 10 or more monomeric units) are cellulose and hemicellulose, pectins, gums and mucilages, inulin, polydextrose, and resistant dextrins. Resistant starches are generally classified into five types: types 1 and 2 are naturally resistant to digestion in the small intestine, and types 3, 4, and 5 become resistant to digestion after heat or chemical treatment. The subgroup of associated substances includes lignin, chitin, and wax, and they are associated to polysaccharides, which have structural functions in foods [5]. Fruits, along with vegetables and grains, are the main sources of dietary fiber, which is present in higher concentrations in the fruit by-product than in the edible fraction [2]. The main fibers found in fruit by-products are cellulose, hemicellulose, pectin, gums, and lignin [3].

Prebiotics are substrates that can be metabolized by the gut microbiota and selectively stimulate the growth or activity of health-promoting gut bacteria [15]. Therefore, dietary fibers or other compounds can be classified as prebiotics if they are metabolized by beneficial bacteria present in the hindgut microbiota, but this property must be confirmed in well-designed clinical trials. Rezende *et al.* [5] highlighted that the term "dietary fiber" cannot be used as a synonym for "prebiotic compound," as not all fibers are prebiotics, and proposed a classification of fiber by prebiotic categories, as following: prebiotic dietary fibers, dietary fibers candidate to prebiotic, and dietary fibers not recognized as prebiotic. Currently, only fructans (FOS, oligofructose, and inulin) and galactans are well documented in the literature as prebiotics. Other fermentable carbohydrates are considered candidates to prebiotic based on *in vitro* and preclinical studies, such as MOS, XOS, galactosides, polydextrose and pectins. In turn, there are dietary fibers such as resistant starches 1 and 5, cellulose and those of the subgroup "associated substances" that have not been recognized as prebiotics. However, in an *in vitro* fecal fermentation study, cellulose, hemicellulose, and lignin beneficially changed the gut microbiota composition [16]. Thus, dietary fibers currently considered candidates to prebiotics and other fibers not yet recognized as prebiotic could be considered as prebiotics in the future if their prebiotic effects are confirmed in clinical trials.

Fermentation of prebiotic fibers by gut bacterial results predominantly in the production of SCFA, such as acetic, propionic, and butyric acid, but other carboxylic acids can also be produced, including lactic and succinic acid [17]. These metabolites are responsible for several beneficial effects on gut and metabolic health, including reduction of colonic pH, energy supply for colonocytes, increased mucus production, maintenance of the enteric barrier integrity, improved anti-inflammatory status, decreased growth of potential pathogens, improved lipid metabolism, increased leptin production and satiety, and prevention of dysbiosis and weight gain [17–19].

2.2.2 Phenolic Compounds

Phenolic compounds are secondary plant metabolites, characterized by the presence of aromatic rings with one or more hydroxyl groups in their chemical structure, ranging from a single molecule to a complex polymer of high molecular weight. Fruits and vegetables, including their by-products, are major dietary sources of phenolic compounds. In general, the concentration of phenolic compounds in by-products is higher than in their edible parts [2].

According to the phenol-explorer database [20], phenolic compounds can be classified into flavonoids, phenolic acids, stilbenes, lignans, other polyphenols, and non-phenolic metabolites. The structure of flavonoids is comprised of two aromatic rings (A and B), linked through a three-carbon chain (C), usually in the form of a heterocyclic ring. Different substitution patterns in ring C allow flavonoids to have several subclasses [21], which includes anthocyanins (glucosides of cyanidin, malvidin, delphinidin, and pelargonidin), chalcones (butein, xanthohumol), dihydrochalcones (phloretin, phloridzin), dihydroflavonols (dihydromyricetin, dihydroquercetin), flavanols (catechin, epicatechin, procyanidin, and theaflavin), flavanones (hesperidin, naringenin), flavones (apigenin, luteolin), flavonols (kaempferol, myricetin, quercetin), and isoflavonoids (daidzein, genistein). Phenolic acids are usually classified into hydroxybenzoic (ellagic acid, gallic acid), hydroxycinnamic (caffeic acid, ferulic acid), hydroxyphenylacetic (phenylacetic acid, homovanillic acid), hydroxyphenylpropanoic (dihydrocaffeic acid, dihydroferulic acid), and hydroxyphenylpentanoic acids (3-hydroxyphenylvaleric acid). The main stilbenes are resveratrol and pterostilbene. The lignans subgroup includes enterodiol, enterolactone, schisandrin, and sesamin. Other polyphenols are classified into 14 types: alkylmethoxyphenols, alkylphenols, curcuminoids, furanocoumarins, hydroxybenzaldehydes, hydroxybenzoketones, hydroxycinnamaldehydes, hydroxycoumarins, hydroxyphenylpropenes, methoxyphenols, naphtoquinones, phenolic terpenes, tyrosols, and other polyphenols. Major non-phenolic metabolites are glycine and vanilloylglycine [20].

Phenolic compounds have antioxidant, anti-inflammatory [22], anti-obesogenic, and antilipidemic properties [23], and can positively modulate the gut microbiota [6], thus, being considered candidates to prebiotics. It is estimated that a small portion of phenolic compounds is absorbed in small intestine (5–10%), and a large percentage (90-95%) arrive intact in the colon, where they interact with the gut microbiota producing metabolites, which are better absorbed [24, 25]. Methylation, sulfation, glucuronidation are the major metabolic pathways of phenolic compounds in gut and liver tissues [26]. There is evidence from preclinical studies that some phenolic compounds, especially catechins, anthocyanins, and proanthocyanidins, can stimulate the growth of beneficial gut bacteria and increase the production of metabolites [6].

A wide range of phenolic metabolites with different structures and potential functions can be produced by the gut bacteria. For example, equol, a metabolite of soy isoflavones, has estrogenic and antioxidant activity [27]. Flavonoids like epicatechin, catechin, and quercetin generate hydroxyphenylacetic and hydroxyphenylpropionic acids. The urolithins, metabolites derived from ellagitannins of berries and pomegranate, have beneficial effects on different tissues including improvement of intestinal barrier integrity, antioxidant and anti-inflammatory properties, and neuroprotection [28]. However, there are still few *in vivo* studies investigating changes in the gut microbiota composition by polyphenol intake, the metabolites generated from phenolic compounds and their health benefits.

2.3 Fruit By-Products From Agro-Industry Processing

2.3.1 Production by Agro-Industry and Processing as Food Ingredient

Approximately 40% to 50% of the fruits and vegetables are wasted worldwide, and fruit processing is responsible for the most part of the generated residues [29]. The environmental impact of this wasting is massive, but social impact is even bigger if we consider the eventual destination of fruit processing residues to mitigating malnutrition and hunger in developing countries [30]. In addition, there are economic and social benefits of using all fractions generated in fruit processing, including the wastes, converting them in value-added ingredients for the development of novel products by food, pharmacy, and cosmetology industries. By using the fruit waste generated in the agro-industry, the sustainability of food chain is warranted, and the circular economy is promoted [21].

Fruit by-products generated from food industries need to be processed prior to their scientific or technologic use, as they are usually rich in moisture and nutrients. First, the by-products are cut in smaller sizes and cleaned in running water. Thereafter, the fruit wastes are submitted to a sanitization process (e.g., sodium hypochlorite and bleaching), and finally, a dehydration process, preferentially freeze-drying to preserve their properties. The drying step, in which the water content is usually reduced to less than 12% (Table 2.1), is one of the main processes applied for the preservation of the nutritional, functional and microbiological quality of the fruit by-products [32, 34]. Subsequently, the freeze-dried fruit by-products are ground into a fine powder. Finally, the powder is packed in vacuum-sealed (for oxidation protection) and opaque (for light protection) packaging, and stored at low temperatures, i.e., under refrigeration or freezing [18]. The moisture in freeze-dried powder of fruit by-products can vary from 12% to almost 0% (Table 2.1). This relatively high range is an important issue as the content of nutrients and bioactive compounds in foods and their by-products vary according to their moisture concentration.

2.3.2 Dietary Fiber and Phenolic Compounds Composition

Some fruit by-products have remarkable contents of dietary fiber (upper 50 g of total dietary fiber/100 g), such as acerola, grape, guava, and jaboticaba pomaces; passion fruit shell, and pineapple peel (Table 2.1). Apple, banana, mango, melon, and pomegranate peels, as well as blueberry, cashew apple, and persimmon pomaces, and orange albedo, have a concentration of total dietary fiber that varies from 17 to 49 g per 100 g (Table 2.1). In the most fruit by-products, the content of insoluble dietary fiber (10–83 g/100 g) is higher than the content of soluble dietary fiber (2–58 g/100 g), but grape and jaboticaba pomaces have remarkable content of soluble dietary fiber, approximately 60 g per 100 g (Table 2.1). This high concentration of soluble fiber indicates potential prebiotic properties, since soluble fiber is highly fermentable by gut microbiota, and promotes growth of beneficial bacteria [43]. The main subgroups of fibers found in fruit by-products are non-starch polysaccharides, and associated substances (Figure 2.1).

The content of total phenolic compounds in freeze-dried fruit by-products vary into a wide range, from 121 to 8,220 mg gallic acid equivalents (GAE) per 100 g (Table 2.1). Beyond the intrinsic chemical properties of each material, differences on their polyphenol concentrations could be related to the moisture of freeze-dried by-product, fruit genetic variety, geographic location, seasonal variation, stage of maturation, and extraction conditions of these compounds [21]. Fruit by-products, especially cashew apple, grape, guava, and jaboticaba pomaces, as well as pineapple and pomegranate peels, showed the highest total content of phenolic compounds (Table 2.1). Flavonoids (quercetin, isoquercitin, kaempferol, myricetin, catechins, hesperidin, rutin, and malvidin, delphinidin, and cyanidin glucosides), and phenolic acids (gallic acid, caffeic acid, ferulic acid, and *p*-coumaric acid) are major phenolic compounds reported in freeze-dried fruit by-products (Table 2.1, Figure 2.1). Among the phenolic compounds, flavonoids have been the most investigated in relation to their potential prebiotic effects [6].

2.4 Prebiotic Potential of Fruit By-Products

2.4.1 *In Vitro* and *Ex Vivo* Studies

In vitro studies may provide results in a short period at lower costs. Thus, in the last decade, a range of fruit by-products has been investigated for their ability to increase probiotic viable cell count in *in vitro* monoculture assays (Table 2.2). Generally, these substrates (carbon sources) are tested

Table 2.1 Moisture, dietary fiber, and phenolic compound composition of fruit by-products.

Composition (g/100 g)		Fruit by-product (freeze-dried powder)							
		Acerola pomace	Apple peel	Blueberry pomace	Banana peel	Cashew apple pomace	Grape pomace	Guava pomace	Jaboticaba pomace
Moisture		3.3	0.6	4.5	0.5	2.0–6.2	–	2.0–9.8	8.3–12.0
Dietary fibers	total	48-69	19	32	42	35-49	68	45–83	85
	soluble	8-14	–	11	–	2-8	58	3–33	57
	insoluble	34-61	–	21	–	28-48	10	41–83	28
Total phenolic compounds (mg GAE/100 g)		492	485	379	121	802-6,588	6,250	253–1,987	8,220
Major phenolic compounds		quercetin	phlorizin	malvidin delphinidin	isoquercitrin kaempferol	myricetin	caftaric acid malvidin epicatechin	catechin	cyanidin gallic acid hesperidin
Reference		[12, 31–33]	[34, 35]	[36, 37]	[34, 38]	[2, 12, 31, 39]	[40]	[2, 12, 31, 32, 41, 42]	[43, 44]

Composition (g/100 g)		Mango peel	Melon peel	Orange albedo	Passion fruit shell	Persimmon pomace	Pineapple peel	Pomegranate peel
Moisture		0.6	8.0	3.3–7.3	7.1–10.0	4.5	11.6	4.3–9.3
Dietary fibers	total	41-47	44	40	42-65	18	62	17
	soluble	–	4	12	17	9	21	–
	insoluble	–	40	28	48	9	41	–
Total phenolic compounds (mg GAE/100 g)		288-376	296	421	253-384	270	2,787	1,387
Major phenolic compounds		gallic acid rutin	*p*-coumaric acid	hesperidin	rutin epicatechin	gallic acid	ferulic acid	catechins
Reference		[2, 8, 33, 34, 45]	[46, 47]	[11, 42]	[8, 33, 42, 45, 48, 49]	[36, 50]	[2, 45, 51]	[52–54]

–: aspect not addressed in the study; shell: mesocarp plus peel; GAE: gallic acid equivalents.

in comparison with other standard sources like monosaccharides, mainly glucose, and prebiotics, including inulin and FOS. Further, the growth of probiotics is evaluated compared to viable cell count of other intestinal bacteria. The main genera of microorganisms recognized as prebiotic targets are *Lactobacillus* and *Bifidobacterium*, in addition to *Roseburia*, *Eubacterium*, *Faecalibacterium*, and *Akkermansia*, among others [4]. *Lactobacillus* and *Bifidobacterium* are the genera most evaluated in *in vitro* studies (Table 2.2). These genera are among the most used probiotics by the food industry and have been studied for their importance to human gut microbiota and beneficial effects to the host health.

Some studies evaluated the selective stimulation of growth and/or metabolism of probiotics by viable cell count of other intestinal bacteria, especially *Escherichia coli* (Table 2.2). These counts usually are represented by growth curves or by prebiotic activity score (PAS), which indicates the ability of a substrate (FOS or fruit by-products) to promote the growth of probiotic relative to the other intestinal bacteria and relative to non-prebiotic substrate (glucose) [56]. In the presence of *L. acidophilus*, jaboticaba pomace was found to have a similar PAS to FOS; however, orange albedo elicited a higher PAS than FOS (Table 2.2). The synergistic effect between dietary fibers, mainly soluble fibers, and phenolic compounds could explain the high prebiotic potential of these food by-products [11, 43].

The positive prebiotic activity of the fruit by-products also is reflected by lower pH values and increased organic acid production (lactate, acetate, propionate, and butyrate) in growth media as indicative of metabolic response of the tested strains (Table 2.2). The production of different organic acids is associated to the fermentation of energy source substrates, such as carbohydrates and polyphenols, increasing bacterial growth. Lactate is the main end product of *Lactobacillus* metabolism, in addition to being one of the main end products of *Bifidobacterium* metabolism. Acetate is produced from carbohydrate fermentation by most enteric bacteria, especially *Bifidobacterium* [32, 43]. Increased propionate production in studies with cactus pear and pineapple peels could be related to the growth of *Pediococcus pentosaceus* (Table 2.2), bacteria with probiotic potential [51].

Ex vivo models have greater similarity, including greater complexity, with human conditions, which improves predictions of gut microbiota metabolism, in addition to being fast and viable [57]. Among the studies presented in Table 2.3, three performed genomic DNA extraction by 16S rRNA sequencing analysis [36, 47, 59]. This method allows identifying the growth of new bacteria, such as the *Veillonella*, *Odoribacter*, and *Butyricimonas* families, which are still little explored, and may be an

Table 2.2 Selected *in vitro* studies on prebiotic potential of fruit by-products.

Fruit by-product	Monoculture assay with probiotics and other bacteria	Outcomes on monoculture medium		Ref.
		Bacterial modulation	Metabolites production	
Acerola pomace	*L. acidophilus* LA-5, *L. fermentum* PCC	↑ all *Lactobacillus* and *Bifidobacterium* strains	–	[33]
	L. reuteri RC-14, *L. rhamnosus* GR-1, LGG			
	L. paracasei subsp. *paracasei* 431, F19			
	B. lactis BB-12, *B. longum* BB-46			
	B. infantis BB-02			
	E. coli ATCC 8739	↓ pathogenic strains		
	Cl. perfringens ATCC 13124			
Acerola pomace	*L. acidophilus* LA-05	↑ all tested strains	↑ lactate, acetate	[32]
	L. casei L-26			
	B. animalis subsp. *lactis* BB-12			
Apple peel	*L. rhamnosus* LGG	↑ all tested strains	–	[34]
	L. casei			
	B. lactis BB-12			
Banana peel	*L. rhamnosus* LGG	↑ all tested strains	–	[34]
	L. casei			
	B. lactis BB-12			

(*Continued*)

Table 2.2 Selected *in vitro* studies on prebiotic potential of fruit by-products. (*Continued*)

Fruit by-product	Monoculture assay with probiotics and other bacteria	Outcomes on monoculture medium		Ref.
		Bacterial modulation	**Metabolites production**	
Cashew apple pomace	*L. acidophilus* LA-05	↑ all *Lactobacillus* strains	↑ lactate, acetate PAS: FOS = 0.95; by-product = 0.89, $p < 0.05$	[39]
	L. casei L-26			
	L. paracasei L-10			
	enteric mixture	↓ pathogenic strains		
Cactus pear peel	*L. rhamnosus* GG	↑ *Lactobacillus* and *Pediococcus* strains	↑ lactate, acetate, propionate	[51]
	Pediococcus pentosaceus UAM22			
	E. coli K12	↓ pathogenic strains		
Guava pomace	*Lactobacillus* spp. LACTO-5	↑ *Lactobacillus* strain	–	[41]
	E. coli ATCC 25922	↓ pathogenic strain		
Guava pomace	*L. acidophilus* LA-05	↑ all *Lactobacillus* and *Bifidobacterium* strains	↑ lactate, acetate	[32]
	L. casei L-26			
	B. animalis subsp. *lactis* BB-12			
Jaboticaba pomace	*L. acidophilus* LA-05	↑ all *Lactobacillus* and *Bifidobacterium* strains	↑ lactate, acetate PAS: FOS = 0.81; by-product = 0.86, $p > 0.05$	[43]
	L. casei L-26			
	B. lactis BB-12			
	enteric mixture	↓ pathogenic strains		

(*Continued*)

Table 2.2 Selected *in vitro* studies on prebiotic potential of fruit by-products. (*Continued*)

Fruit by-product	Monoculture assay with probiotics and other bacteria	Outcomes on monoculture medium		Ref.
		Bacterial modulation	**Metabolites production**	
Jerivá pulp	*L. casei* BGP93	↑ all *Lactobacillus* and *Bifidobacterium* strains	↑ acetate	[55]
	L. acidophilus LA3			
	B. lactis BLC1			
	E. coli CDC 055	↓ pathogenic strain		
Macaúba pulp	*L. casei* BGP93	↑ *L. acidophilus, B. lactis*	↑ lactate, total SCFA	[55]
	L. acidophilus LA3			
	B. lactis BLC1			
	E. coli CDC 055	↓ pathogenic strain		
Mango peel	*L. rhamnosus* LGG	↑ all *Lactobacillus* and *Bifidobacterium* strains	–	[34]
	L. casei			
	B. lactis BB-12			
Orange albedo	*L. acidophilus* LA-05	↑ all *Lactobacillus* and *Bifidobacterium* strains	↑ lactate, acetate PAS: FOS = 0.4; by-product = 1.8, $p < 0.05$	[11]
	B. animalis subsp. *lactis* BB-12			
	enteric mixture	↓ pathogenic strains		
Pineapple peel	*L. rhamnosus* GG	↑ *Lactobacillus* and *Pediococcus* strains	↑ lactate, acetate, propionate	[51]
	P. pentosaceus UAM22			
	E. coli K12	↓ pathogenic strain		

Ref.: reference; ↑: positive effect; ↓: negative effect; —: aspect not addressed in the study; SCFA: short-chain fatty acids; enteric mixture: *E. coli* ATCC 8739 + *E. coli* ATCC 11775 or 25922; PAS: prebiotic activity score; FOS: fructooligosaccharides.

effective tool for analysis of the next generations of probiotics. The other studies [10, 32, 58, 60] used the fluorescent *in situ* hybridization coupled to flow cytometry (FISH-FC) method, which identifies the growth of a limited number of bacteria, not allowing the visualization of new genera. Thus, there is a predominance of *Bifidobacterium* and *Lactobacillus* as outcome of bacterial modulation in the cited studies (Table 2.3).

In addition to the beneficial bacteria modulation, the *ex vivo* studies have also reported an increase in SCFA production (Table 2.3). These findings indicate a great prebiotic potential of fruit by-products. Nonetheless, the design of *ex vivo* studies could be improved by carrying out a wider analysis of gut bacteria genera or species, as well as a more precise stratification according to age, sex, and nutritional status of the donors, given that these are factors influencing the microbiota profile [57].

2.4.2 *In Vivo* Studies

Preclinical experiments simulate the complex interaction of different processes of digestion, absorption, modulation of the gut microbiota, and production of metabolites in mammalian. Thus, animal models are widely used to predict benefits to human health, including prebiotic effects [57]. For instance, administration of grape pomace and peach peel extracts to mice resulted in the growth of *Akkermansia*, a potential and promising probiotic candidate [13, 61]. Further, a decrease in the *Enterobacteriaceae*

Table 2.3 Selected *ex vivo* studies on prebiotic activity of fruit by-products.

Fruit by-product	Study design (pooled human fecal inoculum)	Outcomes on fecal inoculum		Ref.
		Bacterial modulation	Metabolites production	
Acerola pomace	4 healthy adult donors (2 men, 2 women, 25–40 years old)	↑ *Bifidobacterium*, *Lactobacillus*, *Enterococcus*, *E. rectale*, *C. coccoides*	↑ SCFA	[32]
Blueberry pomace	7 healthy adult donors (3 men, 4 women)	↑ *Ruminococcaceae*, *Lactobacillus*	–	[36]

(*Continued*)

Table 2.3 Selected *ex vivo* studies on prebiotic activity of fruit by-products. (*Continued*)

Fruit by-product	Study design (pooled human fecal inoculum)	Outcomes on fecal inoculum		Ref.
		Bacterial modulation	Metabolites production	
Cashew apple pomace	4 healthy adult donors (2 men, 2 women, 25–40 years old)	↑ *Lactobacillus, Enterococcus, Bifidobacterium, Bacteroides, Prevotella, E. rectale, C. coccoides, C. histolyticum*	↑ SCFA	[58]
Guava pomace	4 healthy adult donors (2 men, 2 women, 25–40 years old)	↑ *Bifidobacterium, E. rectale, C. coccoides*	↑ SCFA	[32]
Jaboticaba pomace	4 healthy adult donors (2 men, 2 women, 18–39 years old)	↑ *Lactobacillus, Enterococcus, Bifidobacterium, Bacteroides, Prevotella, E. rectale, C. coccoides, C. histolyticum*	↑ SCFA	[10]
Melon peel (extract)	5 healthy adult donors (men and women)	↑ *Bifidobacterium, Lactobacillus*	↑ SCFA	[47]
Pineapple stems and peels	5 healthy adult donors (sex not known)	↑ *Bifidobacterium, Lactobacillus*	↑ SCFA	[59]
Persimmon pomace	7 healthy adult donors (3 men, 4 women)	↑ *Streptococcus, Veillonella, Odoribacter, Butyricimonas*	–	[36]
Pomegranate pomace (extract)	3 healthy adult donors (sex not known)	↑ *Bifidobacterium, Lactobacillus*	↑ SCFA	[60]

Ref.: reference; ↑: positive effect; ↓: negative effect; –: aspect not addressed in the study; SCFA: short-chain fatty acids.

family, abundant in pathogenic bacteria, was observed in Wistar rats supplemented with guava pomace [12]. As in the *in vitro* and *ex vivo* studies (Tables 2.2 and 2.3), SCFA production was observed in most *animal* studies (Table 2.4). Altogether, preclinical studies demonstrate that supplementation with fruit by-products results in beneficial microbiota modulation, decreased serum and liver lipid accumulation, and attenuation of weight gain and inflammatory response (Figure 2.1).

Although findings are promising, it is important to note that preclinical studies to date are still very heterogeneous regarding the experimental model, route of administration, type of food by-product, dose, and length of the experiment, which limits conclusions on the prebiotic effect of the fruit by-products. In addition, some experimental designs are at risk of bias because fecal samples collected from the cage rather than the colon were used for analysis, as SCFA losses have been observed in the feces when compared to cecal content [13, 44, 61].

To the best of our knowledge, there is only one published clinical trial on fruit residue supplementation and microbiota modulation [62]. This is a pilot study that evaluated the effects of grape pomace extract supplementation (1.4 g per day) in 10 healthy women for 21 days. Serum concentrations of glucose, uric acid, albumin, hepatic metabolism enzymes, cholesterol, triglycerides, proteins, basal hormones, and the inflammatory marker TNF-α were evaluated. Only blood fasting glucose was significantly reduced (from 89.00 ± 1.84 to 84.50 ± 2.61 mg/dL, $p < 0.05$) after the grape by-product supplementation. Overall, supplementation did not lead to significant changes in gut bacterial populations or fecal and urinary phenolic metabolite contents. However, significant changes ($p < 0.05$) were observed in the profiles of short- and medium-chain fatty acids [62].

2.5 Gaps and Perspectives for Further Research

Fruit by-products or waste residues are generated during the manufacturing process of juices, jams, canned fruit, and dehydrated fruit, among others. These fruit by-products are rich in dietary fiber and phenolic compounds, which are very beneficial to human health. While fruit by-products can be incorporated into supplements and other food products for additional nutritional value and health benefits, tons of fruit by-products are not recovered and instead disposed in landfills, which have a significant negative economic impact. As the world is striving for long-term sustainable development, research on the health benefits of fruit by-products has increased in recent years.

Table 2.4 Selected *in vivo* studies on prebiotic effect of fruit by-products.

Fruit by-product (dose, administration via)	Study design (animal model)	Outcomes on gut microbiota		Ref.
		Bacterial modulation	Metabolites production	
Acerola pomace (400 mg/kg BW, intragastric)	Female adult Wistar rats (n=8/group) Dyslipidaemic diet (HFD and high-energy) 28 days, cc	↑ *Lactobacillus*, *Bifidobacterium*	↑ SCFA	[12]
Cashew apple pomace (400 mg/kg BW, intragastric)	Female adult Wistar rats (n=8/group) Dyslipidaemic diet (high-lipid and high-energy, Rhoster®) 28 days, cc	↓ *Enterobacteriacea*	↔ SCFA	[12]
Grape pomace-extract (200 mg/kg BW, intragastric)	C57BL/6 adult mice (n=10/group) Control group (D12450B diet) 7 days, feces	↑ *Akkermansia*	–	[61]
Guava pomace (400 mg/kg BW, intragastric)	Female adult Wistar rats (n=8/group) Dyslipidaemic diet (high-lipid and high-energy) 28 days, cc	↑ *Lactobacillus*, *Bifidobacterium*	↔ SCFA	[12]

(*Continued*)

Table 2.4 Selected *in vivo* studies on prebiotic effect of fruit by-products

Fruit by-product (dose, administration via)	Study design (animal model)	Outcomes on gut microbiota		Ref.
		Bacterial modulation	Metabolites production	
Jaboticaba peel (AIN93-M diet with freeze-dried jabuticaba peel, 12% moisture)	C57BL/6 adult mice (n=6/group) Control group (AIN-93M diet) HFD (AIN-93M with adjusted 35% from lard) 13 weeks, feces	↑ *Muribaculaceae, Lachnospiraceae*	↑ SCFA	[44]
Passion fruit shell (AIN-93M diet with 50% of cellulose content was replaced by fiber from passion fruit shell)	Male Wistar rats (n=4/group) AIN-93M diet 15 days, cc	↔ *Lactobacillus, Bifidobacterium*	↑ butyrate, acetate ↔ propionate	[48]
Peach peel (300 mg/kg BW, intragastric)	Female ICR mice (n=10/group) Normal control (15% kcal from fat) HFD (60% kcal from fat) 12 weeks, feces	↑ *Lactobacillus, Lachnospiraceae, Prevotellaceae, Alloprevotella, Akkermansia, Roseburia, Ruminococcus* Abundance and diversity of gut microbiota were modulated	↑ SCFA	[13]

Ref.: reference; ↑: positive effect; ↓: negative effect; ↔: no change; – : aspect not addressed in the study; BW: body weight; cc: cecum content; SCFA: short-chain fatty acids; HFD: high fat diet; shell: mesocarp plus peel.

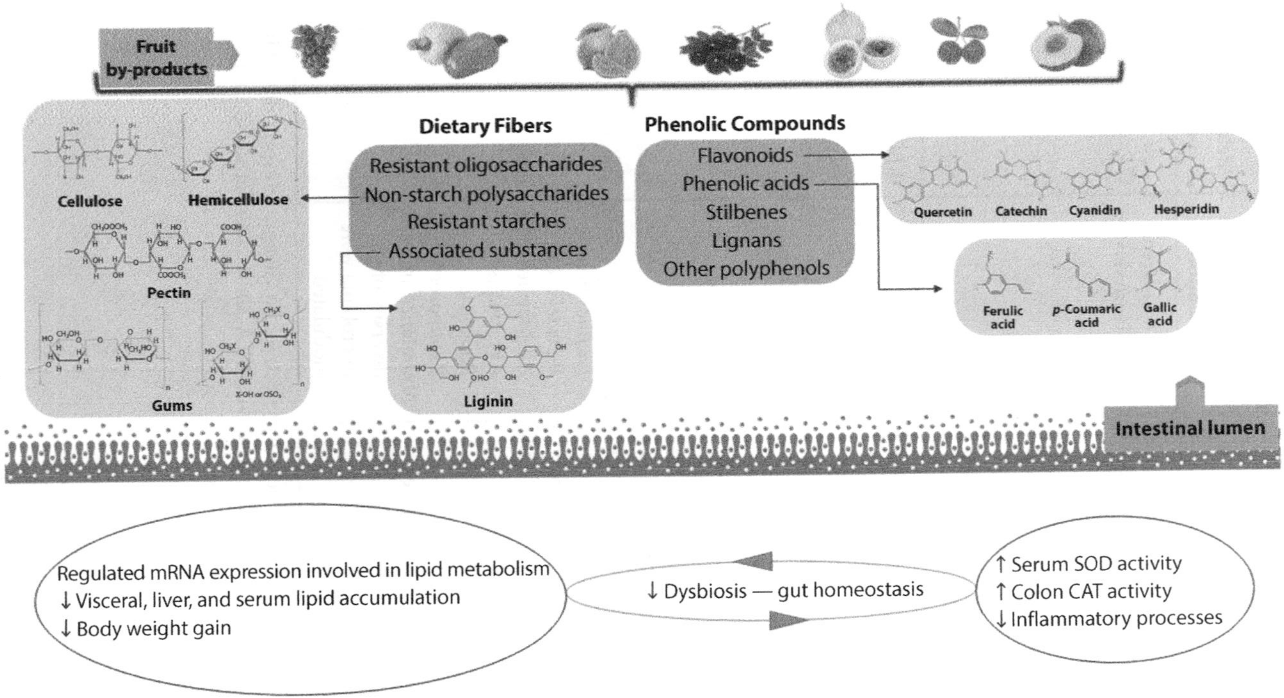

Figure 2.1 Potential prebiotic compounds in fruit by-products and their metabolic benefits as described in preclinical studies. ↓: decreased; ↑: increased; SOD: superoxide dismutase; CAT: catalase.

Dietary fiber and phenolic compounds have great prebiotic potential [5, 6]. Since gut dysbiosis has been linked to the development of several chronic diseases [63], fruit by-products have been explored for their ability to modulate the gut microbiota. Accumulating evidence from *in vitro*, *ex vivo*, and animal studies indicate that fruit by-products (e.g., acerola, guava, jaboticaba pomace) favorably modulate the gut microbiota as indicated by an increase in beneficial bacteria, which may be accompanied by reduction of pathogenic bacteria. Most of the studies also noted an increase in the production of colonic SCFA (Tables 2.2–2.4). To date, only one clinical trial has been conducted to examine the health effects of short-term supplementation with grape pomace extract in healthy women [62]. While no changes were observed in intestinal microbial population, a significant increase in the production of SFCA, as well as a significant reduction in fasting glucose levels were noted [62]. These findings are promising but need to be confirmed in robust clinical trials. Therefore, randomized, double-blind, placebo-controlled crossover clinical trials in healthy and diseased populations are warranted to confirm in humans the prebiotic effects of fruit by-products, and explore the association of the changes in gut microbiota with functional and systemic health outcomes.

2.6 Conclusion

Fruit by-products are potential modulators of beneficial gut microbiota, as they are abundant sources of prebiotic compounds. Grape, guava, and jaboticaba pomaces are highlighted since they have high contents of soluble fibers and phenolic compounds, as well as cashew apple pomace, and pineapple and pomegranate peels as remarkable sources of phenolics. Evidence from preclinical studies indicate that the supplementation with fruit by-products may elicit health benefits via modulation of gut microbiota, such as reduction in serum and liver lipid accumulation, inflammatory process, and in weight gain. Although food by-products have been shown to have prebiotic potential, given their high content of dietary fiber and phenolics, more clinical trials are needed to confirm the prebiotic properties of fruit by-products in humans.

References

1. Majerska, J., Michalska, A., Figiel, A., A review of new directions in managing fruit and vegetable processing by-products. *Trends Food Sci. Technol.*, 88, 207, 2019.

2. Silva, L.M.R., Figueiredo, E.A.T., Ricardo, N.M.P.S., Vieira, I.G.P., Figueiredo, R.W., Brasil, I.M. *et al.*, Quantification of bioactive compounds in pulps and by-products of tropical fruits from Brazil. *Food Chem.*, 143, 398, 2014.
3. Chaouch, M.A. and Benvenuti, S., The role of fruit by-products as bioactive compounds for intestinal health. *Foods*, 9, 1716, 2020.
4. Gibson, G.R., Hutkins, R., Sanders, M.E., Prescott, S.L., Reimer, R.A., Salminen, S.J. *et al.*, Expert consensus document: The International Scientific Association for Probiotics and Prebiotics (ISAPP) consensus statement on the definition and scope of prebiotics. *Nat. Rev. Gastroenterol. Hepatol.*, 14, 491, 2017.
5. Rezende, E.S.V., Lima, G.C., Naves, M.M.V., Dietary fibers as beneficial microbiota modulators: A proposal classification by prebiotic categories. *Nutrition*, 89, 111217, 2021.
6. Alves-Santos, A.M., Sugizaki, C.S.A., Lima, G.C., Naves, M.M.V., Prebiotic effect of dietary polyphenols: A systematic review. *J. Funct. Foods*, 74, 104169, 2020.
7. Fan, Y. and Pedersen, O., Gut microbiota in human metabolic health and disease. *Nat. Rev. Microbiol.*, 19, 55, 2021.
8. Albuquerque, M.A.C., Levit, R., Beres, C., Bedani, R., LeBlanc, A.M., Saad, S.M.I. *et al.*, Tropical fruit by-products water extracts as sources of soluble fibres and phenolic compounds with potential antioxidant, antiinflammatory, and functional properties. *J. Funct. Foods*, 52, 724, 2019.
9. Campos, D.A., Gómez-García, R., Vilas-Boas, A.A., Madureira, A.R., Pintado, M.M., Management of fruit industrial by-products - a case study on circular economy approach. *Molecules*, 25, 320, 2020.
10. Massa, N.M.L., Oliveira, S.P.A., Rodrigues, N.P.A., Menezes, F.N.D.D., Lima, M.S., Magnani, M., Souza, E.L., *In vitro* colonic fermentation and potential prebiotic properties of pre-digested jabuticaba (*Myrciaria jaboticaba* (Vell.) Berg) by-products. *Food Chem.*, 388, 133003, 2022.
11. Rezende, E.S.V., Lima, G.C., Lima, M.S., Coelho, A.S.G., Naves, M.M.V., Prebiotic potential of isolated commercial dietary fibres compared to orange albedo in *Lactobacillus* and *Bifidobacterium* species. *Bioact. Carbohydr. Diet. Fibre*, 28, 100316, 2022.
12. Batista, K.S., Alves, A.F., Lima, M.D.S., Silva, L.A., Lins, P.P., Gomes, J.A.S. *et al.*, Beneficial effects of consumption of acerola, cashew or guava processing by-products on intestinal health and lipid metabolism in dyslipidaemic female Wistar rats. *Br. J. Nutr.*, 119, 30, 2018.
13. Kan, J., Chen, C., Huo, T., Xie, W., Hui, Y., Liu, J., Jin, C., Polyphenolic-enriched peach peels extract regulates lipid metabolism and improves the gut microbiota composition in high fat diet-fed mice. *J. Funct. Foods*, 72, 104082, 2020.
14. Stephen, A.M., Champ, M.M.J., Cloran, S.J., Fleith, M., Van Lieshout, L., Mejborn, H. *et al.*, Dietary fibre in Europe: Current state of knowledge on

definitions, sources, recommendations, intakes and relationships to health. *Nutr. Res. Rev.*, 30, 149, 2017.
15. Lordan, C., Thapa, D., Ross, R.P., Cotter, P.D., Potential for enriching next-generation health-promoting gut bacteria through prebiotics and other dietary components. *Gut Microbes*, 11, 1, 2020.
16. Ge, Q., Li, H., Zheng, Z., Yang, K., Li, P., Xiao, Z. *et al.*, *In vitro* fecal fermentation characteristics of bamboo insoluble dietary fiber and its impacts on human gut microbiota. *Food Res. Int.*, 156, 111173, 2022.
17. Williams, B., Grant, L., Gidley, M., Mikkelsen, D., Gut fermentation of dietary fibres: Physico-chemistry of plant cell walls and implications for health. *Int. J. Mol. Sci.*, 18, 2203, 2017.
18. Oliveira, F.L., Arruda, T.Y.P., Morzelle, M.C., Pereira, A.P.A., Casarotti, S.N., Fruit by-products as potential prebiotics and promising functional ingredients to produce fermented milk. *Food Res. Int.*, 161, 111841, 2022.
19. Sugizaki, C.S.A. and Naves, M.M.V., Potential prebiotic properties of nuts and edible seeds and their relationship to obesity. *Nutrients*, 10, 1645, 2018.
20. Rothwell, J.A., Perez-Jimenez, J., Neveu, V., Medina-Remon, A., M'Hiri, N., Garcia-Lobato, P. *et al.*, Phenol-Explorer 3.0: A major update of the Phenol-Explorer database to incorporate data on the effects of food processing on polyphenol content. *Database*, 2013, bat070, 2013.
21. Trigo, J.P., Alexandre, E.M.C., Saraiva, J.A., Pintado, M.E., High value-added compounds from fruit and vegetable by-products - characterization, bioactivities, and application in the development of novel food products. *Crit. Rev. Food Sci. Nutr.*, 60, 1388, 2020.
22. Rudrapal, M., Khairnar, S.J., Khan, J., Dukhyil, A.B., Ansari, M.A., Alomary, M.N. *et al.*, Dietary polyphenols and their role in oxidative stress-induced human diseases: Insights into protective effects, antioxidant potentials and mechanism(s) of action. *Front. Pharmacol.*, 13, 806470, 2022.
23. Singh, M., Thrimawithana, T., Shukla, R., Adhikari, B., Managing obesity through natural polyphenols: A review. *Future Foods*, 1–2, 100002, 2020.
24. Rodríguez-Daza, M.C., Pulido-Mateos, E.C., Lupien-Meilleur, J., Guyonnet, D., Desjardins, Y., Roy, D., Polyphenol-mediated gut microbiota modulation: Toward prebiotics and further. *Front. Nutr.*, 8, 689456, 2021.
25. Shortt, C., Hasselwander, O., Meynier, A., Nauta, A., Fernández, E.N., Putz, P. *et al.*, Systematic review of the effects of the intestinal microbiota on selected nutrients and non-nutrients. *Eur. J. Nutr.*, 57, 25, 2018.
26. García-Villalba, R., González-Sarrías, A., Giménez-Bastida, J.A., Selma, M.V., Espín, J.C., Tomás-Barberán, F.A., Metabolism of dietary (poly)phenols by the gut microbiota, in: *Comprehensive Gut Microbiota*, M. Glibetic (Ed.), pp. 149–175, Elsevier, Amsterdam, 2022.
27. Mayo, B., Vázquez, L., Flórez, A.B., Equol: A bacterial metabolite from the daidzein isoflavone and its presumed beneficial health effects. *Nutrients*, 11, 2231, 2019.

28. Marhuenda-Muñoz, M., Laveriano-Santos, E.P., Tresserra-Rimbau, A., Lamuela-Raventós, R.M., Martínez-Huélamo, M., Vallverdú-Queralt, A., Microbial phenolic metabolites: Which molecules actually have an effect on human health? *Nutrients*, 11, 2725, 2019.
29. FAO - Food and Agriculture Organization of the United Nations, *Global initiative on food loss and waste reduction*, FAO, Rome, 2015.
30. Torres-León, C., Ramírez-Guzman, N., Londoño-Hernandez, L., Martinez-Medina, G.A., Díaz-Herrera, R., Navarro-Macias, V. *et al.*, Food waste and by-products: An opportunity to minimize malnutrition and hunger in developing countries. *Front. Sustain. Food Syst.*, 2, 52, 2018.
31. Araújo, C.M., Sampaio, K.B., Menezes, F.N.D.D., Almeida, E.T.C., Lima, M.S., Viera, V.B. *et al.*, Protective effects of tropical fruit processing coproducts on probiotic lactobacillus strains during freeze-drying and storage. *Microorganisms*, 8, 96, 2020.
32. Menezes, F.N.D.D., Melo, F.H.C., Vieira, A.R.S., Almeida, E.T.C., Lima, M.S., Aquino, J.S. *et al.*, Acerola (*Malpighia glabra* L.) and guava (*Psidium guayaba* L.) industrial processing by-products stimulate probiotic *Lactobacillus* and *Bifidobacterium* growth and induce beneficial changes in colonic microbiota. *J. Appl. Microbiol.*, 130, 1323, 2021.
33. Vieira, A.D.S., Bedani, R., Albuquerque, M.A.C., Biscola, V., Saad, S.M.I., The impact of fruit and soybean by-products and amaranth on the growth of probiotic and starter microorganisms. *Food Res. Int.*, 97, 356, 2017.
34. Zahid, H.F., Ranadheera, C.S., Fang, Z., Ajlouni, S., Utilization of mango, apple and banana fruit peels as prebiotics and functional ingredients. *Agriculture*, 11, 584, 2021.
35. Du, G., Zhu, Y., Wang, X., Zhang, J., Tian, C., Liu, L., Meng, Y., Guo, Y., Phenolic composition of apple products and by-products based on cold pressing technology. *J. Food Sci. Technol.*, 56, 1389, 2019.
36. Bas-Bellver, C., Andrés, C., Seguí, L., Barrera, C., Jiménez-Hernández, N., Artacho, A. *et al.*, Valorization of persimmon and blueberry by-products to obtain functional powders: *In vitro* digestion and fermentation by gut microbiota. *J. Agric. Food Chem.*, 68, 8080, 2020.
37. Herrera-Balandrano, D.D., Chai, Z., Beta, T., Feng, J., Huang, W., Blueberry anthocyanins: An updated review on approaches to enhancing their bioavailability. *Trends Food Sci. Technol.*, 118, 808, 2021.
38. Avram, I., Gatea, F., Vamanu, E., Functional compounds from banana peel used to decrease oxidative stress effects. *Processes*, 10, 248, 2022.
39. Duarte, F.N.D., Rodrigues, J.B., Lima, M.C., Lima, M.S., Pacheco, M.T.B., Pintado, M.M.E. *et al.*, Potential prebiotic properties of cashew apple (*Anacardium occidentale* L.) agro-industrial by-product on *Lactobacillus* species. *J. Sci. Food Agric.*, 97, 3712, 2017.
40. Martínez-Meza, Y., Perez-Jimenez, J., Rocha-Guzman, N.E., Rodríguez-García, M.E., Alonzo-Macías, M., Reynoso-Camacho, R., Modification on

the polyphenols and dietary fiber content of grape pomace by instant controlled pressure drop. *Food Chem.*, 360, 130035, 2021.

41. Lim, S.Y., Tham, P.Y., Lim, H.Y.L., Heng, W.S., Chang, Y.P., Potential functional by-products from guava puree processing. *J. Food Sci.*, 83, 1522, 2018.
42. Casarotti, S.N., Borgonovi, T.F., Batista, C.L.F.M., Penna, A.L.B., Guava, orange and passion fruit by-products: Characterization and its impacts on kinetics of acidification and properties of probiotic fermented products. *LWT - Food Sci. Technol.*, 98, 69, 2018.
43. Massa, N.M.L., Menezes, F.N.D.D., Albuquerque, T.M.R., Oliveira, S.P.A., Lima, M.S., Magnani, M. *et al.*, Effects of digested jabuticaba (*Myrciaria jaboticaba* (Vell.) Berg) by-product on growth and metabolism of *Lactobacillus* and *Bifidobacterium* indicate prebiotic properties. *LWT - Food Sci. Technol.*, 131, 109766, 2020.
44. Loubet Filho, P.S., Baseggio, A.M., Vuolo, M.M., Reguengo, L.M., Biasoto, A.C.T., Correa, L.C. *et al.*, Gut microbiota modulation by jabuticaba peel and its effect on glucose metabolism via inflammatory signaling. *Curr. Res. Food Sci.*, 5, 382, 2022.
45. Selani, M.M., Bianchini, A., Ratnayake, W.S., Flores, R.A., Massarioli, A.P., Alencar, S.M. *et al.*, Physicochemical, functional and antioxidant properties of tropical fruits co-products. *Plant Foods Hum. Nutr.*, 71, 137, 2016.
46. Gómez-García, R., Campos, D.A., Oliveira, A., Aguilar, C.N., Madureira, A.R., Pintado, M., A chemical valorisation of melon peels towards functional food ingredients: Bioactives profile and antioxidant properties. *Food Chem.*, 335, 127579, 2021.
47. Gómez-García, R., Sánchez-Gutiérrez, M., Freitas-Costa, C., Vilas-Boas, A.A., Campos, D.A., Aguilar, C.N., Ana R. Madureira, A.R., Manuela Pintado, M., Prebiotic effect, bioactive compounds and antioxidant capacity of melon peel (*Cucumis melo* L. *inodorus*) flour subjected to *in vitro* gastrointestinal digestion and human faecal fermentation. *Food Res. Int.*, 154, 111045, 2022.
48. Silva, J.K., Cazarin, C.B.B., Bogusz Junior, S., Augusto, F., Marostica Junior, M.R., Passion fruit (*Passiflora edulis*) peel increases colonic production of short-chain fatty acids in Wistar rats. *LWT - Food Sci. Technol.*, 59, 1252, 2014.
49. Souza, C.B., Saad, S.M.I., Venema, K., Lean and obese microbiota: Differences in *in vitro* fermentation of food-by-products. *Benef. Microbes*, 12, 397, 2021.
50. Lucas-González, R., Fernández-López, J., Pérez-Álvarez, J.A., Viuda-Martos, M., Effect of particle size on phytochemical composition and antioxidant properties of two persimmon flours from *Diospyros kaki* Thunb. vars. 'Rojo Brillante' and 'Triumph' co-products. *J. Sci. Food Agric.*, 98, 504, 2018.
51. Diaz-Vela, J., Totosaus, A., Cruz-Guerrero, A.E., Perez-Chabela, M.L., *In vitro* evaluation of the fermentation of added-value agroindustrial by-products: Cactus pear (*Opuntia ficus-indica* L.) peel and pineapple (*Ananas comosus*) peel as functional ingredients. *Int. J. Food Sci. Technol.*, 48, 1460, 2013.

52. Ismail, T., Akhtar, S., Riaz, M., Ismail, A., Effect of pomegranate peel supplementation on nutritional, organoleptic and stability properties of cookies. *Int. J. Food Sci. Nutr.*, 65, 661, 2014.
53. Ismail, T., Akhtar, S., Riaz, M., Hameed, A., Afzal, K., Sheikh, A.S., Oxidative and microbial stability of pomegranate peel extracts and bagasse supplemented cookies. *J. Food Qual.*, 39, 658, 2016.
54. Omer, H.A.A., Abdel-Magid, S.S., Awadalla, I.M., Nutritional and chemical evaluation of dried pomegranate (*Punica granatum* L.) peels and studying the impact of level of inclusion in ration formulation on productive performance of growing Ossimi lambs. *Bull. Natl. Res. Cent.*, 43, 182, 2019.
55. Andrade, A.C., Marinho, J.F.U., Souza, A.C., Tavares, T.S., Dias, D.R., Schwan, R.F. *et al.*, Prebiotic potential of pulp and kernel cake from jerivá (*Syagrus romanzoffiana*) and macaúba palm fruits (*Acrocomia aculeata*). *Food Res. Int.*, 136, 109595, 2020.
56. Huebner, J., Wehling, R.L., Hutkins, R.W., Functional activity of commercial prebiotics. *Int. Dairy J.*, 17, 770, 2007.
57. Xu, Y., Sherestha, N., Préat, V., Beloqui, A., An overview of *in vitro*, *ex vivo* and *in vivo* models for studying the transport of drugs across intestinal barriers. *Adv. Drug Deliv. Rev.*, 175, 113795, 2021.
58. Menezes, F.N.D.D., Almeida, E.T.C., Vieira, A.R.S., Aquino, J.S., Lima, M.S., Magnani, M. *et al.*, Impact of cashew (*Anacardium occidentale* L.) by-product on composition and metabolic activity of human colonic microbiota *in vitro* indicates prebiotic properties. *Curr. Microbiol.*, 78, 2264, 2021.
59. Campos, D.A., Coscueta, E.R., Vilas-Boas, A.A., Silva, S., Teixeira, J.A., Pastrana, L.M., Pintado, M.M., Impact of functional flours from pineapple by-products on human intestinal microbiota. *J. Funct. Foods*, 67, 103830, 2020.
60. Bialonska, D., Ramnani, P., Kasimsetty, S.G., Muntha, K.R., Gibson, G.R., Ferreira, D., The influence of pomegranate by-product and punicalagins on selected groups of human intestinal microbiota. *Int. J. Food Microbiol.*, 140, 175, 2010.
61. Lu, F., Liu, F., Zhou, Q., Hu, X., Zhang, Y., Effects of grape pomace and seed polyphenol extracts on the recovery of gut microbiota after antibiotic treatment in high-fat-diet-fed mice. *Food Sci. Nutr.*, 7, 2897, 2019.
62. Gil-Sánchez, I., Esteban-Fernández, A., Llano, D.G., Sanz-Buenhombre, M., Guadarrana, A., Salazar, N. *et al.*, Supplementation with grape pomace in healthy women: Changes in biochemical parameters, gut microbiota and related metabolic biomarkers. *J. Funct. Foods*, 45, 34, 2018.
63. Chen, Y., Zhou, J., Wang, L., Role and mechanism of gut microbiota in human disease. *Front. Cell. Infect. Microbiol.*, 11, 625913, 2021.

3

Valorization of *Actinidia* spp. By-Products and Wastes for Nutraceutical Applications

Ana Margarida Silva[1], Catarina Macedo[1], Piotr Latocha[2], Paulo Costa[3,4], Cristina Delerue-Matos[1] and Francisca Rodrigues[1]*

[1]*REQUIMTE/LAQV, ISEP, Polytechnic of Porto, Rua Dr. António Bernardino de Almeida, Porto, Portugal*
[2]*Department of Environmental Protection and Dendrology, Institute of Horticultural Sciences, Warsaw University of Life Sciences – SGGW, Nowoursynowska, Warsaw, Poland*
[3]*REQUIMTE/UCIBIO, MedTech-Laboratory of Pharmaceutical Technology, Department of Drug Sciences, Faculty of Pharmacy, University of Porto, Rua de Jorge Viterbo Ferreira, Porto, Portugal*
[4]*Associate Laboratory i4HB—Institute for Health and Bioeconomy, Faculty of Pharmacy, University of Porto, Porto, Portugal*

Abstract
Circular economy is an essential point of sustainable development that supports waste minimization and efficient use of resources. The reuse and recovery of by-products and wastes from different fruit species is a world challenge. *Actinidia* spp. is an example of a fruit genus widely disseminated worldwide and coupled with an enormous economic value. This genus is composed by 54 species, of which 2 are of great commercial interest (*A. chinensis* var. *chinensis, A. chinensis* var. *deliciosa,* and *A. arguta*). The growing interest in the cultivation of kiwi leads to the generation of different by-products and wastes (e.g., leaves, stems, damage fruits). In addition, the widespread interest observed has created a demand for the identification of the bioactive compounds present in its by-products and wastes. Different compounds have been identified, such as polyphenols, vitamins, fiber, and phenolic compounds, which have been associated with several beneficial health effects (e.g., antioxidant, anti-inflammatory, antimicrobial, neuroprotective, and anti-obesity effects). Therefore, kiwi by-products and wastes could be valorized as new sources of nutraceuticals ingredients. This chapter aims to explore the pro-healthy properties

**Corresponding author*: francisca.rodrigues@graq.isep.ipp.pt; franciscapintolisboa@gmail.com

Umile Gianfranco Spizzirri (ed.) Nutraceutics from Agri-Food By-Products, (69–116) © 2023 Scrivener Publishing LLC

of *Actinidia* genus by-products and wastes, reviewing the most recent studies and overview the challenges for its validation as new nutraceutical ingredients.

Keywords: Actinidia by-products, phenolics, valorization, sustainability, circular economy

3.1 Introduction

Although the genus *Actinidia* contains 54 species, only few of them are tasty fruits and show economic potential [1]. Currently, the most important of these species are two varieties of *A. chinensis*—var. *deliciosa* and var. *chinensis*—called as kiwifruit. Until recently, both varieties were commonly treated as separate species (*A. chinensis* and *A. deliciosa*). However, the development of the world kiwifruit industry is related almost entirely to the first one. The history of the commercial kiwifruit cultivation is only about two hundred years old. Kiwifruit is amongst the best examples of the successful domestication and commercialization of a new crop. The homeland of *Actinidia* is China, but until the beginning of the 20th century, it had not been commercially cultivated there. The spectacular success of the kiwifruit industry is associated with New Zealand and, which is unique, with only one cultivar—'Hayward.' The first plantations in the world were established in New Zealand and the first fruit produced was sold on the domestic market in the 1930s. The first fruit export under the trade name "Chinese gooseberry" took place in the 1950s. However, the dynamic increase in the interest in this fruit occurred after the fruit's trade name had been changed to kiwifruit, which resembled New Zealand's iconic native bird - the kiwi. By the 1970s, kiwifruit had already been produced outside New Zealand in many countries with a temperate climate, such as France, Italy, South America, and the United States of America (USA) [2]. Thanks to the cultivation of kiwifruit in both hemispheres, the fruit was available on the market all year round, which contributed to a further increase in the interest in this fruit. Subsequently, kiwifruit became one of the newest fruit crops to gain international commercial importance.

At the beginning of the 21st century, the production of the second variety (*A. chinensis* var. *chinensis*) began, and a completely different species—*A. arguta*—was commercialized. Its fruit, called kiwiberry, differs significantly from kiwifruit. The fruit is smaller in size (~10 times) and has edible delicate skin. Although the global production of kiwiberry is growing dynamically and fruit is increasingly entering store shelves, it is still at an early stage. In 2015, the total global area under kiwiberry cultivation was about 1640 ha and the estimated total production was about

1,600 tonnes [3], but since, then many plantations (including over 2,300 ha in China) have been established, and the 2018 production in China alone reached estimated 1,500 tonnes [4]. Due to the low post-harvest durability of the fruit and their limited transport possibilities, the production is rather focused on local markets. The advantage of this species, however, is that it is much more resistant to frost and can therefore be grown in areas that are too cold for kiwi cultivation.

Differently than at the beginning, the future of the global kiwifruit category has been linked to the successful development of new cultivars. The market for the production of *Actinidia* fruit seems to be prospective as several other species show significant economic potential, e.g., *A. eriantha*, *A. kolomikta*, and *A. polygama* [5], but their commercialization is progressing very slowly.

Actinidia fruit has been identified as exceptionally nutritious, with significant health-promoting properties [6, 7]. While the kiwifruit is generally consumed fresh, its uses have changed over time with trends in food consumption. Over time, kiwifruit has been used as the base in processed products, such as jams, preserves, fruit juice, wine, nectar, and pulp products for use as fruit topping, dried fruit, and even candy [8].

Starting from the 1970s, the world production of kiwifruit has been growing steadily but with varying strength, reaching about 800,000 tonnes in 1990, while in the next twenty years this production doubled [9]. In 2012, the world kiwifruit production amounted to about 2 million tonnes, about 90 percent of which was provided by 10 main producers – China, which became the leader, followed by Italy, New Zealand, Chile, Greece, France, Japan, Iran, the USA, and Spain. At that time, Italy and New Zealand were the main exporters of kiwifruit (over 60 percent of global trade in fresh kiwifruit in total). According to Ward and Courtney [9], the value of the global kiwifruit industry in 2012 was estimated to be around US $ 1.9 billion. In New Zealand, kiwifruit has become an essential component of the country's economy, where nearly about 2,500 farmers were associated with kiwifruit production and they harvested about 3.7 billion kiwifruits per year and marketed them in about 55 countries under the brand name Zespri—one of the most recognized kiwi brands in the world [10, 11]. Over the last decade, New Zealand's kiwifruit exports have increased by 176 percent from under $ 1 billion to over $ 2.5 billion in 2020. The fruit is mainly exported to European Union (EU) countries, Japan and China, accounting for 70 percent of all New Zealand kiwifruit exports [12].

The global kiwifruit market keeps growing. According to FAOSTAT [13], the global volume of kiwifruit production increases reached 4.41 million tonnes in 2020, worth a value of approximately US $ 4,448,493 thousand, with China (2,230.0 thousand tonnes harvested annually) dominating the market, followed by New Zealand (624.9 thousand tonnes), Italy (521.5 thousand tonnes), Greece (307.4 thousand tonnes) and Chile (158.9 thousand tonnes). The remaining 7 countries produce about 654 thousand tonnes of kiwifruit annually. Most of the Chinese kiwifruit production is for domestic consumption, whereas New Zealand, Chile, Greece, and Italy produce kiwifruit mainly for export [14].

Despite the significant growth in the global volume over the past 40 years, kiwifruit is still considered to be a relatively minor crop. However, is easily recognized by consumers due to its original appearance. Currently, kiwifruit is the sixth most valuable fruit crop in the world, after citrus, apples, table grapes, peaches/nectarines, and pears [10].

In recent years, there is a lot of public awareness and interest in protecting the environment through reducing waste or by-products, recycling items, and reusing products. The agricultural industry is deeply committed to this cause as demonstrated by the increasing amount of published research. Published studies point to the high nutritional value of *Actinidia* spp. discarded unmarketable fruit as well as leaves, flowers, and shoots (Figure 3.1). Containing valuable compounds, this plant's parts could be used as food additives or as ingredients for pharmaceutic or cosmetic industries [8, 15–17]. *Actinidia* is a very vigorous climber that can produce a remarkable amount of leaf and shoot biomass during the summer and winter pruning. The fruit that is not marketable due to inadequate size, cosmetic blemishes, or physical defects may be used in infusions and decoctions as well [18]. In the next sections, the bioactive

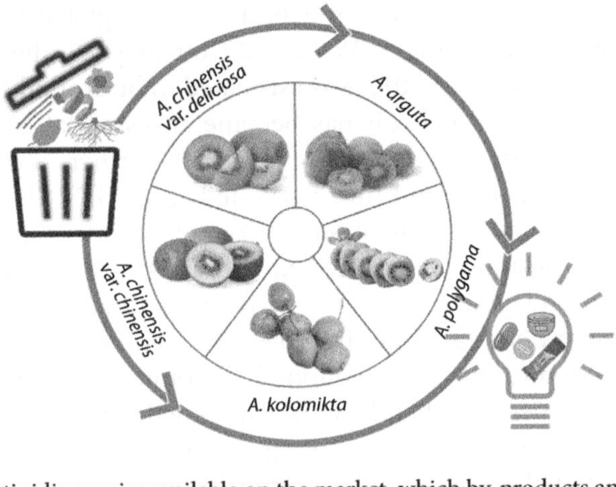

Figure 3.1 Actinidia species available on the market, which by-products and wastes have healthy benefits.

composition, health benefits and potential applications of these residues will be deeply explored. In this paper, for simplicity, we distinguished the two varieties of kiwi as two species - *A. chinensis* and *A. deliciosa.*

3.2 Bioactive Composition and Health Benefits of *Actinidia* spp. By-Products and Wastes

The global kiwi production generates enormous amounts of residues and wastes, being estimated that among 4.41 million tonnes of kiwi (2020 production) more than 1 million tons is converted into waste [15]. Fruits losses are principal composed by fruits that are not acceptable for commercialization due to damages, inadequate sizes or absence of comply with standard aspects established [18]. However, other by-products are produced during processing, such as skin, seeds, leaves, shoots, flowers, and roots [8, 15]. These by-products are characterized by high levels of bioactive compounds, being an actual challenge their valorization for food, pharmaceutical and cosmetic industries [15]. Therefore, this section will explore the bioactive composition of the different *Actinidia* fruits and associated by-products.

3.2.1 *A. chinensis var. deliciosa*

A. chinensis var. *deliciosa*, also known as green kiwifruit, is consumed in many parts of the world due to the health benefits associated [19]. According to the United States Department of Agriculture, green kiwifruit present 83.9 g of water/100 g of fruit and is composed by different minerals, such as potassium (198 mg/100 g of fresh weight (fw), calcium (35 mg/100 g fw), phosphorus (34 mg/100 g fw), magnesium (16 mg/100 g fw), and sodium (5 mg/100 g fw) [20]. Per 100 g of fruit, almost 75 mg of vitamin C are available [20]. Also, β-carotene is the main carotenoid available in the fruit (52 μg/100 g fw) [20]. The main soluble sugars present are glucose and fructose, whereas sucrose appears in smaller amounts [21]. In different varieties of *A. chinensis* var. *deliciosa*, *myo*-inositol varied between 0.029 and 0.177 g/100 g fw [21]. Forty-seven volatile and semi-volatile compounds were identified in the fruit, being distributed in chemical classes, namely 7 alcohols (e.g.:1-pentanol), 15 aldehydes (octanal), 3 ketones (1-penten-3-one), 11 esters (methylpropionate), 6 monoterpenes (limonene), 3 norisoprenoids (β-cyclocitral), and 2 hydrocarbon compounds (toluene) [22].

Latocha *et al.* studied the chemical composition of six *Actinidia* species, including *A. chinensis* var. *deliciosa* [23]. According to the authors, the total polyphenol content (TPC) of kiwifruit was 76.5 mg of gallic acid equivalents (GAE)/ 100 g fw [23]. Tannic acid and 2,5-dihydroxybenzoic acid were the principal phenolic compounds identified in *A. chinensis* var. *deliciosa* fruits, respectively, 4.63 and 2.86 mg/100 g fw [23]. The *in-vitro* antioxidant activity of the fruits was evaluated by the 2,2-diphenyl-2-picrylhydrazyl hydrate (DPPH) radical scavenging assay, achieving a value of 0.98 mg of ascorbic acid equivalent (AAE)/g fw [23]. In another study, 2,2′-azino-bis-(3-ethylbenzothiazoline-6-sulphonic acid) (ABTS) and DPPH radicals scavenging assays were applied to evaluate the antioxidant profile of two different genotypes (*A. chinensis* var. *deliciosa* cv. Qinmei and *A. chinensis* var. *deliciosa* cv. Hayward) [24]. The ABTS values ranged between 346 and 800 μM trolox, while the DPPH results were between 952 and 1600 μM trolox [24].

Fiorentino *et al.* also assessed the phytochemical composition of green kiwifruits [25]. In this study, the pulps were infused with ethanol during 5 days, being followed by infusion in diethyl ether for 5 days and hexane for 5 days [25]. The authors identified 27 phytochemicals, reporting compounds, such as δ-tocomonoenol, tocopherol, sterols, ursolic acid, caffeic acid glucosyl derivatives, chlorogenic acid, coumarin derivatives, kaempferol, quercetin, and rutin [25]. An *ex-vivo* study demonstrated that the consumption of green kiwifruits decreases the H_2O_2-induced DNA breakage in lymphocytes, exhibiting high antioxidant activity [26]. Based on these findings, the effect of kiwifruit on large intestine was assessed in healthy rats [27]. The animals were divided into two groups (n = 8/group) that received 10% of green kiwifruit or control diet (10 % glucose solution) for 4 or 6 weeks [27]. The consumption of green kiwifruit for 6 weeks leads to differences in the expression regulation of β-defensins, increasing the length of colon crypts, modifying the resident microbiota composition by increasing the total beneficial bacteria, which contributed to the optimal digestive health [27]. In another study, ovariectomized mice were used to evaluate the effect of *A. chinensis* var. *deliciosa* fruits on bone resorption [28]. After 9 weeks of kiwi intake, the authors observed a decrease of the serum C-terminal telopeptide of type I collagen levels, the mRNA expression and the bone resorption [28]. Furthermore, an *in-vivo* study performed by Duttaroy and Jørgensen associated the consumption of two or three kiwi fruits per day with the reduction of platelet aggregation and plasma lipid levels [29]. Thirty volunteers were recruited and distributed in two groups; a group received two green kiwifruits per day in the first period and three green kiwifruits per day in the second period and the other group took two and three kiwi fruit per day during all period, indistinctly (28 days) [29]. The trial shown that the consumption of two or three kiwifruits per day for 28 days increased the plasma levels of antioxidant activity (evaluated by FRAP assay) as well as the plasma levels of vitamin C (from 65.92 μM at day 0 to 102 μM and from 68 μM to 92.38 μM after consumption of two and three green kiwifruits, respectively) [29]. In addition, a significant decrease of the triglyceride concentrations was observed as well as the inhibition of the ADP-induced platelet aggregation [29]. A similar study was performed by Chang and Liu, which evaluated the effects of consumption of two green kiwifruits per day for 8 weeks on the antioxidant activity and the lipidic profile [30]. Forty-three subjects who had hyperlipidemia participated in this trial [30]. After 8 weeks, it was observed an increase of the high-density lipoprotein cholesterol (HDL-C) and a decrease of the low-density lipoprotein cholesterol (LDL-C)/HDL-C ratio and total cholesterol (TC)/HDL-C ratio [30]. Regarding the antioxidant activity, the concentration of vitamin C increased after 4 weeks of green kiwifruit consumption (from 50.2 μmol/L to 59.2 μmol/L), while the vitamin E concentration augmented after 8 weeks (from 43.6 μmol/L to 63.5 μmol/L) and the malondialdehyde + 4-hydroxy-2(E)-nonenal (MDA + 4-HNE) concentration decreased (from 17.2 μM to 5.67 μM), demonstrating the benefits of green kiwifruits intake [30].

During green kiwifruit preparation, the peels are removed, generating high amounts of residues. Annually, wastes and by-products from kiwifruits can account for up to 30% of the total kiwifruit crop and processing [31]. Almost 13% of the content of green kiwifruit peel is protein [32]. The peels are composed by different minerals, such as magnesium, potassium, calcium, sodium, and phosphorus [32]. In a study conducted by Asker Mohsen and Ahmed Habiba, the TPC and total flavonoid content (TFC) of four different peel extracts (prepared with 100 mL of distilled water, 80% ethanol, 80 % methanol (MeOH) and 80% acetone, respectively) were evaluated [32]. The results demonstrated that the TPC varied between 14.68 and 24.54 mg/g of dry weight (dw), while the TFC ranged from 11.69 to 24.47 mg/g dw for the ethanol and acetone extracts, respectively [32]. In addition, the authors selected the extract prepared with 80% acetone to characterize the phenolic profile [32]. The compounds identified were mainly syringic acid (197.72 mg/100 g dw), chyrsin (120.40 mg/100 g dw), quercetin (84.45 mg/100 g dw), catechin (26.66 mg/100 g dw), gallic acid (25.62 mg/100 g dw), and ellagic acid (17.79 mg/100 g dw) [32]. In another study, catechin, epicatechin, caffeic acid, and sinapic acid were reported in green kiwifruit peels [33]. The peels extracted with ultrasound-assisted extraction (UAE) were also evaluated regarding the phenolic profile, revealing the presence of epicatechin, quercetin-3-*O*-glucoside and quercetin-3-*O*-rhamnoside [34]. In what concern to the *in-vitro* antiradical activity, Asker Mohsen and Ahmed Habiba evaluated the potential through DPPH and ABTS assays [32]. The IC_{50} value for the DPPH assay was 54.85 μg/mL and was reported for the kiwifruit peel extract with 80 % acetone, while for the ABTS assay the acetone extract showed the highest antiradical activity (IC_{50} = 22.02 μg/mL) [32]. In the same way, the antioxidant power of the ultrasound-assisted extract of green kiwifruit peels was assessed through the capacity to inhibit the formation of thiobarbituric acid reactive substances (TBARS) and the oxidative hemolysis (OxHLIA). The IC_{50} values of 0.37 mg/mL and 1.01 mg/mL, respectively, highlight the antioxidant capacity of the phenolic compounds presents in peels [34]. Moreover, peels exhibited antibacterial and antifungal potential against gram-positive (*Bacillus subtilis* and *Staphylococcus aureus*) and gram-negative bacteria (*Escherichia coli, Enterobacter cloacae,* and *Pseudomonas aeruginosa*), yeasts (*Saccharomyces cerevisiae* and *Candida albicans*) and filamentous fungus (*Aspergillus flavus*) [32, 34]. The *in-vitro* anti-inflammatory activity of *A. chinensis* var. *deliciosa* peels was also evaluated by D'Eliseo *et al.* [35]. The extract prepared with ethanol/acidified water (pH 3.2; 70:30 (*v/v*)) presented as main polyphenol procyanidins (5.42 mg/ g of peel) [35]. The peel extract can inhibit inflammatory mediators, such as interleukins (IL-6 and IL-1β) and tumor necrosis factor (TNF-α), achieving inhibitory percentages of 81%, 68% and 63%, respectively [35]. In another study, peels from *A. chinensis* var. *deliciosa* and *A. chinensis* var. *chinensis* were extracted with 95% ethanol at 50°C for 3 days and their effects in mice sleep were evaluated [36]. The animals were administrated with 125, 250, 500, or 1000 mg/kg of extracts and a decrease of the sleep latency and an increase of the sleep duration was observed [36]. In the studies performed by Priyadarshi *et al.* and Li *et al.*, the green kiwifruit peels were incorporated in nanoparticles [37, 38]. Due to the polyphenol composition, the peels extracts were used as capping agents for sulfur nanoparticles (SNPs), leading to higher dispersion and stability, as well as better functionality [37]. In addition, the SNPs revealed antimicrobial potential against *E. coli* and *L. monocytogenes* [37]. In a study of Li *et al.*, zinc oxide nanoparticles were synthesized using peel extracts and the formulation exhibited bactericidal activity against *S. aureus*, as well as inhibitory activity of the colon cancer cells (HCT116) [38].

According to the traditional Chinese medicine, the roots of *A. chinensis* var. *deliciosa* are described as a "drug" for the treatment of various diseases, particularly hepatitis, edema, rheumatoid arthritis, or even cancer [39]. β-sitosterol, stearic acid, aesculetin, vanillic acid, protocatechuic acid, and tachioside are some of the phenolics identified in the roots extracts [39]. Bai *et al.* optimized the extraction time (min), ethanol fraction (%), liquid:solid ratio (volume per mass) and microwave power (W) of microwave-assisted extraction (MAE) of green kiwifruit roots [40]. The optimal conditions determined were 72.67%, 362.12 W, 15:1 (volume per mass), 30 min, that led to an extraction of 84.96% of triterpenoids [40].

The leaves of *A. chinensis* var. *deliciosa* are an excellent source of phenolic compounds with antioxidant activity [41]. The main compounds identified in leaves extracts were quercitrin and rutin, but compounds, such as proantocyanidin, quinic acid, myricitrin, quercetin, kaempferol, and triterpene acid-*O*-hexoside were also present [41, 42]. The authors also identified carotenoids, namely β-carotene and lutein [42]. Henriques *et al.* prepared a leaves decoction with distilled water for 10 minutes and evaluated the cytotoxicity on intestinal cell lines (Caco-2) as well as the permeation capacity [43]. Only rutin and quercitrin were able to permeate the monolayer of Caco-2 cells, being detected 0.09% and 0.14%, respectively, in the basolateral chamber [43]. The acetylcholinesterase (AChE) inhibitory capacity was also determined, achieving an IC_{50} of 0.56 mg/mL [43]. The anti-hyperglycemic effect of *A. chinensis* var. *deliciosa* leaves was investigated by Shirosaki *et al.* through *in-vivo* studies in mice [44]. The authors evaluated the postprandial absorption of carbohydrates and demonstrated that the administration of green kiwifruit leaves extract with soluble starch or sucrose suppressed the blood glucose level after 30 minutes [44]. In addition, the α-amylase and α-glucosidase inhibitory activities were evaluated *in-vitro*, with leaves extracts prepared with MeOH achieving an IC_{50} value of 42.9 and 33.9 mg/mL, respectively [44].

In what concerns to *A. chinensis* var. *deliciosa* flowers, few studies have been performed, reporting their richness in terpene volatile compounds, dominated by (*E,E*)-α-farnesene and germacrene [45], and containing 12% of volatile ester, 6% to 9% of aromatic and 9% of straight-chain alcohols [46].

3.2.2 *A. chinensis var. chinensis*

A. chinensis var. *chinensis*, commonly known as golden kiwifruit, is a very nutritional and tasty fruit traditionally consumed fresh, in juices, yogurt, wine, milk beverage, and vinegar [47]. The branches and leaves, as well as vines and roots, have been described as a rich source of bioactive compounds with human health benefits [47].

A. chinensis var. *chinensis* fruit is constituted by high amounts of phenolic compounds, triterpenes, vitamin C and E [47]. The vitamin C content vary between 108.5 and 161.33 g/100 g fw [48]. Vitamins E (1.40 mg/100 g fw), folate (31 μg/100 g fw) and minerals, such as potassium (315 mg/100 g fw), and magnesium (12 mg/100 g fw), are also present [47, 49]. The citric acid ranges from 0.8 to 1.8 g/100 g fw, while the malic acid varied between 0.1 and 0.5 g/100 g fw [50]. Regarding the total sugar content, fructose and glucose are the main constituents, respectively ranging between 5.68 g and 5.80 g/100 g fw and 5.20 to 5.28 g/100 g fw [48]. The main carotenoids present are α-carotene, β-carotene, lutein and zeaxanthin [48]. In what concerns to the TPC, the golden kiwifruits present a value of 115.76 mg GAE/100 g fw [51]. Zuo *et al.* screened the *in-vitro* antioxidant capacity of golden fruits extracted with ethanol (60 %) through an ultrasonic apparatus, using a ratio of 1:5 (*w/v*). The authors reported IC_{50} values of 13.97 and 45.60 mg/mL, respectively, for the DPPH and ABTS radicals' assays [51]. Epicatechin, procyanidin B2, quercetin-3-*O*-rhamnoside, gallic acid, chlorogenic acid, neochlorogenic acid, and caffeic acid were the principal compounds identified in *A. chinensis* var. *chinensis* fruits, being probably responsible for the antiradical results [52]. Nie *et al.* studied the inhibitory effects on digestive enzymes responsible for the control of hyperlipidemia, obesity, and diabetes, reporting inhibitory IC_{50} values on pancreatic and lipase α-glucosidase that ranged from 1.93 to 3.23 mg/mL and IC_{50} values that varied from 48.02 to 70.1 mg/mL, respectively [52]. In the same way, Eady *et al.* studied the impact of golden kiwifruit on the gastrointestinal health [53]. The study was based on the daily consumption of three fruits to alleviate constipation and improve gastrointestinal discomfort and enrolled 32 volunteers [53]. For 4 weeks, the participants consumed either three golden kiwifruit or 14.75 g Metamucil® (used as positive control) [53]. The trial compared the effect of kiwifruit with a baseline period of no consumption and demonstrated that the fruit consumption conducted to an increase on the number of complete spontaneous bowel movements, comparatively to the baseline period and the Metamucil® consumption [53]. The daily consumption of three golden kiwifruits improved the symptoms of gastrointestinal discomfort in mildly constipated adults [53]. Furthermore, the *in-vivo* antioxidant protection of golden kiwifruit was evaluated by Brevik and colleagues on 24 human healthy volunteers [54]. The volunteers received a supplementation with golden kiwifruit in a cross-over study for 4 weeks and venous blood samples were collected before, during and after tests to determine the levels of vitamin C, carotenoids, antioxidant activity and lipid oxidation [54]. The supplementation with two golden kiwifruits per day for 4 weeks increased the content of vitamin C and, consequently, decreased the H_2O_2-induced DNA damages [54]. The consumption of one fruit per day for 4 weeks reduced the blood platelet aggregation and the plasma triglyceride levels, without affecting the cholesterol levels [54]. Moreover *A. chinensis* var. *chinensis* fruits modulated the immune system through decrease the common infections, such as colds and influenza [49].

Regarding the peels of *A. chinensis* var. *chinensis*, extracts prepared with absolute MeOH (1:20; *w/v*) at 37°C present phenolic compounds such as catechin, epicatechin, quercetin, epigallocatechin, isoquercetin, and gallic acid, being catechin and quercetin the principal ones [55]. The TPC was 12.8 mg GAE/g dw, while the *in-vitro* antiradical activities determined by the DPPH and ABTS assays resulted in inhibitions of 81.5% and 90.0%, respectively, for the concentration of 60 μg/mL [55]. Alim *et al.* concluded that the golden kiwi peels presented an excellent antibacterial activity against *E. coli*, *L. monocytogenes* and *S. aureus*, and can also inhibit the proliferation of human liver cancer HepG2 cells (IC_{50} = 170.0 μg/mL) [55].

The roots of *A. chinensis var chinensis* are commonly used in traditional Chinese medicine [47], with different compounds isolated from them, such as epicatechin, epiafzelechin, benzylthioprocyanidol B2, emodin and β-sitosterol [56]. The main essential oils present in roots are dodecane, octane, decane, paeonal, camphor, *n*-decanoic acid, 4-methyldodecane, undecane, linoleic acid and linalool oxide [57]. Fang *et al.* investigated the anticancer activity on hepatocellular carcinoma of golden kiwi root extracts prepared by decoction, heating the roots in water at 100°C for 1 hour [58]. The cells were used to study the effect of roots on proliferation, migration, invasion, and apoptosis, followed by *in-vivo* assays in mice [58]. The root extracts inhibited the hepatocellular carcinoma cell proliferation to values inferior to 20%, while the increasing concentration of golden kiwi roots significantly decreased the distal-less homeobox 2 mRNA expression [58]. The *in-vivo* study was performed in the subcutaneous and orthotropic cell tumors using a mouse model. The animals were divided in two groups ($n = 6$), one administered with the roots extract and the other with distilled water (control) [58]. The results shown that golden kiwi roots inhibited the hepatocellular carcinoma growth, decreased the lung and intrahepatic metastasis, and reduced the mice weight and the rates of ascites in the xenograft tumor model [58]. In another study, the extracted triterpenoids from *A. chinensis* var. *chinensis* roots revealed antitumor effects against five human cancer cell lines, namely hepatocellular carcinoma, human lung cancer, human breast carcinoma, human ovarian carcinoma, and human cervical cancer cell lines, with IC_{50} values ranging between 10.30 and 94.45 μM [59]. Likewise, the roots of golden kiwi were investigated regarding the inhibition of cholesterol metabolism in hepatocellular carcinoma [60]. The results demonstrated that the roots extracts decreased the hepatocellular carcinoma cell viability (5 mg/mL in LM3 and HepG2 cells, and 30 mg/mL in HL-7702 cells) and inhibited the intracellular cholesterol levels through upregulation of PCSK9 and LDL receptor expression [60].

The leaves of *A. chinensis* var. *chinensis* were also investigated regarding composition and antioxidant activity [61]. Lv *et al.* employed the UAE to evaluate the influence of parameters such as power, duration, temperature, and solvent:solid ratio on the proanthocyanidins content extracted from leaves [61]. According to the authors, the optimal extract was obtained at 70 °C, with 40% of amplitude, 30 mL/g (solvent:solid ratio) during 15 min [61]. The optimal extract achieved IC_{50} values of 10.88 and 13.67 µg/mL, respectively, for the ABTS and DPPH assays [61]. In what concerns to the proanthocyanidins content, the optimal extract revealed a value of 122.19 mg procyanidin/g dw [61]. Phenolic compounds such as catechin, epicatechin, procyanidin and their derivatives, rutin, quercetin, isoquercitin and benzoyl glucuronide were identified in the extracts, being quercetin the major one (18.617 %) [61]. In addition, leaf extract at 50 µg/mL significantly reduced the viability of HepG-2 and Caco-2 cells [61]. In another study, golden kiwi leaves were extracted with 80% MeOH originating an extract rich in sinensin, naringin, rutin, quercitrin, kaempferol, myricitrin, isoquercitrin, and hydroxybenzoic acid [62]. The leaves also presented antibacterial activity, namely against *Streptococcus pyogenes* and *P. aeruginosa* [63].

Regarding *A. chinensis var chinensis* flowers, 48 volatile compounds were identified, particularly octanal, nonanal, linalool, decanal, 2-phenylethyl acetate, tetradecane, geranyl acetone, pentadecane, hexadecane, heptadecane, octadecane and nonadecane [64]. In another study, Tatsuka *et al.* identified 87 volatile components by capillary gas chromatography and gas chromatography-mass spectrometry, being 32 carbonyls, 24 alcohols, 23 hydrocarbons, 6 esters, 1 acid, and 2 miscellaneous compounds [65].

3.2.3 *Actinidia arguta*

A. arguta fruit, commonly known as kiwiberry, is characterized by a sensitive and delicate skin, associated with a sweet and intense flavor and an aromatic taste [66]. Despite the health benefits associated with its consumption, many residues and by-products are generated, such as pomace, skins, or leaves [67]. The rich composition of kiwiberry in bioactive compounds has been deeply reported by different authors. Wojdyło *et al.* revealed that *A. arguta* fruits have more than 6000 mg of phenolic compounds/100 g dw [19]. The main constituents are flavonols, anthocyanins, quercetin, kaempferol, quinic acid, luteolin glucuronide, and myristin [18, 19]. In addition, kiwiberry fruits are a good source of vitamin C, with values ranging between 76.09 and 282.56 mg/100 g fw [19]. The rich composition underlines the health benefits of the fruit consumption, such as antioxidant capacity and antidiabetic effect through the inhibition of α-amylase and α-glucosidase [19]. The benefits of this fruit also involve the gastrointestinal and hypercholesterolemia prevention as well as anti-cancer effects [6].

In the same line, Nishimura *et al.* studied the *in-vivo* effect of kiwiberry juice as anti-mutagenic [68]. The authors used a mouse ear model and divided the animals in four groups ($n = 8$): water (as control), kiwiberry juice in different concentrations and 2-amino-3,8-dimethylimidazo[4,5-f]quinoxaline (MeIQx) (as positive control) [68]. The results demonstrated that the topical application of juice inhibited the 12-*O*-tetradecanoylphorbol-13-acetate (TPA)-induced edema of mouse ear, reporting that the application of juice led to a decrease of 48.2 % in the mice ears inflammation [68]. The authors also evaluated the formation of DNA adducts in mice's liver and verified that a significant decrease after the administration of juice compared with group without juice, revealing the antimutagenicity potential of the kiwiberry juice [68]. In another study, Ha *et al.* evaluated the effect of kiwiberry fruits extracts prepared with 80% ethanol at 60°C for 2 h in the cognitive function [69]. Firstly, the authors evaluated the inhibitory effect against AChE enzyme, reporting a good inhibitory capacity (IC_{50} = 53 µg/mL) [69]. The extract was also tested against the H_2O_2-induced oxidative stress on neuronal cells (PC 12 cell line), showing a potent inhibitory effect (41%) on the intracellular oxidative stress [69]. Finally, *in-vivo* assays were performed in mice. The animals were administrated with kiwiberry fruit extract at concentrations of 5, 10, and 20 mg/kg body weight (bw) and after 3 weeks were injected with trimethyltin (TMT) (a neurotoxic compound) [69]. The *in-vivo* studies demonstrated that the extract at 20 mg/kg bw was able to decrease the AChE activity (around 10 %), increasing the superoxide dismutase (SOD) activity (2.72 U/mg protein), decreasing the ratio between oxidized glutathione dimer (GSSG) and glutathione peroxidase (GSH) (8.20%) and the lipid peroxidation (approximately 18%) [69]. Thus, the kiwiberry fruit can be a possible ingredient to prevent neurodegenerative disorders. Inclusively, in China dried kiwiberry fruits are commercialized in traditional herbal medicine markets as DA-9102 [70]. In a study conducted by Choi *et al.*, male hairless rats were used to evaluate spontaneous dermatitis induced by magnesium deficiency. The animals daily received DA-9102 (100 mg/kg) via oral route for 16 days [70]. The treatment inhibited the erythematous pruritic rash, the erosion, the hemorrhage and decreased the thickening of the epidermis and dermis, demonstrating efficacy and safety in the systemic treatment of atopic dermatitis-like skin disease [70].

The leaves of *A. arguta* are another by-product generated in high quantities during fruit maturation [71]. Various studies showed that leaves contain a high number of bioactive compounds [8, 16, 66, 67, 71–74]. Regarding vitamin C content, Tan *et al.* demonstrated that young apical leaves had highest vitamin C content (7.47 mg/g), comparatively with fruits [72]. In addition, phenolic acids are the main class of phenolics identified in leaves, followed by flavonols [67]. The neochlorogenic acid, a phenolic acid, is the principal compound quantified in leaves (145.39 mg/g of leaves) [73]. Bioactive compounds, such as chlorogenic acid, gallic acid, catechin, epicatechin, kaempferol-3-*O*-glucoside, isorhamnetin-3-*O*-rutinoside, and rutin are other examples of bioactive compounds present in *A. arguta* leaves extracts [16, 66, 67, 73]. *In-vitro* studies concluded about the great antioxidant/ antiradical potential of *A. arguta* leaves extracts through FRAP, DPPH$^{\cdot}$ and ABTS scavenging assays [16, 66, 67, 71, 74]. The highest capacity was obtained by the extract prepared by Almeida *et al.* with water and ethanol during 1 h, at 50 °C through conventional extraction (IC_{50} = 53.95 μg/mL for DPPH and 6628.42 μmol of ferrous sulphate/g dw for FRAP assays), followed by the hydroalcoholic extract obtained by Silva *et al.* through MAE (IC_{50} = 131.58 μg/mL for ABTS assay) [16, 74]. In the same way, the *A. arguta* leaves extracts exhibited a high capacity to scavenge reactive oxygen and nitrogen species (ROS and RNS, respectively), with IC_{50} values of 29.10, 1.87, and 1.18 μg/mL against superoxide anion radical ($O_2^{\cdot-}$), hypochlorous acid (HOCl), and nitric oxide ($^{\cdot}$NO) [74]. A great antidiabetic effect has been also reported [75]. Kwon *et al.* evaluated the *in-vitro* antidiabetic effect based on the α-glucosidase inhibition by *A. arguta* leaves extracted with water, 50% of ethanol, and 100% of ethanol [75]. The results showed that the extracts prepared with 50% of ethanol exhibited the highest inhibitory activity on α-glucosidase, namely 61.6 % [75]. Pinoresinol diglucoside and fertaric acid were the compounds identified as potential inhibitors of α-glucosidase activity [75]. Yoo *et al.* also screened the potential effect of kiwiberry leaves against hyperglycemic caused by diabetes [76]. The authors performed *in-vivo* studies with five groups of mice (a control group, a group with streptozotocin, a group with streptozotocin + metformin 100 mg/kg bw, a group with streptozotocin + kiwiberry leaves 100 mg/kg bw and a group with streptozotocin + kiwiberry leaves 200 mg/kg bw) and observed that the group treated with streptozotocin and kiwiberry leaves at concentration of 200 mg/kg bw maintained the body weight, reduced the fasting glucose levels, increased the content of SOD and decreased the content of MDA [76]. In addition, the leaves of *A. arguta* can contribute to prevent neuronal diseases, with AChE inhibitory activity of almost 30 % [74]. The *A. arguta* leaves extracted with ethanol (75 %) also presented anti-inflammatory effects through nitric oxide synthase suppression, being caffeoylthreonic acid, a caffeic acid derivative, the responsible compound for this activity [77].

More recently, Silva *et al.* studied the skin application effects of *A. arguta* leaves extracts [78]. The study revealed that the hydroalcoholic extract of

A. arguta obtained by MAE presented an elastase inhibition of 65.62% and a hyaluronidase inhibition of 54.64 %, enzymes enrolled in the of skin aging process [78]. Furthermore, the *ex-vivo* skin permeation assay through Franz cells (coupled to human skin) showed a good permeability of the bioactive compounds, namely 1-feruloylquinic acid, 5-caffeoylquinic acid, and rutin [78]. The extract was classified as non-irritating for skin and eye applications and safety for topical application by *in-vivo* Patch test performed in human volunteers [78].

A. arguta stems are another by-product that have been associated with bioactive compounds, such as catechin, epicatechin, luteolin, caffeic acid, chicoric acid, chlorophyll a and β-carotene [79, 80]. The total phenolic content of stems varied between 32.87 and 557.23 mg GAE/g of extract [81]. The DPPH radical scavenging activity assay achieved IC_{50} of 14.28 and 296.31 μg/mL for ethyl acetate and *n*-butanol extracts, respectively [81]. Stems also have a good α-glucosidase inhibitory potential, having a positive impact on diseases such as diabetes [81]. Takano *et al.* extracted stems with MeOH and assessed the extract effect on bone marrow cell proliferation [80]. The bone marrow cells were isolated from femurs of mice that were first administrated orally with MeOH extract of *A. arguta* stems at a dose of 100 mg/kg per day for 5 consecutive days. The authors observed an increase in the number of myeloid colonies and IL-3, as well as a decrease on the bone marrow functions, probably due to the presence of catechin and epicatechin [80]. The anti-inflammatory activity of *A. arguta* stems were investigated by Kim *et al.* [82]. The anti-inflammatory capacity was proved due to the significant reduction of nitric oxide production, lipopolysaccharide-induced mRNA levels of cytokines (including IL-1β and TNF-α) and blocking of the signaling pathways of nuclear factor (NF)-κB and mitogen-activated protein kinases (MAPKs) [82].

The roots of *A. arguta* are commonly used in traditional medicine due to its benefic properties for obesity [83–85]. Jang *et al.* isolated six pentacyclic triterpenes, as ursolic acid and betulinic acid, to evaluate the inhibitory capacity of pancreatic lipase [84]. The results showed that the ursolic acid has achieved the highest inhibitory potential (IC_{50} = 9.45 μg/mL) [84]. Thereafter, the anti-lipase effect of ursolic acid isolated from *A. arguta* roots was investigated in Male Wistar rats [85]. The ursolic acid, administrated at concentrations of 50 and 100 mg/kg bw, decreased the plasma triacylglycerol levels and inhibited the phosphodiesterase product, also inducing the release of free fatty acids in fat cells by lipolysis [85]. In another study, Teng *et al.* reported the antinociceptive and anti-inflammatory effects of *A. arguta* roots [83]. Roots were extracted with ethanol and *in-vivo* studies were performed in Wistar rats through chemically and thermally induced nociceptive pain, while carrageenan induced paw edema [83]. The extract was able to inhibit the acetic acid-induced writhing and the carrageenan induced paw oedema, while increased the phases of the formalin-induced pain response [83].

3.2.4 *Actinidia kolomikta*

A. kolomikta is a fruit with multiple benefits, being described as a healthy food [86]. The fruit present a TPC of 430.03 mg/100 g fw, a TFC of 69.05 mg of catechin equivalents/100 g fw, and a total vitamin C of 211.12 mg/100 g fw [51]. Regarding the antiradical capacity, the IC_{50} values of the DPPH and ABTS assays were, respectively, 1.35 and 5.68 mg/mL [51]. In addition, the fruit extract inhibited the HT29 and HepG2 cells growth, presenting an $EC_{50} > 75$ mg/mL [51]. The *A. kolomikta* fruits also have capacity to inhibit the TNF-α, ciclooxigenase-2 (COX-2) and IL-6 expression [86].

In what concerns to *A. kolomikta* leaves, different phenolic compounds are present, with particular focus on epicatechin, catechin, procyanidin, apigenin, caffeic acid, chlorogenic acid, neochlorogenic acid, kaempferol, quercetin, rutin, and isorhamnetin [87]. Six flavonoids, namely kaempferol-7-*O*-α-L-rhamnoside, kaempferol-3-*O*-β-D-galactoside, quercetin-3-*O*-β-D-glucoside, isorhamnetin-3-*O*-β-D-glucoside, kaempferol-7-*O*-α-L-rhamnosyl-3-*O*-rutinoside, and kaempferol-3-*O*-[α-rhamnopyranosyl-(1→3)-α-L-rhamnopyranosyl-(1→6)-α-D-galactopyranoside], have been isolated from leaves [88]. Yuan *et al.* studied the *in-vitro* and *in-vivo* antidiabetic capacity of *A. kolomikta* roots and leaves extracted with water and ethanol [89]. The roots ethanolic extract showed the highest inhibitory activity against α-glucosidase, decreasing *in-vivo* the serum uric acid and the blood urea nitrogen, as well as the total cholesterol, triglycerides and high-density lipoproteins [89]. The authors also observed that the aqueous leaves extract can prevent the blood glucose increase, without causing a hypoglycemic state [89]. In the same way, Hu *et al.* investigated the antidiabetic effect of the aqueous extract of *A. kolomikta* roots [90]. The results demonstrated that the extract has a dose dependent inhibition of yeast α-glycosidase activity, probably due to the phenolic composition that is rich in phenolic acids, phenols, phenyl propanols, flavonoids, minor flavonoids, quinines, and tannins [90]. In addition, the ethanolic extract of *A. kolomikta* roots exhibited anti-hyperglycemic effects and inhibitory capacity against α-glucosidase, improving the insulin release [91]. The ethanolic root extract is characterized by high contents of phenolic and flavonoid compounds, with a TPC of 80.49 mg GAE/g and a

TFC of 430.69 mg of rutin equivalents/g [91]. The *A. kolomikta* roots are also associated with high antioxidant levels and inhibitory capacity against DLD-1 colon cancer cells proliferation [92]. Therefore, the roots extracts may reduce the radical oxygen species levels and protect macrophages from damages by immunomodulatory activation [93].

3.2.5 *Actinidia polygama*

A. polygama is another specie of *Actinidia* genus with extremely interesting properties for medicinal application. The fruit is characterized by the presence of vitamins C and A, tannins, proteins, and organic acids [94]. New triterpenoids were recently identified in *A. polygama* fruits, namely 2α,3α,23,24-tetrahydroxyurs-12-en-28-oic acid, 2α,3α-dihydroxyurs-12-en-24-al-28-oic acid, 2α,3α,24-trihydroxyurs-12-ene-23,28-dioic acid, 2α,3α,23-trihydroxyursa-12,20(30)-dien-28-oic acid, and 3β-(trans-*p*-coumaroyloxy)-2α,23-dihydroxy-urs-12-en-28-oic acid [95]. The main polysaccharides present are maltose, glucose, and fructose [94]. This fruit is known as a medicinal herb in Korea, being employed to treat diseases such as pain, rheumatic arthritis, and inflammation [96]. In fact, the fruit extracts have anti-inflammatory and anti-asthmatic effects, demonstrated by the inhibitory effects on airway hyperresponsiveness, namely through the reduction of leukocytes and eosinophils as well as IL-4, IL-5, IL-13, and IgE levels [96]. The *A. polygama* fruit extracts can also activate the mRNA expression of COX-2, TNF-α, IL-6, and IL-1β [94].

In-vivo studies performed in mice with high-fat-diet-induced obesity allowed to evaluate the anti-obesity effect of *A. polygama* dried fruits extracted with 70% ethanol [97]. The results demonstrated that the fruit extract had the capacity to reduce the body weight gain, the adipose tissue accumulation, the adipocyte size and the serum triglyceride [97]. *In-vitro* studies also proved that the fruit extracts have heat-stable agents with inhibitory capacity against promyelocytic leukemia (HL-60) cell line [98]. In addition, Kim *et al.* evaluated the anti-photoaging effect of *A. polygama* fruits extracted with 30% ethanol or water at 85°C for 6 h [99]. Both extracts, when administrated to hairless female mice, were able to suppress the degradation of collagen in the dorsal skin of ultraviolet B (UVB)-irradiated mice, inhibit the UVB-induced matrix metalloproteinases (MMPs) 1, 3, and 9 and protect the UVB-induced collagen type 1 down-regulation [99].

The flowers of *A. polygama* showed the presence of 11 volatile compounds, being linalool, 2-hexenal, 2-hexen-1-ol, and β-citronellol the main constituents [100]. Unfortunately, few studies reported the effect of *A. polygama* by-products and residues.

Khromykh *et al.* studied the antioxidant and antimicrobial activities of fruits and leaves of *A. polygama*, reporting a TPC that varied between 136.2 and 291.9 mg GAE/100 g fw, respectively, for the pulp and the peel of *A. polygama* [101]. In addition, the total antioxidant capacity was 736.2 and 1269.1 mg AAE/100 g fw for pulp and peel, respectively [101]. Regarding leaves, the TPC was 1138.9 mg GAE/100 g fw, while the antioxidant activity was 2487.4 mg of AAE/100 g fw [101]. The leaves extracts revealed a high inhibitory capacity against *P. aeruginosa* and *E. coli* [101].

In Table 3.1 summarizes health benefits associated with bioactive compounds from *Actinidia* sp. fruits and by-products as described above.

3.3 Products Available on the Market With Actinidia Fruits and By-Products—An Overview

Actinidia fruits and its by-products and residues are rich in compounds with biological activities that are of huge interest for food, cosmetic and pharmaceutical industries, which is justified by the exhibited pharmacologic properties, including antidiabetic [102, 103], antifungal [104], antiviral [105], anti-constipation [106, 107], anti-hypertensive [108], anti-inflammatory [109, 110], antimicrobial [63, 111], anti-tumor [59, 112], anti-ulcer [113], antioxidant activity [114, 115], hypoglycemic [108], hypolipidemic [109], hepatoprotective [116], and sedative-hypnotic [36] effects.

Currently, there is a large variety of products with *Actinidia* in their constitution on the market. The products available on the market that contain in their composition *Actinidia* fruits, by-products or residues are listed in the following subsections.

3.3.1 Food

The consumption of kiwifruit is associated with several digestive benefits [7]. In addition to the ingestion motivated by its nutritional value, namely its richness in vitamins and minerals previously mentioned, kiwifruit consumption provides digestive benefits, both for healthy people and people with constipation or other gastrointestinal disorders, including irritable bowel syndrome [7]. There are two main compounds responsible for these

Table 3.1 Bioactive compounds present in *Actinidia* species fruits and by-products and their potential associated benefits to human health.

Actinidia species	Fruits or by-products parts	Bioactive compounds	Health benefits associated	References
A. chinensis var. *deliciosa* (green kiwifruit)	Fruits	Vitamin C Vitamin E Potassium Calcium Phosphorus Magnesium Sodium β-carotene Myo-inositol Tannic acid 2,5-dihydroxybenzoic acid δ-tocomonoenol Tocopherol Sterols Ursolic acid Caffeic acid glucosyl derivatives Chlorogenic acid Coumarin derivatives Kaempferol Quercetin	Antioxidant activity Protect DNA breakage in lymphocytes Bone resorption Improve digestive health through increase the length of colon crypts and microbiota composition Increase the plasma levels of vitamin C Decrease of the triglyceride concentrations Reduction of platelet aggregation Increase of the HDL-C and a decrease of the LDL-C/ HDL-C ratio and TC/ HDL-C ratio	[20, 22–25, 27–30]

(Continued)

Table 3.1 Bioactive compounds present in Actinidia species fruits and by-products and their potential associated benefits to human health. (*Continued*)

Actinidia species	Fruits or by-products parts	Bioactive compounds	Health benefits associated	References
		Rutin Semi-volatile compounds - Alcohols; - Aldehydes; - Ketones; - Esters; - Monoterpenes; - Norisoprenoids; - Hydrocarbon compounds		
	Peels	Syringic acid Chyrsin Catechin Gallic acid Ellagic acid Catechin Epicatechin Caffeic acid Sinapic acid Quercetin-3-*O*-glucoside Quercetin-3-*O*-rhamnoside Procyanidins	Antioxidant activity Antibacterial and antifungal activities Anti-inflammatory activity Improve sleep duration capping action of sulfur nanoparticles	[32–38]

(*Continued*)

Table 3.1 Bioactive compounds present in Actinidia species fruits and by-products and their potential associated benefits to human health. (*Continued*)

Actinidia species	Fruits or by-products parts	Bioactive compounds	Health benefits associated	References
	Roots	β-sitosterol Stearic acid Aesculetin Vanillic acid Protocatechuic acid Tachioside Tripernoids	Treatment for hepatitis, edema, rheumatoid arthritis, and cancer	[39, 40]
	Leaves	Quercitrin Quercetin Rutin Proantocyanidin Quinic acid Myricitrin Kaempferol Triterpene acid-*O*-hexoside β-carotene Lutein	Antioxidant activity AChE inhibitory capacity Anti-hyperglycemic activity Inhibitory activities of α-amylase and α-glucosidase	[41–44]

(*Continued*)

Table 3.1 Bioactive compounds present in Actinidia species fruits and by-products and their potential associated benefits to human health. (*Continued*)

Actinidia species	Fruits or by-products parts	Bioactive compounds	Health benefits associated	References
	Flowers	(*E*,*E*)-α-farnesene Germacrene Volatile ester Aromatic and straight-chain alcohols		[45, 46]

(*Continued*)

Table 3.1 Bioactive compounds present in Actinidia species fruits and by-products and their potential associated benefits to human health. (*Continued*)

Actinidia species	Fruits or by-products parts	Bioactive compounds	Health benefits associated	References
A. chinensis var chinensis (golden kiwifruit)	Fruits	Vitamin C Vitamin E Potassium Magnesium Triterpenes Citric acid Malic acid α-carotene β-carotene Lutein Zeaxanthin Epicatechin Procyanidin B2 Quercetin-3-*O*-rhamnoside Gallic acid Chlorogenic acid, Neochlorogenic acid Caffeic acid	Antioxidant activity Inhibitory activities of pancreatic and lipase α-glucosidase Improve the gastrointestinal discomfort Increase the plasma levels of vitamin C Protect the H_2O_2-induced DNA damages Reduce cholesterol levels Improve immune system	[47–54]

(*Continued*)

Table 3.1 Bioactive compounds present in Actinidia species fruits and by-products and their potential associated benefits to human health. (*Continued*)

Actinidia species	Fruits or by-products parts	Bioactive compounds	Health benefits associated	References
	Peels	Catechin Epicatechin Quercetin Epigallocatechin Isoquercetin Gallic acid	Antioxidant activity Antibacterial activity	[55]

(*Continued*)

Table 3.1 Bioactive compounds present in Actinidia species fruits and by-products and their potential associated benefits to human health. (*Continued*)

Actinidia species	Fruits or by-products parts	Bioactive compounds	Health benefits associated	References
	Roots	Epicatechin Epiafzelechin Benzylthioprocyanidol B2 Emodin β-sitosterol Dodecane Octane Decane Paeonal Camphor *n*-decanoic acid 4-methyldodecane Undecane Linoleic acid Linalool oxide Triterpenoids	Anticancer activity against hepatocellular carcinoma, human lung cancer, human breast carcinoma, human ovarian carcinoma, and human cervical cancer Reduce cholesterol levels	[56–60]

(*Continued*)

Table 3.1 Bioactive compounds present in Actinidia species fruits and by-products and their potential associated benefits to human health. (*Continued*)

Actinidia species	Fruits or by-products parts	Bioactive compounds	Health benefits associated	References
	Leaves	Catechin Epicatechin Procyanidin Rutin Quercetin Isoquercitin Benzoyl glucuronide Sinensin Naringin Quercitrin Kaempferol Myricitrin Isoquercitrin Hydroxybenzoic acid	Antioxidant activity Antibacterial activity	[61–63]
	Flowers	Volatile compounds - Carbonyls; - Alcohols; - Hydrocarbons; - Esters; - Acids; - Miscellaneous compounds		[64, 65]

(*Continued*)

Table 3.1 Bioactive compounds present in Actinidia species fruits and by-products and their potential associated benefits to human health. (*Continued*)

***Actinidia* species**	**Fruits or by-products parts**	**Bioactive compounds**	**Health benefits associated**	**References**
A. arguta (kiwiberry)	Fruits	Vitamin C Flavonols Anthocyanins Quercetin Kaempferol Quinic acid Luteolin glucuronide Myristin	Antioxidant activity Antidiabetic activity Inhibitory activities of α-amylase and α-glucosidase Antimutagenicity potential Improve gastrointestinal function AChE inhibitory capacity Reduce lipid peroxidation Prevent neurodegenerative diseases Prevent and treat atopic dermatitis	[19, 68–70]

(*Continued*)

Table 3.1 Bioactive compounds present in Actinidia species fruits and by-products and their potential associated benefits to human health. (*Continued*)

Actinidia species	Fruits or by-products parts	Bioactive compounds	Health benefits associated	References
	Leaves	Vitamin C Neochlorogenic acid Chlorogenic acid Gallic acid Catechin Epicatechin Kaempferol-3-*O*-glucoside Isorhamnetin-3-*O*-rutinoside Rutin Pinoresinol diglucoside Fertaric acid Caffeoylthreonic acid Caffeic acid derivative 1-feruloylquinic acid 5-caffeoylquinic acid	Antioxidant activity Antidiabetic activity Anti-inflammatory activity Inhibitory activity of α-glucosidase Prevent neuronal diseases AChE inhibitory capacity Decrease glucose levels Skin anti-aging potential	[16, 66, 67, 71–78]

(*Continued*)

Table 3.1 Bioactive compounds present in Actinidia species fruits and by-products and their potential associated benefits to human health. (*Continued*)

Actinidia species	Fruits or by-products parts	Bioactive compounds	Health benefits associated	References
	Stems	Catechin Epicatechin Luteolin Caffeic acid Chicoric acid Chlorophyll a β-carotene	Antioxidant activity Anti-inflammatory activity Inhibitory activity of α-glucosidase Reduce the bone marrow functions	[79–82]
	Roots	Pentacyclic triterpenes	Antinociceptive activity Anti-inflammatory activity Prevent obesity	[83–85]
A. kolomikta	Fruit	Polyphenols Flavonoids Vitamin C	Antioxidant activity Anti-inflammatory activity Inhibitory activitis of TNF-α, COX-2 and IL-6 expressions	[51, 86]

(*Continued*)

Table 3.1 Bioactive compounds present in Actinidia species fruits and by-products and their potential associated benefits to human health. (*Continued*)

Actinidia species	Fruits or by-products parts	Bioactive compounds	Health benefits associated	References
	Leaves	Epicatechin Catechin Procyanidin Apigenin Caffeic acid Chlorogenic acid Neochlorogenic acid Kaempferol Quercetin Rutin Isorhamnetin	Prevent the blood glucose increase	[87–89]
	Roots	Phenolic acids Phenols Phenyl propanols Flavonoids Quinines Tannins	Antioxidant activity Antidiabetic activity Anti-hyperglycemic activity Reduce cholesterol levels Decrease serum uric acid and the blood urea nitrogen Inhibitory activity of α-glucosidase	[89–92]

(*Continued*)

Table 3.1 Bioactive compounds present in Actinidia species fruits and by-products and their potential associated benefits to human health. (*Continued*)

Actinidia species	Fruits or by-products parts	Bioactive compounds	Health benefits associated	References
A. polygama	Fruits	Vitamin A Vitamin C Tannins Triterpenoids	Anti-inflammatory activity Anti-asthmatic activity Prevent and treat pain, rheumatic arthritis, and inflammation Prevent obesity Anti-photoaging activity Antioxidant activity	[94–97, 99, 101]
	Flowers	Volatile compounds - Linalool - 2-hexenal - 2-hexen-1-ol - β-citronellol		[100]
	Peels	Polyphenols	Antioxidant activity Antimicrobial activity	[101]
	Leaves	Polyphenols	Antioxidant activity Antimicrobial activity	[101]

gastrointestinal effects: (i) actinidin, a proteolytic enzyme that facilitates the digestion of protein-rich foods, such as meat; and (ii) the high amount of fibers present in kiwifruit. The soluble and insoluble fibers of kiwifruit add bulk, increase water retention in the colon and change faecal consistency, while soluble fiber act as substrate for the microflora in the gastrointestinal tract. As a result, there is a decrease in the transit time and a reduction of the abdominal discomfort [7].

Kiwifruit is consumed all over the world not only raw, but also processed forms. These products often maintain the aroma profile as well as the beneficial nutritional components of the kiwifruit. Presently, a wide range of kiwifruit processed products can be found in the market (Table 3.2), namely yogurts, cookies, jams, jellies, fruit concentrates, juices, chocolates, syrups, energy gels and bars, baby food (fruit puree), and vinegar. Nevertheless, no food products that use *Actinidia* residues or by-products in their production were observed on the market.

3.3.2 Cosmetics

The value of the beauty and personal care market worldwide reached US$534 billion in 2022 and is expected to exceed US$670 billion by 2026 [117]. In such a demanding and dynamic market, it is essential to develop innovative products that can respond to the growing awareness of the potential toxicity and safety of ingredients by consumers, who are increasingly looking for fragrance-free products made with natural ingredients.

Considering all the properties mentioned above, *Actinidia* emerges as a multifunctional natural ingredient with potential to be incorporated as active ingredient into cosmetics. Currently, a considerable number of cosmetics with *Actinidia* in its composition can be found on the market (Table 3.3). Most of the cosmetic products identified are intended for skin care, which is in line with the data revealed in 2022 that showed skin care as approximately 27.23% of the global beauty and personal care market [117].

According to the cosmetic companies, skin care products containing kiwi extract in its formulation have a revitalizing and rejuvenating effect on the skin, with soothing properties. These properties may be explained by the polysaccharides present in *A. chinensis* var. *chinensis* extract, as its effects on cell proliferation via enhanced growth factor receptors, energy production and collagen synthesis were already observed in keratinocytes, fibroblasts and three-dimensional skin equivalents [118]. The high content of ascorbic acid present in *A. chinensis* var. *chinensis* fruit may also be responsible for the activities of these products, since it has important biological functions that are relevant to skin health, such as promoting the

Table 3.2 Commercialization of *Actinidia* spp. fruit processed food products.

Kiwi-based product	Example of producing brand or company	Product name
Dehydrated	Nutraj	Premium Dried Kiwi
Jam	Aródama	Kiwi-apple-banana marmalade
Jelly	Aloha From Oregon	Kiwi pepper jelly
Fruit concentrate	Fabbri	Delipaste Kiwi
Yogurt	Activia	0% Strawberry & Kiwi Yogurt
Cookies	Hijas del Sol	Oat Bran Cookies with Kiwi
Juice	Snapple	Kiwi Strawberry Juice
Chocolate	Supremely Gourmet	Kiwifruit chocolates
Syrup	Le Sirop de Monin	Monin Kiwi
Energy gels	Just Loading	Energy Gels Fresa y Kiwi
Energetic bar	Nutrinovex	Glucobar Kiwi
Fruit puree (baby food)	Holle	Kiwi Koala Pear & banana with kiwi
Powder	Green Heart Store	Kiwi Powder
Vinegar	Goulter's	Organic Kiwifruit Vinegar

Table 3.3 Cosmetic products commercially available containing *A. chinensis* var. *chinensis* fruit or by-products/wastes.

	Cosmetic	Product name	Brand	Active ingredient form
Face	Anti-aging cream	Rénergie Multi-Lift	Lancôme	Fruit water
		Age-defying Hydro Jelly Kiwi	Revuele	Fruit extract
		Sisleyouth Anti-Pollution	Sisley	Fruit extract
		Multi-active Nuit	Clarins	Fruit extract
		Facial cream anti-aging dark spot prevention	Radiance Cosmetics	Fruit water
	Anti-aging serum	Age-defying Intense Serum Kiwi	Revuele	Fruit extract
		Expert +	Revuele	Fruit juice
		Hydra Therapy	Revuele	Fruit juice
		Double Serum	Clarins	Fruit extract
		Anti-aging Dark Spot Prevention	Radiance Cosmetics	Fruit water
	Brightening serum	Brightening Peel Serum Papaya	Revuele	Fruit juice
		Gold kiwi brightening serum	Holika holika	Seed oil
	Brightening cream	Gold Kiwi Brightening Sleeping Cream	Holika holika	Seed oil Fruit extract
	Moisturizer oil	Kiwi Seed	Organic pure oil	Seed oil

(Continued)

Table 3.3 Cosmetic products commercially available containing *A. chinensis* var. *chinensis* fruit or by-products/wastes. (*Continued*)

	Cosmetic	Product name	Brand	Active ingredient form
	Facial cleanser	Age-defying Face Wash Kiwi	Revuele	Fruit extract
		Just Squalane Cleanser	Revox	Fruit juice
		Healing Tea Garden White Tea	The Saem	Fruit extract
		Face Cleansing Water Coconut Oil	Eco U	Fruit extract
	Face mask	Natural Mask Sheet Golden Kiwi	The Saem	Fruit extract
		Kiwi & Yogurt Face Mask	SYS	Fruit extract
		Botanic Treat	Marion	Fruit extract
		Natural Super Face Mask	Fito Bomb	Fruit juice
	Face primer	Clean ID Vitamin C Fresh Glow	Catrice	Fruit extract
		Energizing Prime & Care Spray	Catrice	Fruit extract
	Face milk	Milky boost	Clarins	Fruit extract
		Brightening milk whitening effect	Byphasse	Fruit extract
	Face scrub	Lemon + kiwi face scrub	Fancy Handy	Seeds
		Smooth Sugar Clear Scrub	L'Oréal Paris	Seeds
	Makeup setting mist	Fix It Forget It	Lancôme	Fruit extract

(*Continued*)

Table 3.3 Cosmetic products commercially available containing *A. chinensis* var. *chinensis* fruit or by-products/wastes. (*Continued*)

	Cosmetic	Product name	Brand	Active ingredient form
	Foundation	Forever Better Skin	Wibo	Fruit extract
		Phyto-teint nude	Sisley	Fruit extract
		Phyto-hydra teint	Sisley	Fruit extract
Hair	Shampoo	Smiling Kiwi Mild Shampoo	Freshly Cosmetics	Fruit water
		Shampoo Kiwi Fruit	Glatten Professionals	Fruit extract
		Volume Shampoo Chamomile Kiwi	Eubiona	Fruit extract
		Kiwi Lime Squeeze	Alberto VO5	Fruit extract
		Kiwi Dry Shampoo	Centifolia	Fruit extract Seeds
	Hair mask	Multivitamin Mask	Occun	Fruit extract
	Hair serum	Bed Head Control Freak	Tigi	Fruit extract
Eyes	Eye serum	Superfood cooling eye serum	Dr Botanicals	Seed oil
		Doble Serum Eye	Clarins	Fruit extract
	Eye patches	Liquid Eye Patches Kiwi	Revuele	Fruit extract
Body	Body cream	Body Firming	Clarins	Fruit extract
		Sugar Crush Moisturizing Body Buttercream	Soap & Glory	Fruit water
	Body wash	Sugar Crush Body Wash	Soap & Glory	Fruit extract
Nails	Nail polish	Clean & Strong	Essence	Fruit extract

collagen formation and providing the antioxidant protection to the skin [119].

The use of kiwi extract in brightening creams and serums is due to the presence of proanthocyanidins such as procyanidins, procyanidins gallate, and prodelphinidins, which strongly inhibit tyrosinase activity, the enzyme responsible to produce melanin [120]. Concerning eye cosmetics, the same benefits are advertised, namely the rejuvenating effect. The purpose of the use of *A. chinensis* var. *chinensis* seed oil in one of the eye serums is related to its ability to lock in moisture. As for hair cosmetics, the use of kiwi extract is closely related to its richness in ascorbic acid, which, according to the cosmetic companies, strengthens the hair's natural structure and slows down hair aging.

Interestingly, in all the cosmetic products listed, *A. chinensis* var. *chinensis* was the only specie used, more frequently in the form of fruit extract, juice, or water. Even so, there is an increasing focus on the inclusion of kiwi by-products, such as seeds and seed oil, in the production of some cosmetics.

3.3.3 Nutraceuticals and Pharmaceuticals

Several research studies have been conducted to explore the pharmacologic profile of kiwi. Although no evidence was found on the use of kiwifruit for pharmaceutical purposes, it is widely used for nutraceutical approaches. According to the producing companies, the benefits of consuming kiwi-based nutraceuticals are very similar to consuming the fresh fruit: support bowel regularity and digestive system health, decrease bloating and flatulence, and facilitate the digestion of proteins. The commercialized products available on the market are listed in Table 3.4. The commercially available products have different active substances, such as actinidin, the soluble and insoluble fibers naturally present in kiwi, or kiwi extract, rich in fibers, sugars, and enzymes. The products are divided into two large groups: those that facilitate the digestion and intestinal transit and the ones that have a detoxifying effect. No products were found that made use of waste or by-products from the kiwi production industry.

3.4 Conclusion and Future Prospects

The importance of the *Actinidia* fruits is undeniable, both for its economic value and for the health benefits already proven by several studies. Not only the fruits are rich in pro-health compounds but also the residues and by-products, namely the skin, seeds, leaves, roots, and flowers, proved

Table 3.4 Nutraceutical products commercially available containing *Actinidia* spp. fruit.

Product name	Producing brand or company	Kiwi-derived ingredient
Kiwi Bio-Boost Kids	Lifestream	Livaux® (kiwifruit extract)
Kiwi Gold Biotics	Lifestream	Livaux®
Pro-15	Hyperbiotics	Fruit extract
Sene + Kiwi	Biokygen	Fruit extract
Kiwisoothe	AFT Pharmaceuticals	Actinidin
Actinidina p200	Salvena	Actinidin
Kiwizyme	Viridian	Actinidin
Kiwi and Hibisco Fibers	Biolax	Kiwi fruit
Drenax 500 Detox & Depur	B. Green	Kiwi fruit
Multibiotic Microbiome Support	Together	Kiwi fruit
Blevit L	Ordesa	Kiwi powder
Greens Blend, Detox & Digest	Amazing Grass	Actazin® (kiwifruit powder)
Gut Connection - Digestive Balance	Country Life	Actazin®

to be an interesting source of bioactive compounds, such as carotenoids, polysaccharides, minerals, polyphenols, triterpenes, vitamins, and volatile substances. These compounds confer to the fruits interesting pharmacological properties such as antidiabetic, anti-constipation, anti-hypertensive, anti-inflammatory, antioxidant, antimicrobial, and anti-tumor activity, and hypoglycemic, hypolipidemic, hepatoprotective effects, among others.

While the kiwifruits are used as food, the wastes and by-products have great potential to be used in food, cosmetic, nutraceutical, and pharmaceutical industries. The wastes and by-products resulting from the cultivation and processing of these fruits can be transformed and incorporated into new products. Its subsequent introduction into the market favors a circular economy model that contributes to the reduction of the environmental impact in the generation of additional economic profits.

Nevertheless, more human studies are still needed to fully understand the effect of biologically active compounds present on these fruits, by-products and wastes on the human body, in order to be able to use them isolated or synergistically with other compounds to successfully prevent and combat diseases that affect humans.

Acknowledgments

This work received financial support from project EXPL/BAA-GR/0663/2021—Kiwi4Health—Exploring the Eco-Innovative Re-Use of Kiwiberry, supported by national funds by FCT/MCTES and by the projects UIDB/50006/2020 and UIDP/50006/2020 through national funds. This work was also financed by national funds from FCT - Fundação para a Ciência e a Tecnologia, I.P., in the scope of the project UIDP/04378/2020 and UIDB/04378/2020 of the Research Unit on Applied Molecular Biosciences - UCIBIO and the project LA/P/0140/2020 of the Associate Laboratory Institute for Health and Bioeconomy i4HB.

Ana Margarida Silva and Catarina Macedo are thankful for their Ph.D. grant (SFRH/BD/144994/2019 and 2022.12642.BD, respectively) financed by POPH-QREN and subsidized by the European Science Foundation and Ministério da Ciência, Tecnologia e Ensino Superior. Francisca Rodrigues (CEECIND/01886/2020) is thankful for her contract financed by FCT/MCTES—CEEC Individual Program Contract.

References

1. Huang, H., *The genus actinidia, a world monograph*, Science Press, Beijing, China, 2014.
2. Ferguson, A. and Bollard, E., Domestication of the kiwifruit, in: *Kiwifruit: Science and Management*, I.J. Warrington and G.C. Weston (Eds.), pp. 165–246, Ray Richards Publisher in association with the New Zealand Society for Horticultural Science, Auckland, 1990.
3. Latocha, P., Vereecke, D., Debersaques, F., Kiwiberry commercial production-what stage are we at?, in: *IX International Symposium on Kiwifruit*, Porto, Portugal, pp. 559–564, 2017.
4. Latocha, P., Debersaques, F., Hale, I., *Actinidia arguta* (Kiwiberry): Botany, production, genetics, nutritional value, and postharvest handling, in: *Horticultural Reviews*, I. Warrington, (Ed.), pp. 37–152, John Wiley & Sons, Inc, Chichester, UK, New Zealand, 2021.
5. Ferguson, A. and Seal, A., Kiwifruit, in: *Temperate Fruit Crop Breeding - Germplasm to Genomics*, J.F. Hancock (Ed.), pp. 235–264, Springer Dordrecht, Amsterdam, Netherlands, 2008.

6. Latocha, P., The nutritional and health benefits of kiwiberry (*Actinidia arguta*)–A review. *Plant Foods Hum. Nutr.*, 72, 4, 325–334, 2017.
7. Richardson, D.P., Ansell, J., Drummond, L.N., The nutritional and health attributes of kiwifruit: A review. *Eur. J. Nutr.*, 57, 8, 2659–2676, 2018.
8. Silva, A.M., Costa, P.C., Delerue-Matos, C., Latocha, P., Rodrigues, F., Extraordinary composition of *Actinidia arguta* by-products as skin ingredients: A new challenge for cosmetic and medical skincare industries. *Trends Food Sci. Technol.*, 116, 842–853, 2021.
9. Ward, C. and Courtney, D., Chapter one - kiwifruit: Taking its place in the global fruit bowl, in: *Advances in Food and Nutrition Research*, M. Boland and P.J. Moughan (Eds.), pp. 1–14, Academic Press, Amsterdam, Netherlands, 2013.
10. Guroo, I., Wani, S.A., Wani, S.M., Ahmad, M., Mir, S.A., A review of production and processing of kiwifruit. *J. Food Process. Technol.*, 8, 10, 1000699, 2017.
11. Scrimgeour, F., Hughes, W., Kumar, V., *The economic contribution of kiwifruit industry expansion to the Bay of Plenty, Northland and New Zealand economies*, Inst. Business Research, Uni. Waikato, p. 1, 2017.
12. McIndoe, C., *Kiwifruit leading the charge for horticulture*, 2021, https://berl.co.nz/economic-insights/kiwifruit-leading-charge-horticulture.
13. Food and Agriculture Organization (FAO), *Value of agricultural production*, 2022, https://www.fao.org/faostat/en/#data/QV.
14. Ferguson, A.R., World economic importance, in: *The Kiwifruit Genome*, R. Testolin, H.-W. Huang, A.R. Ferguson (Eds.), pp. 37–42, Springer International Publishing, Cham, 2016.
15. Chamorro, F., Carpena, M., Fraga-Corral, M., Echave, J., Rajoka, M.S.R., Barba, F.J., Cao, H., Xiao, J., Prieto, M.A., Simal-Gandara, J., Valorization of kiwi agricultural waste and industry by-products by recovering bioactive compounds and applications as food additives: A circular economy model. *Food Chem.*, 370, 131315, 2022.
16. Almeida, D., Pinto, D., Santos, J., Vinha, A.F., Palmeira, J., Ferreira, H.N., Rodrigues, F., Oliveira, M.B.P.P., Hardy kiwifruit leaves (*Actinidia arguta*): An extraordinary source of value-added compounds for food industry. *Food Chem.*, 259, 113–121, 218.
17. Sanz, V., López-Hortas, L., Torres, M.D., Domínguez, H., Trends in kiwifruit and by-products valorization. *Trends Food Sci. Technol.*, 107, 401–414, 2020.
18. Silva, A.M., Pinto, D., Fernandes, I., Albuquerque, T.G., Costa, H.S., Freitas, V., Rodrigues, F., Oliveira, M.B.P.P., Infusions and decoctions of dehydrated fruits of *Actinidia arguta* and *Actinidia deliciosa*: Bioactivity, radical scavenging activity and effects on cells viability. *Food Chem.*, 289, 625–534, 2019.
19. Wojdyło, A., Nowicka, P., Oszmiański, J., Golis, T., Phytochemical compounds and biological effects of Actinidia fruits. *J. Funct. Foods*, 30, 194–202, 2017.

20. Department of Agriculture of United States, *Kiwi fruit, raw*, 2020, https://fdc.nal.usda.gov/fdc-app.html#/food-details/1102667/nutrients.
21. Nishiyama, I., Fukuda, T., Shimohashi, A., Oota, T., Sugar and organic acid composition in the fruit juice of different actinidia varieties. *Food Sci. Technol. Res.*, 14, 1, 67–73, 2008.
22. Mota, L.M., Aguiar, A., Ferreira, I.M., Guedes de Pinho, P., Volatile profiling of kiwifruits (*Actinidia deliciosa* 'Hayward') evaluated by HS-SPME and GC-IT/MS: Influence of ripening, training system and storage. *Food Bioproc. Tech.*, 5, 8, 3115–3128, 2012.
23. Latocha, P., Krupa, T., Wołosiak, R., Worobiej, E., Wilczak, J., Antioxidant activity and chemical difference in fruit of different Actinidia sp. *Int. J. Food Sci. Nutr.*, 61, 4, 381–394, 2010.
24. Du, G., Li, M., Ma, F., Liang, D., Antioxidant capacity and the relationship with polyphenol and vitamin C in Actinidia fruits. *Food Chem.*, 113, 2, 557–562, 2009.
25. Fiorentino, A., D'Abrosca, B., Pacifico, S., Mastellone, C., Scognamiglio, M., Monaco, P., Identification and assessment of antioxidant capacity of phytochemicals from kiwi fruits. *J. Agric. Food Chem.*, 57, 10, 4148–4155, 2009.
26. Collins, B.H., Horská, A., Hotten, P.M., Riddoch, C., Collins, A.R., Kiwifruit protects against oxidative DNA damage in human cells and *in vitro*. *Nutr. Cancer*, 39, 1, 148–153, 2001.
27. Paturi, G., Butts, C.A., Bentley-Hewitt, K.L., Ansell, J., Influence of green and gold kiwifruit on indices of large bowel function in healthy rats. *J. Food Sci.*, 79, 8, H1611–H1620, 2014.
28. Katsumata, S., Wolber, F.M., Tadaishi, M., Tousen, Y., Ishimi, Y., Kruger, M.C., Effect of kiwifruit on bone resorption in ovariectomized mice. *J. Nutr. Sci. Vitaminol.*, 61, 4, 332–337, 2015.
29. Duttaroy, A.K. and Jørgensen, A., Effects of kiwi fruit consumption on platelet aggregation and plasma lipids in healthy human volunteers. *Platelets*, 15, 5, 287–292, 2004.
30. Chang, W.-H. and Liu, J.-F., Effects of kiwifruit consumption on serum lipid profiles and antioxidative status in hyperlipidemic subjects. *Int. J. Food Sci. Nutr.*, 60, 8, 709–716, 2009.
31. Đilas, S., Čanadanović-Brunet, J., Ćetković, G., By-products of fruits processing as a source of phytochemicals. *Chem. Ind. Chem. Eng. Q.*, 15, 4, 191–202, 2009.
32. Asker Mohsen, S. and Ahmed Habiba, A., Active constituents of kiwi (*Actinidia deliciosa* Planch) peels and their biological activities as antioxidant, antimicrobial and anticancer. *Res. J. Chem. Environ.*, 22, 9, 52–59, 2018.
33. Cairone, F., Garzoli, S., Menghini, L., Simonetti, G., Casadei, M.A., Di Muzio, L., Cesa, S., Valorization of kiwi peels: Fractionation, bioactives analyses and hypotheses on complete peels recycle. *Foods*, 11, 4, 589, 2022.
34. Giordano, M., Pinela, J., Dias, M.I., Calhelha, R.C., Stojković, D., Soković, M., Tavares, D., Cánepa, A.L., Ferreira, I.C.F.R., Caleja, C., Barros, L.,

Ultrasound-assisted extraction of flavonoids from kiwi peel: Process optimization and bioactivity assessment. *Appl. Sci.*, 11, 14, 6416, 2021.
35. D'Eliseo, D., Pannucci, E., Bernini, R., Campo, M., Romani, A., Santi, L., Velotti, F., *In vitro* studies on anti-inflammatory activities of kiwifruit peel extract in human THP-1 monocytes. *J. Ethnopharmacol.*, 233, 41–46, 2019.
36. Yang, H., Lee, Y.C., Han, K.S., Singh, H., Yoon, M., Park, J.H., Cho, C.-W., Cho, S., Green and gold kiwifruit peel ethanol extracts potentiate pentobarbital-induced sleep in mice via a GABAergic mechanism. *Food Chem.*, 136, 1, 160–163, 2013.
37. Priyadarshi, R., Roy, S., Rhim, J.-W., Enhanced functionality of green synthesized sulfur nanoparticles using kiwifruit (*Actinidia deliciosa*) peel polyphenols as capping agents. *J. Nanostructure Chem.*, 12, 3, 389–399, 2022.
38. Li, J., Li, Y., Wu, H., Naraginti, S., Wu, Y., Facile synthesis of ZnO nanoparticles by *Actinidia* deliciosa fruit peel extract: Bactericidal, anticancer and detoxification properties. *Environ. Res.*, 200, 111433, 2021.
39. Shastri, K.V., Bhatia, V., Parikh, P.R., Chaphekar, V.N., *Actinidia deliciosa*: A review. *Int. J. Pharm. Sci. Res.*, 3, 10, 3543, 2012.
40. Bai, X., Qiu, A., Guan, J., Optimization of microwave-assisted extraction of antihepatotoxic triterpenoid from *Actinidia* deliciosa root and its comparison with conventional extraction methods. *Food Technol. Biotechnol.*, 45, 2, 174–180, 2007.
41. Henriques, J., Ribeiro, M.J., Falé, P.L., Pacheco, R., Ascensão, L., Florêncio, M.H., Serralheiro, M.L.M., Valorization of kiwifruit production: Leaves of the pruning branches of *Actinidia deliciosa* as a promising source of polyphenols. *Eur. Food Res. Technol.*, 243, 8, 1343–1353, 2017.
42. Chawla, H., Parle, M., Yadav, M., Medicinal potential and phytopharmacology of *Actnidia deleciosa. Int. J. Exp. Pharm.*, 6, 20–25, 2016.
43. Henriques, J., Fale, P.L., Pacheco, R., Florêncio, M.H., Serralheiro, M.L., Phenolic compounds from *Actinidia deliciosa* leaves: Caco-2 permeability, enzyme inhibitory activity and cell protein profile studies. *J. King Saud Univ. Sci.*, 30, 4, 513–518, 2018.
44. Shirosaki, M., Koyama, T., Yazawa, K., Anti-hyperglycemic activity of kiwifruit leaf (*Actinidia deliciosa*) in mice. *Biosci. Biotechnol. Biochem.*, 72, 4, 1099–1102, 2008.
45. Nieuwenhuizen, N.J., Wang, M.Y., Matich, A.J., Green, S.A., Chen, X., Yauk, Y.K., Beuning, L.L., Nagegowda, D.A., Dudareva, N., Atkinson, R.G., Two terpene synthases are responsible for the major sesquiterpenes emitted from the flowers of kiwifruit (*Actinidia deliciosa*). *J. Exp. Bot.*, 60, 11, 3203–3219, 2009.
46. Crowhurst, R.N., Gleave, A.P., MacRae, E.A., Ampomah-Dwamena, C., Atkinson, R.G., Beuning, L.L., Bulley, S.M., Chagne, D., Marsh, K.B., Matich, A.J., Montefiori, M., Newcomb, R.D., Schaffer, R.J., Usadel, B., Allan, A.C., Boldingh, H.L., Bowen, J.H., Davy, M.W., Eckloff, R., Fergusson, A.R., Fraser, L.G., Hellen, R.P., Laing, W.A., Analysis of expressed sequence tags from

Actinidia: Applications of a cross species EST database for gene discovery in the areas of flavor, health, color and ripening. *BMC Genom.*, 9, 1, 351, 2008.
47. He, X., Fang, J., Chen, X., Zhao, Z., Li, Y., Meng, Y., Huang, L., *Actinidia chinensis* Planch.: A review of chemistry and pharmacology. *Front. Pharmacol.*, 10, 1236, 2019.
48. Sivakumaran, S., Huffman, L., Sivakumaran, S., Drummond, L., The nutritional composition of Zespri® SunGold kiwifruit and Zespri® sweet green kiwifruit. *Food Chem.*, 238, 195–202, 2018.
49. Skinner, M.A., Loh, J.M., Hunter, D.C., Zhang, J., Gold kiwifruit (*Actinidia chinensis* 'Hort16A') for immune support. *Proc. Nutr. Soc.*, 70, 2, 276–280, 2011.
50. Esti, M., Messia, M.C., Bertocchi, P., Sinesio, F., Moneta, E., Nicotra, A., Fantechi, P., Palleschi, G., Chemical compounds and sensory assessment of kiwifruit (*Actinidia chinensis* (Planch.) var. chinensis): Electrochemical and multivariate analyses. *Food Chem.*, 61, 3, 293–300, 1998.
51. Zuo, L.-L., Wang, Z.Y., Fan, Z.L., Tian, S.Q., Liu, J.R., Evaluation of antioxidant and antiproliferative properties of three Actinidia (*Actinidia kolomikta, Actinidia arguta, Actinidia chinensis)* extracts *in vitro*. *Int. J. Mol. Sci.*, 13, 5, 5506–5518, 2012.
52. Nie, X.-R., Li, H.Y., Wei, S.Y., Han, Q.H., Zhao, L., Zhang, Q., Li, S.-Q., Quin, W., Wu, D.T., Changes of phenolic compounds, antioxidant capacities, and inhibitory effects on digestive enzymes of kiwifruits (*Actinidia chinensis*) during maturation. *J. Food Meas. Charact.*, 14, 3, 1765–1774, 2020.
53. Eady, S.L., Wallace, A.J., Butts, C.A., Hedderley, D., Drummond, L., Ansell, J., Gearry, R.B., The effect of 'Zesy002'kiwifruit (*Actinidia chinensis var. chinensis*) on gut health function: A randomised cross-over clinical trial. *J. Nutr. Sci.*, 8, 1–11, 2019.
54. Brevik, A., Gaivão, I., Medin, T., Jørgenesen, A., Piasek, A., Elilasson, J., Karlsen, A., Blomhoff, R., Veggan, T., Duttaroy, A.K., Collins, A.R., Supplementation of a western diet with golden kiwifruits *(Actinidia chinensis var.'Hort 16A':)* effects on biomarkers of oxidation damage and antioxidant protection. *Nutr. J.*, 10, 1, 1–9, 2011.
55. Alim, A., Li, T., Nisar, T., Ren, D., Zhai, X., Pang, Y., Yang, X., Antioxidant, antimicrobial, and antiproliferative activity-based comparative study of peel and flesh polyphenols from *Actinidia chinensis*. *Food Nutr. Res.*, 63, 1–10, 2019.
56. Chang, J. and Case, R., Cytotoxic phenolic constituents from the root of *Actinidia chinensis*. *Planta Med.*, 71, 10, 955–959, 2005.
57. Yu, A.-N., Tian, D.T., Qu, W.Y., Tan, Z.W., Song, X.J., Chemical composition of the essential oil. *Chem. Nat. Compd.*, 45, 1, 108, 2009.
58. Fang, T., Fang, Y., Xu, X., He, M., Zhao, Z., Huang, P., Yuan, F., Guo, M., Yang, B., Xia, J., *Actinidia chinensis* Planch root extract attenuates proliferation and metastasis of hepatocellular carcinoma by inhibiting epithelial-mesenchymal transition. *J. Ethnopharmacol.*, 231, 474–485, 2019.

59. Wei, L.B., Ma, S.Y., Liu, H.X., Huang, C.S., Liao, N., Cytotoxic triterpenoids from roots of *Actinidia chinensis. Chem. Biodivers.*, 15, 2, e1700454, 2018.
60. He, M., Hou, J., Wang, L., Zheng, M., Fang, T., Wang, X., Xia, J., *Actinidia chinensis* Planch root extract inhibits cholesterol metabolism in hepatocellular carcinoma through upregulation of PCSK9. *Oncotarget*, 8, 26, 42136, 2017.
61. Lv, J.-M., Gouda, M., Zhu, Y.Y., Ye, X.Q., Chen, J.C., Ultrasound-assisted extraction optimization of proanthocyanidins from kiwi (*Actinidia chinensis)* leaves and evaluation of its antioxidant activity. *Antioxidants*, 10, 8, 1317, 2021.
62. Bai, H.-H., Xia, T.S., Jiang, Y.P., Zhao, Z.X., Gou, X.J., Xin, H.L., Chemical constituents and antioxidant activity of leaves of *Actinidia chinensis. Chem. Nat. Compd.*, 58, 1, 132–134, 2022.
63. Basile, A., Vuotto, M.L., Violante, U., Sorbo, S., Martone, G., Castaldo-Cobianchi, R., Antibacterial activity in *Actinidia chinensis, Feijoa sellowiana* and *Aberia caffra. Int. J. Antimicrob. Agents*, 8, 3, 199–203, 1997.
64. Twidle, A., Barker, D., Seal, A.G., Fedrizzi, B., Suckling, D., Identification of floral volatiles and pollinator responses in kiwifruit cultivars, *Actinidia chinensis var. chinensis. J. Chem. Ecol.*, 44, 4, 406–415, 2018.
65. Tatsuka, K., Suekane, S., Sakai, Y., Sumitani, H., Volatile constituents of kiwi fruit flowers: Simultaneous distillation and extraction versus headspace sampling. *J. Agric. Food Chem.*, 38, 12, 2176–2180, 1990.
66. Silva, A.M., Luís, A.S., Moreira, M.M., Ferraz, R., Brezo-Borjan, T., Švarc-Gajić, J., Costa, P.C., Delerue-Matos, C., Rodrigues, F., Influence of temperature on the subcritical water extraction of *Actinidia* arguta leaves: A screening of pro-healthy compounds. *Sustain. Chem. Pharm.*, 25, 100593, 2022.
67. Silva, A.M., Pinto, D., Moreira, M.M., Costa, P.C., Delerue-Matos, C., Rodrigues, F., Valorization of kiwiberry leaves recovered by ultrasound-assisted extraction for skin application: A response surface methodology approach. *Antioxidants*, 11, 4, 763, 2022.
68. Nishimura, M., Okimasu, Y., Miyake, N., Tada, M., Hida, R., Negishi, T., Arimoto-Kobayashi, S., Inhibitory effect of *Actinidia arguta on* mutagenesis, inflammation and two-stage mouse skin tumorigenesis. *Genes Environ.*, 38, 1, 25–25, 2016.
69. Ha, J.S., Jin, D.E., Park, S.K., Park, C.H., Seung, T.W., Bae, D.W., Kim, D.-O., Heo, H.J., Antiamnesic effect of *Actinidia* arguta extract intake in a mouse model of TMT-induced learning and memory dysfunction. *Evid. Based Complement. Altern. Med.*, 2015, 1–13, 2015.
70. Choi, J.J., Park, B., Kim, D.H., Pyo, M.Y., Choi, S., Son, M., Jin, M., Blockade of atopic dermatitis-like skin lesions by DA-9102, a natural medicine isolated from *Actinidia arguta,* in the Mg-deficiency induced dermatitis model of hairless rats. *Exp. Biol. Med.*, 233, 8, 1026–1034, 2008.
71. Marangi, F., Pinto, D., de Francisco, L., Alves, R.C., Puga, H., Sut, S., Dall'Acqua, S., Rodrigues, F., Oliveira, M.B.P.P., Hardy kiwi leaves extracted

by multi-frequency multimode modulated technology: A sustainable and promising by-product for industry. *Food Res. Int.*, 112, 184–191, 2018.

72. Tan, C., Wang, Z., Feng, X., Irfan, M., Liu, C., Identification of bioactive compounds and antioxidant activity in leaves and fruits of Actinidia arguta accessions from northeastern China, Preprints, 2020.
73. Cyboran, S., Oszmiański, J., Kleszczyńska, H., Modification of the properties of biological membrane and its protection against oxidation by *Actinidia* arguta leaf extract. *Chem.-Biol. Interact.*, 222, 50–59, 2014.
74. Silva, A.M., Pinto, D., Fernandes, I., d.Freitas, V., d.L.L.Cádiz-Gurrea, M., Costa, P., Delerue-Matos, C., Rodrigues, F., An insight into kiwiberry leaf valorization: Phenolic composition, bioactivity and health benefits. *Molecules*, 26, 8, 2314, 2021.
75. Kwon, D., Kim, G.D., Kang, W., Park, J.E., Kim, S.H., Choe, E., Kim, J.-I., Auh, J.H., Pinoresinol diglucoside is screened as a putative α-glucosidase inhibiting compound in *Actinidia arguta* leaves. *J. Korean Soc. Appl. Biol.*, 57, 4, 473–479, 2014.
76. Yoo, S.K., Kang, J.Y., Lee, U., Park, S.K., Kim, J.M., Han, H.J., Kim, D.O., Heo, H.J., Improving effect of *Actinidia arguta* leaf on hyperglycemia-induced cognitive dysfunction. *J. Funct. Foods*, 76, 104315, 2021.
77. Kim, G.-D., Lee, J.Y., Auh, J.-H., Metabolomic screening of anti-inflammatory compounds from the leaves of *Actinidia arguta* (Hardy Kiwi). *Foods*, 8, 2, 47, 2019.
78. Silva, A.M., Garcia, J., Dall'Acqua, S., Costa, P., Delerue-Matos, C., Rodrigues, F., Eco-friendly insights on kiwiberry leaves valorization through *in-vitro* and *in-vivo studies*. *Ind. Crops Prod.*, 184, 115090, 2022.
79. Kim, J. and Choe, E., Improvement of the lipid oxidative stability of soybean oil-inwater emulsion by addition of daraesoon (shoot of *Actinidia arguta*) and samnamul (shoot of *Aruncus dioicus*) extract. *Food Sci. Biotechnol.*, 26, 1, 113–119, 2017.
80. Takano, F., Tanaka, T., Tsukamoto, E., Yahagi, N., Fushiya, S., Isolation of (+)-catechin and (-)-epicatechin from *Actinidia arguta* as bone marrow cell proliferation promoting compounds. *Planta Med.*, 69, 04, 321–326, 2003.
81. Lee, J., Sowndhararajan, K., Kim, M., Kim, J., Kim, D., Kim, S., Kim, G.-Y., Kim, S., Jhoo, J.W., Antioxidant, inhibition of α-glucosidase and suppression of nitric oxide production in LPS-induced murine macrophages by different fractions of *Actinidia arguta* stem. *Saudi J. Biol. Sci.*, 21, 6, 532–538, 2014.
82. Kim, H.-Y., Hwang, K.W., Park, S.-Y., Extracts of *Actinidia arguta* stems inhibited LPS-induced inflammatory responses through nuclear factor–κB pathway in raw 264.7 cells. *Nutr. Res.*, 34, 11, 1008–1016, 2014.
83. Teng, K., Ruan, H.-S., Zhang, H.-F., Flavonoid and saponin rich fractions of kiwi roots (*Actinidia* arguta (Sieb. et Zucc.) Planch) with antinociceptive and anti-inflammatory effects. *Afr. J. Pharm. Pharmacol.*, 7, 35, 2445–2451, 2013.

84. Jang, D.S., Lee, G.Y., Kim, J., Lee, Y.M., Kim, J.M., Kim, Y.S., Kim, J.S., A new pancreatic lipase inhibitor isolated from the roots of *Actinidia arguta*. *Arch. Pharm. Res.*, 31, 5, 666–670, 2008.
85. Kim, J., Jang, D.S., Kim, H., Kim, J.S., Anti-lipase and lipolytic activities of ursolic acid isolated from the roots of *Actinidia arguta*. *Arch. Pharm. Res.*, 32, 7, 983–987, 2009.
86. Ye, C., Jin, M., Jin, C., Jin, L., Sun, J., Ma, Y.J., Zhou, W., Li, G., Inhibitory effects of chemical constituents from *Actinidia kolomikta on* LPS-induced inflammatory responses. *Rev. Bras. Farmacogn.*, 30, 1, 127–131, 2020.
87. Česonienė, L., Štreimikytė, P., Liaudanskas, M., Žvikas, V., Viškelis, P., Viškelis, J., Daubaras, R., Berries and leaves of *Actinidia kolomikta* (Rupr. & Maxim.) Maxim.: A source of phenolic compounds. *Plants*, 11, 2, 147, 2022.
88. Lu, J., Jin, Y., Liu, G., Zhu, N., Gui, M., Yu, A., Li, X., Flavonoids from the leaves of *Actinidia kolomikta*. *Chem. Nat. Compd.*, 46, 2, 205–208, 2010.
89. Hu, X. and Zhang, Z., *In vitro* and *in vivo* anti-diabetic activity of extracts from *Actinidia kolomikta*. *Int. J. Biol.*, 6, 3, 1–10, 2014.
90. Hu, X., Li, S., Wang, L., Zhu, D., Wang, Y., Li, Y., Yang, Y., Zhang, Z., Cheng, D., Anti-diabetic activities of aqueous extract from *Actinidia kolomikta* root against [alpha]-glucosidase. *J. Pharmacogn. Phytochem.*, 2, 4, 53–57, 2013.
91. Hu, X., Cheng, D., Wang, L., Li, S., Wang, Y., Li, K., Yang, Y., Zhang, Z., Evaluation of anti-hyperglycemic effect of *Actinidia kolomikta* (Maxim. et Rur.) Maxim. root extract. *Pak. J. Pharm. Sci.*, 28, 1135–1140, 2015.
92. Guan, D., Zhang, Z., Yang, Y., Sugiura, N., Hu, H., Xing, G., Liu, J., Zu, X., Zhang, Y., Antioxidant and antitumor activities of water extracts from the root of *Actinidia kolomikta*. *Exp. Ther. Med.*, 2, 1, 33–39, 2011.
93. Guan, D., Zhang, Z., Yang, Y., Xing, G., Liu, J., Immunomodulatory activity of polysaccharide from the roots of *Actinidia kolomikta on* macrophages. *Int. J. Biol.*, 3, 2, 3, 2011.
94. Eo, H.J., Shin, H., Song, J.H., Park, G.H., Immuno-enhancing effects of fruit of *Actinidia* polygama in macrophages. *Food Agric. Immunol.*, 32, 1, 754–765, 2021.
95. Sashida, Y., Ogawa, K., Mori, N., Yamanouchi, T., Triterpenoids from the fruit galls of *Actinidia polygama*. *Phytochem.*, 31, 8, 2801–2804, 1992.
96. Lee, Y.-C., Kim, S.H., Seo, Y.B., Roh, S.S., Lee, J.C., Inhibitory effects of *Actinidia* polygama extract and cyclosporine A on OVA-induced eosinophilia and bronchial hyperresponsiveness in a murine model of asthma. *Int. Immunopharmacol.*, 6, 4, 703–713, 2006.
97. Sung, Y.Y., Yoon, T., Yang, W.K., Moon, B.C., Kim, H.K., Anti-obesity effects of *Actinidia polygama* extract in mice with high-fat diet-induced obesity. *Mol. Med. Rep.*, 7, 2, 396–400, 2013.
98. Yoshizawa, Y., Fukiya, Y., Izumi, Y., Hata, K., Iwashita, J., Murofushi, N., Abe, T., Induction of apoptosis with an extract of *Actinidia polygama* fruit in the promyelocytic leukemia cell line HL-60. *J. Health Sci.*, 48, 4, 303–309, 2002.

99. Kim, Y.J., Lee, J.O., Kim, S.Y., Lee, J.M., Lee, E., Na, J., Yoo, K.H., Park, S.J., Kim, B.J., Effect of *A. polygama* APEE (*Actinidia polygama* ethanol extract) or APWE (*Actinidia polygama* water extract) on wrinkle formation in UVB-irradiated hairless mice. *J. Cosmet. Dermatol.*, 22, 1, 311–319, 2023.
100. Stasiak, A. and Latocha, P., Comparative analysis of volatile compounds in flowers of different actinidia species. *Plants*, 9, 12, 1675, 2020.
101. Khromykh, N., Lykholat, Y.V., Didur, O.O., Sklyar, T.V., Davydov, V.R., Lavrentieva, K.V., Lykholat, T.Y., Phytochemical profiles, antioxidant and antimicrobial activity of *Actinidia polygama* and *A. arguta* fruits and leaves. *Biosyst. Divers.*, 30, 1, 39–45, 2022.
102. Birgani, G.A., Ahangarpour, A., Khorsandi, L., Moghaddam, H.F., Anti-diabetic effect of betulinic acid on streptozotocin-nicotinamide induced diabetic male mouse model. *Braz. J. Pharm. Sci.*, 54, 1–7, 2018.
103. Babu, S. and Jayaraman, S., An update on β-sitosterol: A potential herbal nutraceutical for diabetic management. *Biomed. Pharmacother.*, 131, 110702, 2020.
104. Lahlou, E.H., Hirai, N., Kamo, T., Tsuda, M., Ohigashi, H., Actinidic acid, a new triterpene phytoalexin from unripe kiwi fruit. *Biosci. Biotechnol. Biochem.*, 65, 2, 480–483, 2001.
105. Aucoin, M., Cooley, K., Saunders, P.R., Cardozo, V., Remy, D., Cramer, H., Abad, C.N., Hannan, N., The effect of quercetin on the prevention or treatment of COVID-19 and other respiratory tract infections in humans: A rapid review. *Adv. Integr. Med.*, 7, 4, 247–251, 2020.
106. Chan, A.O., Leung, G., Tong, T., Wong, N.Y., Increasing dietary fiber intake in terms of kiwifruit improves constipation in Chinese patients. *World J. Gastroenterol.*, 13, 35, 4771–5, 2007.
107. Rao, S.S.C. and Brenner, D.M., Efficacy and safety of over-the-counter therapies for chronic constipation: An updated systematic review. *Am. J. Gastroenterol.*, 116, 6, 1156–1181, 2021.
108. Błaszczak, W., Latocha, P., Jeż, M., Wiczkowski, W., The impact of high-pressure processing on the polyphenol profile and anti-glycaemic, anti-hypertensive and anti-cholinergic activities of extracts obtained from kiwiberry (*Actinidia arguta*) fruits. *Food Chem.*, 343, 128421, 2021.
109. Qu, L., Liu, Q., Zhang, Q., Tuo, X., Fan, D., Deng, J., Yang, H., Kiwifruit seed oil prevents obesity by regulating inflammation, thermogenesis, and gut microbiota in high-fat diet-induced obese C57BL/6 mice. *Food Chem. Toxicol.*, 125, 85–94, 2019.
110. Edmunds, S.J., Roy, N.C., Davy, M., Cooney, J.M., Barnett, M.P., Zhu, S., Park, Z., Love, D.R., Laing, W.A., Effects of kiwifruit extracts on colonic gene and protein expression levels in IL-10 gene-deficient mice. *Br. J. Nutr.*, 108, 1, 113–129, 2012.
111. JianJun, D., HaiXia, Y., DaiDi, F., Wei, C., Yan'e, L., Antibacterial activities of polyphenols extract from kiwi fruit (*Actinidia chinensis* Planch.) seeds. *J. Pure Appl. Microbiol.*, 7, 1, 491–496, 2013.

112. Xu, Y.-X., Xiang, Z.B., Jin, Y.S., Shen, Y., Chen, H.S., Two new triterpenoids from the roots of *Actinidia chinensis*. *Fitoterapia*, 81, 7, 920–924, 2010.
113. Aleid, I.S., Alfheeaid, H.A., Aljutaily, T., Alhomaid, R.M., Alharbi, H.F., Althwab, S.A., Abdel-Rahman, H.A., AlGeffari, M.A., Barakat, H., Gastroprotective effects of *spirulina platensis*, golden kiwifruit flesh, and golden kiwifruit peel extracts individually or in combination against indomethacin-induced gastric ulcer in rats. *Nutrients*, 13, 10, 3499, 2021.
114. Iwasawa, H., Morita, E., Yui, S., Yamazaki, M., Anti-oxidant effects of kiwi fruit *in vitro* and *in vivo*. *Biol. Pharm. Bull.*, 34, 1, 128–134, 2011.
115. Hwang, J.-S., Cho, C.H., Baik, M.Y., Park, S.K., Heo, H.J., Cho, Y.S., Kim, D.O., Effects of freeze-drying on antioxidant and anticholinesterase activities in various cultivars of kiwifruit (Actinidia spp.). *Food Sci. Biotechnol.*, 26, 1, 221–228, 2017.
116. El-Demerdash, F.M., Talaat, Y., El-Sayed, R.A., Kang, W., Ghanem, N.F., Hepatoprotective effect of *Actinidia deliciosa* against streptozotocin-induced oxidative stress, apoptosis, and inflammations in rats. *Oxid. Med. Cell. Longev.*, 2022, 1499510, 2022.
117. Statista, *Beauty & personal care – worldwide*, 2022, https://www.statista.com/outlook/cmo/beauty-personal-care/worldwide.
118. Deters, A.M., Schröder, K.R., Hensel, A., Kiwi fruit (*Actinidia chinensis* L.) polysaccharides exert stimulating effects on cell proliferation via enhanced growth factor receptors, energy production, and collagen synthesis of human keratinocytes, fibroblasts, and skin equivalents. *J. Cell. Physiol.*, 202, 3, 717–22, 2005.
119. Pullar, J.M., Carr, A.C., Vissers, M.C.M., The roles of vitamin C in skin health. *Nutrients*, 9, 8, 866, 2017.
120. Chai, W.-M., Shi, Y., Feng, H.L., Xu, L., Xiang, Z.H., Gao, Y.S., Chen, Q.X., Structure characterization and anti-tyrosinase mechanism of polymeric proanthocyanidins fractionated from kiwifruit pericarp. *J. Agric. Food Chem.*, 62, 27, 6382–6389, 2014.

4

Valorization of Grape By-Products: Toward The Circular Economy Concept

Rosa Rodriguez[1*], Germán Mazza[2], María Paula Fabani[3], Yanina Baldán[1], Celia Román[1], Mathias Riveros-Gomez[1], Iside Mut[1] and Anabel Fernandez[1]

[1]Instituto de Ingeniería Química - Facultad de Ingeniería (UNSJ) - Grupo Vinculado al PROBIEN (CONICET-UNCo), San Juan, Argentina
[2]Instituto de Investigación y Desarrollo en Ingeniería de Procesos, Biotecnología y Energías Alternativas, PROBIEN (CONICET-Universidad Nacional del Comahue), Neuquén, Argentina
[3]Instituto de Biotecnología- Facultad de Ingeniería, Universidad Nacional de San Juan, San Juan, Argentina

Abstract
In the challenge of transforming waste into useful products that can be re-used in a circular economy perspective, wine industry can represent a suitable model for the application of its principles, including the valorization of the agricultural and food waste.

In this work, the potentiality of a biorefinery system, based on winery waste and aimed at recovering useful bio-based products, was evaluated. Thus, the stalk transformation into bioenergy vectors, and the grape marc use to obtain a flour to make gluten-free baked were analyzed.

The kinetic analysis of stalk pyrolysis was studied by thermogravimetric analysis at different heating rates (5, 10, and 15°C/min). This parameter does not affect the thermal decomposition behavior and the final value of weight loss is not influenced either.

Additionally, the quality of marc powder changed with drying temperature was evaluated. The drying temperature has a negligible influence on the flour techno-functional properties. This parameter, in the range considered, does not modify the presence of the different organic groups present in the flour obtained. The physicochemical analysis of the gluten-free baked goods made with the premixes

**Corresponding author*: rrodri@unsj.edu.ar

Umile Gianfranco Spizzirri (ed.) Nutraceutics from Agri-Food By-Products, (117–136) © 2023 Scrivener Publishing LLC

showed that the addition of marc flour improves the nutritional composition as its content increases.

***Keywords*:** By-products, winery bio-wastes, grape stalk, grape marc, valorization, piro-gasification, alternative flour

4.1 Introduction

Food and agro-industrial wastes are increasingly problematic with undesirable effects on the environment, economy, and human health. Alternatively, waste valorization for the obtention of energy and chemical products is an important step in the development and deployment of alternative production sources.

To use renewable natural capital and to minimize bio-waste, the concept of circular bio-economy has emerged. This concept proposes a context for using renewable natural capital to convert and manage the land, food, health, and industrial systems, to reach sustainable comfort in synchronization with the environment. In this context, the biorefinery concept is an auspicious prospect for moving up the bio-waste hierarchy through the coupling of the bio-waste and production areas combined in a circular bio-economy. The circular model implementation has been recognized as essential to the achievement of the sustainable development objectives and allows that socio-economic activity to align with acceptable levels of access to natural resources.

According to Watson [1], grape production is higher than 77×10^6 t. The grape is the principal raw material of wine production. Worldwide, Italy, France, Spain, the United States, and Australia are the principal wine producers [2]. In Latin America, Chile, and Argentina are the most important producer. Significant quantities of bio-wastes are generated during wine production, which generally do not have exploitation or economic value and their disposal has a thoughtful negative impact on the environment. The wineries generate principally bio-wastes, such as marc (o pomace) and stalks. The quantity of bio-wastes generated depends on juice extraction level, equipment type, fermentation numbers, and stages of clarification. Kyzas *et al.* [3] reported that the transformation in the wine of 1 t of grape generates 2.7-3.8 kg of solid bio-wastes.

The grape marc is a mixture of skin, pulp, and seeds in variable proportions (25, 55, and 20%, on average, respectively). Its characteristics vary considerably depending on the type of wine produced (red or white), the grape variety, and the type of separation process used. It is compounded by carbohydrates, polyphenols, pectins, heteroxylans, xyloglucan, and cellulose [4]. The grape stalks are composed principally of cellulose, hemicellulose, lignin, tannins, and proteins [5]. These agro-industrial wastes can be utilized as food for human beings and animals, composting, and energy generation. These wastes can be transformed into energy using thermochemical and microbiological processes [6]. Another alternative is the extraction of different added-value compounds such as polyphenols, pectins, pigments, dietary fiber, or platform molecules, such as levulinic acid, oligosaccharides, and hydroxymethylfurfural, among others [7].

The winery wastes can be directly valorized by use as food or by thermochemical processes, but it is also possible to take advantage of multiple-stage processing based on different objectives [8]. Compounds such as polyphenols and pectin can be recovered from waste, before the thermochemical processes, by using adequate extraction methods. The yield of the obtained compounds and their quality depends on the operational conditions of the extraction, such as the type of solvent and process duration, and temperature [9].

Considering the use of winery wastes as a human being's food, it is possible to mention the transformation in flour [10]. Today, consumers request healthy foods. Enrichment of foods with the addition of functional ingredients is a promising way to improve food quality. For this motive, the development of functional flours has an important interest [11].

Considering the thermochemical processes, the piro-gasification transformation produces three main products: biochar, liquid (tar and water) and non-condensable gases with yields that depend on the operating conditions and feedstock properties. Syngas and tar have a high LHV (lower heating value) and can be used for energy recovery. Tar can also be further upgraded to renewable transportation fuels. Biochar can be used as biofuel, a precursor of activated carbons, or as a soil amendment, among other applications, offering an advanced option for the biological sequestration of carbon [12].

To apply the circular economy and biorefinery concepts to grape marc and stalk from a winery, different alternatives are proposed (Figure 4.1).

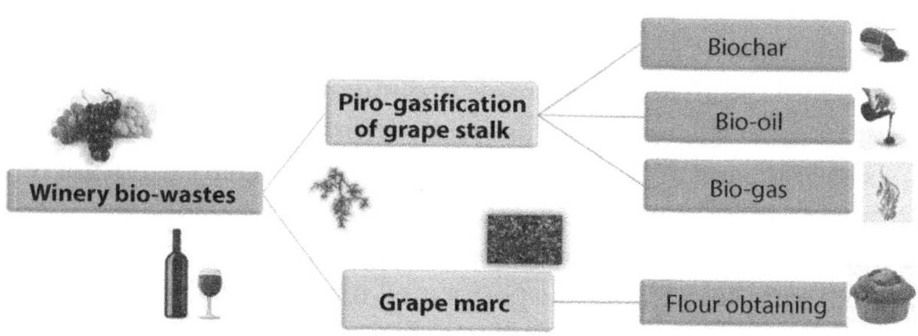

Figure 4.1 Alternatives proposed for this work.

For the grape marc, the biorefinery proposal is the flour production to make gluten-free baked. The piro-gasification process is proposed as a potential transformation of this bio-waste into biochar, bio-oil, and biogas.

4.2 Materials and Methods

4.2.1 Samples

The grapes have different characteristics depending on the climatic conditions, altitude, soil, and varietal. The grape marc (composed of skins and seeds) and stalk samples studied were of the Syrah variety, from the winery of the Sánchez González family, located in the town of Balde de las Chilcas, Department of Valle Fértil, Cuyo Region, in the province of San Juan, Argentina [13, 14]. The samples were collected and stored refrigerated at a temperature of 4±1°C until they are subjected to the corresponding analyses and treatments, before 2 months.

4.2.2 Stalk Characterization

The stalk was analyzed to determine their proximate composition: moisture, volatile matter (VM), fixed carbon (FC), and ash contents, according to ASTM standards methods [16]. The high heating value (HHV) was calculated by correlations [23]. Then, ultimate analysis was achieved using Euro 3000 model elemental analyzer. The lignocellulosic composition (lignin, cellulose, and hemicellulose) was determined by ASTM standard methods [16].

4.2.3 Thermogravimetric Analysis

4.2.3.1 Thermogravimetric Analysis of Pyrolysis

The stalk was used as raw material to be pyrolyzed. Before the thermogravimetric analysis studied biomass was ground and sieved to a size fraction of 0.1 to 0.2 mm. The experiments were performed at three different heating rates of 5, 10, and 15°C/min under an inert (nitrogen) atmosphere with a flow rate of 100 mL/min. TGA-50 Shimadzu microbalance was used and the temperature increased from room temperature to 900°C. Before measuring each sample, a blank was performed under the same conditions for background correction. Three replications were performed, and the reproducibility of the experiments was acceptable.

4.2.3.2 *Thermogravimetric Analysis of Drying*

Grape marc was isothermally dried at two temperatures (55 and 75°C) in a stainless-steel dryer with an internal diameter of 12 cm and a height of 100 cm as described by Baldán *et al.* [15]. Moisture loss was recorded every 30 seconds until a constant weight was reached and the air velocity was 1 m/s and constant.

The dimensionless moisture ratio (MR) was determined from the experimental data of initial moisture, mass loss, and final moisture of the samples [16]. The drying curve (MR vs t) was represented by six semi-theoretical and empirical models [17–22].

4.2.4 Grape Marc Powder (GPP)

4.2.4.1 *Elaboration of GPP*

The samples of marc were ground to a size of 0.10 to 0.20 mm in a stainless-steel knife mill (TecnoDalvo, model TDMC, Argentina). The samples were then stored in the dark in zip lock bags until further use, within 1 month of being collected.

4.2.4.2 *GPP Characterization*

The moisture, acidity, pH, lipids, proteins, crude fiber, and ash content were analyzed in the samples of GPP at 55 and 75°C [24], while total carbohydrate was calculated according to Sousa *et al.* [25]. Then, the total energy was determined by a formula based on the factors of Atwater [26]. Also, lignin and gluten content were quantified by ASTM Method [27] and ELISA Enzyme Immunoassay Method (Kit Ridascreen), respectively. The techno-functional properties of GPP, water holding capacity (WHC) expressed as g water/g dry weight basis (dwb), and oil holding capacity (OHC) (g oil/g dwb) were analyzed according to Garau *et al.* [28]. The values of bulk and compacted density were calculated by Yusof *et al.* [29].

4.2.4.3 *Fourier Transform Infrared Spectroscopy and Scanning Electron Microscopy Analysis*

The Infralum FT-80-FTIR spectrometer was used to obtain FTIR spectra of the GPP samples obtained at 55 and 75°C, respectively. The KBr compressed pellet method was used, and spectra were acquired between 500 and 4000 l/cm [30]. SEM images of the samples of GPP were from a scanning electron microscope (SEM-EDS) [30].

4.2.5 Muffins

4.2.5.1 *Muffin Production Process*

The gluten-free muffins were prepared according to the following recipe: 240 g premix (67% rice flour, 22% cornstarch, and 11% cassava flour), 80 g sugar, 2 eggs, 12 g baking powder, 120 mL oil, 8 g baking soda, 120 mL milk, 12 g xanthan gum, and 2 g salt. Three different premixes were used to make the muffins, replacing 0, 15, and 25% of the rice flour with GPP. The ingredients were mixed until homogeneous preparations were obtained and placed in muffin cups 4 cm in diameter. Subsequently, the premixes were baked for 20 minutes at 180°C. The muffins were then waited until they were at room temperature and stored for 1 to 2 days in Ziploc bags at 20 to 25°C until nutritional and sensory evaluations were carried out.

4.2.5.2 *Sensory Analysis of Muffins*

A group of consumers of gluten-free foods, representative of the population, was selected for a descriptive profile test. Once the study population was identified, a panel of 20 adult male and female jurors was randomly selected to anonymously evaluate the texture and the flavor of the coded muffins. The analysis consisted of filling out a spreadsheet indicating whether or not they would buy the product under study and choosing among four characteristics for each of the attributes (texture and flavor) [31].

4.2.5.3 *Physicochemical Analysis of Muffins*

The volume of the muffins was measured following the work of Maidana *et al.* [13], then they were sliced and subsequently dried at a temperature of 50°C in an oven until a constant weight could be achieved, for approximately 12 hours. Afterward, the dried muffin slices were then ground and analyzed for pH, acidity, moisture, and ash content. In addition, the lipid, crude fiber, and protein contents of the muffins were determined [24].

4.2.6 Statistical Analysis

All analyses were carried out in triplicate ($n = 3$), and results were expressed in terms of mean and standard deviation (SD).

To compare the drying models, the statistical coefficients used to evaluate the fit of the different mathematical models with the experimental data were as follows: Chi-square (χ^2), the sum of squared errors (SSE), and the square root of squared errors (RMSE), selecting those models with the lowest values of χ^2, SSE and RMSE.

Results were analyzed by one-way ANOVA and significant differences between mean values were determined by Tuckey's test ($p < 0.05$) using InfoStat software [16]. Pearson's correlation analysis was used to determine statistical significance.

4.3 Results and Discussions

4.3.1 Stalk Characterization

The stalk had a moisture content of 7.7%, indicating that this biomass can be used in different thermal treatments as a raw material for energy production, without the need for a previous drying process. The MV content was high (66.77%), this value indicates that ignition is easier at low temperatures, improving the thermal process and overall process speed due to the release of gases [23]. The FC was 23.07% dry base (db), while the relation VM/FC ratio was 2.89 indicating a medium reactivity degree of the biomass [23]. Otherwise, the ash content influences costs and the environment, dust emission, and reduction of heat process efficiency. Higher ash content has a negative impact on the HHV value, being 18.12 MJ/kg for this residue. The results of the final analysis showed a high content of %C 64.14, %O 37.54%, %H 5.74, and only traces of nitrogen and sulfur, which proves that the analyzed stalk residues are environmentally eco-friendly. About lignocellulosic biomass content, cellulose, hemicellulose, and lignin contents content was 16.14, 6.18, and 28.11%, respectively, related to the values reported by other authors [32–34].

4.3.2 Thermogravimetric Analysis

4.3.2.1 Thermogravimetric Analysis of Pyrolysis

Figure 4.2 shows the weight loss (TG) and weight loss rates (DTG) at different heating rates. Thermal decomposition occurred in three stages. During the first step, the average weight loss was approximately 10%. It is related to water evaporation and free sugar decomposition or volatilization, taking place in the temperature range of 30 – 200°C [35]. It can be observed that the greatest weight loss (42%) is produced during the second stage (200 - 500°C), where the main components of the bio-waste (hemicellulose, cellulose, and lignin) are decomposed with the temperature

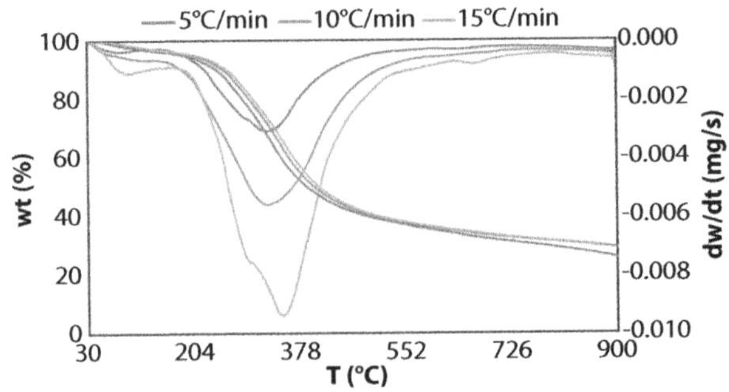

Figure 4.2 Weight loss and weight loss rate at different heating ramps.

increases [36]. Also, different gaseous compounds are released, a fraction of them are condensable and they form the liquid fraction (bio-oil and tar) [37]. The weight loss and corresponding temperature ranges for the third stage were 26% (500–900°C). It can be attributed to the decomposition of complex substances. Lignin is one of these substances and its thermal decomposition range is wider [38].

When the heating rate increased, the peaks of maximum decomposition have no displacement whit respect to temperature. This was because the reaction time is the same in all the cases, indicating that no more energy is required to achieve the same conversion [34, 39].

The influence of the heating ramp on the TG and DTG curves can also be appreciated in Figure 4.2. It can be seen that TG curves shift to the right at a higher heating ramp. The lower peak temperature is achieved for experiments carried out at 15°C/min. This phenomenon might be caused by the time required to approach a potential thermal equilibrium between the gas phase and the solid sample [40]. However, the final weight loss value is not affected by varying the operating heating rate [41].

The biochar yield was 16.77, 9.21, and 24.15% at 5, 10, and 15°C/min, respectively. In this sense, the heating rate had a significant influence on the biochar yield, although no defined trend was observed, which is in agreement with the observations reported by Sun *et al.* [42].

4.3.2.2 Isothermal Drying Curves

Figure 4.3 shows the drying curves for grape marc at 55 and 75°C. The moisture content under different drying conditions was converted into the expression MR. It is observed that increasing the temperature causes a reduction in drying time of up to 43%. This is because higher drying air temperatures lead to an increase in the drying rate [43]. Other authors

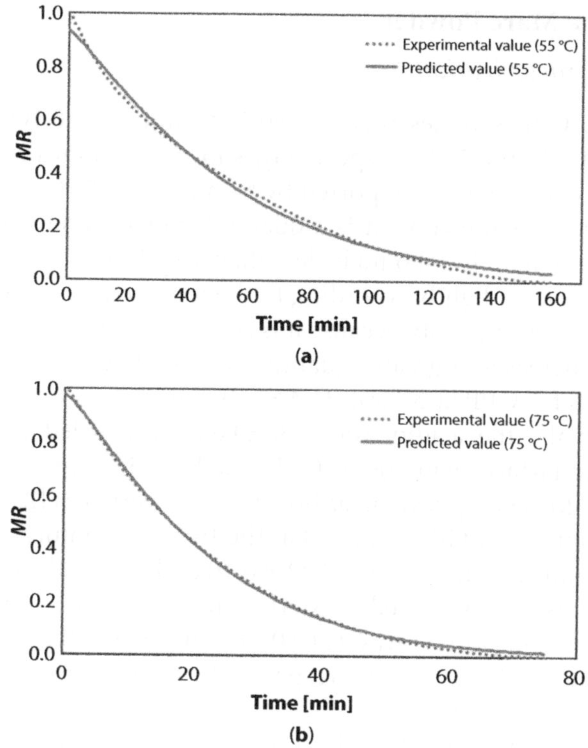

Figure 4.3 Experimental and model approximation drying curves for grape marc at 55°C (a) and 75°C (b).

have identified temperature as the factor that most affects the drying kinetics of food products [44].

The Midilli model [45] best represented the experimental drying curves, obtaining the lowest values for the calculated statistical analysis variables. Figure 4.3 shows the experimental drying curves and the model fit for both temperatures (55 and 75°C). The constant k, corresponding to the decreasing exponential value, depends on the temperature, increasing as this parameter. This effect has been observed in previous work related to fruit drying [46].

D_{eff} values of 3.29×10^{-10} and 7.78×10^{-10} m^2/s were obtained for 55 and 75°C, respectively. As can be seen, the D_{eff} value for 75°C was higher than for 55°C. The difference is related to a shorter drying time and higher convective efficiency. The results obtained were comparable with values reported for other foods 10^{-12} and 10^{-8} m^2/s [47].

4.3.3 Grape Marc Powder

4.3.3.1 *Characterization*

The moisture GPP samples was 7.23±0.34% and 9.22±0.10% for marc dehydrated at 55 and 75°C, respectively, showing significant differences ($p < 0.05$), in coincidence to reported by Sainz *et al.* [49] (pinot noir variety) (10.12%). Moisture content is a quality factor for flour, according to Codex Standard, its value should be less than 15% [48].

The pH increases slightly according to the temperature increases, being associated with lower acidity because it is certain that during drying, as the temperature increases, a greater quantity of volatile compounds are lost. The pH values for GPP at 55 was 3.84±0.10, while at 75°C was 4.18±0.10 ($p > 0.05$), while the acidity was 0.88±0.03 (GPP dried at 55°C) and 0.85±0.01 g of tartaric acid/100 g (GPP dried at 75°C) ($p > 0.05$). Similar values were informed by several authors for both variables [25, 50].

The ash contents of GPP obtained at the two temperatures studied not presenting significant differences (7.22 ± 0.10 and 7.14±0.05 g/100 g dwb). Similar contents were reported in Merlot and Cabernet S. flour by Beres *et al.* [47]. The lipid content was 9.30±0.31 and 9.57±0.80 g/100 g dwb for GPP obtained at 55°C and 75°C, respectively, in accordance as informed for Benitaka variety flour [25].

The protein content in a food is desirable because it contributes to the growth, development, regeneration, and reconstruction of the organism and also helps with the proper functioning of different organs. At 55 and 75°C protein content was 10.04±0.08 and 10.03±0.32 g/100 g dwb, for each temperature, comparable to reported to Cabernet S. flours (10.25±4.89 g/100 g dwb) [51]. About the crude fiber content, their value becomes from 17.55±0.75 to 23.11±1.64 g/100 g dwb for GPP obtained at 55 and 75°C, respectively, being similar to those reported by other authors [25, 52].

Total carbohydrates and total energy were calculated. The values for GPP obtained at 55°C were 50.53 g/100 g dwb and 326.24 kcal/100 g dwb, while at 75°C were 40.09 g/100 g dwb, and 286.62 kcal/100 g dwb respectively, not showing significant differences between temperatures.

The lignin content is related to beneficial physiological effects on the body (laxation, reduced blood cholesterol, and glucose) [53]. In the GPP samples, the lignin content decreased from 46.70±1.8 to 44.96±2.6 g/100 g dwb when the drying temperature increments ($p>0.05$). Similar values were informed for Pinot Noir marc flour (42.53 g/100 g dwb) [51].

Finally, the gluten content of GPP samples was quantified and the values were < 5 mg of gluten/kg of GPP, being, according to ANMAT regulation, a gluten-free food [54].

The techno-functional properties of GPP were measured. The WHC values were 1.72±0.13 for GPP dried at 55 and 1.86±0.02 g water/g dwb for GPP dried at 75°C. The OHC for GPP obtained at 55 and 75°C was 0.95±0.03 and 1.25±0.03 g oil/g dwb, respectively. The WHC and OHC values were greater in the GPP dried at 75°C, related to the augment of the protein and crude fiber content [53]. Finally, apparent and compact density was determined in the two samples. The apparent density values were 0.45 ± 0.01 g/mL for both samples and the compacted density was similar: 0.62±0.01 (GPP dried at 55°C) and 0.60±0.01 g/mL (GPP dried at 75°C).

4.3.4 Muffins

4.3.4.1 Muffin Production Process

The drying temperature to obtain GPP was 75°C, this decision was made based on the following points: (i) The D_{eff} was higher at 75 than 55°C, which leads to about half drying time (Figure 4.3); (ii) crude fiber percentage is higher and carbohydrate is lower for the GPP obtained by drying at this temperature; and (iii) the techno-functional properties were better for this GPP.

The muffins (Figure 4.4) were prepared with the three different premixes. After cooking the muffins were physicochemically analyzed and subjected to sensory analysis as described below.

Some general characteristics of muffins can be appreciated in Table 4.1. It is possible to observe a reduction of volume and mass, and an increase in specific volume when adding the GPP, the same behavior has been reported by other authors in breadsticks [55], bread [56, 57], vegan muffins [58], and cookies [59].

4.3.4.2 Sensory Analysis of Muffins

The 20 panelists randomly selected as described above, analyzed the texture and flavor attributes of the muffins prepared. Concerning flavor, the

Table 4.1 General characteristics of muffins.

Characteristics	Muffin 1	Muffin 2	Muffin 3
Mass (g)	47.65±0.56	47.59±0.42	40.53±2.06
Volume (cm^3)	92.33±5.51	88.35±6.32	76.42±0.77
Specific volume (cm^3/g)	1.90±0.12	1.88±0.16	1.90±0.08

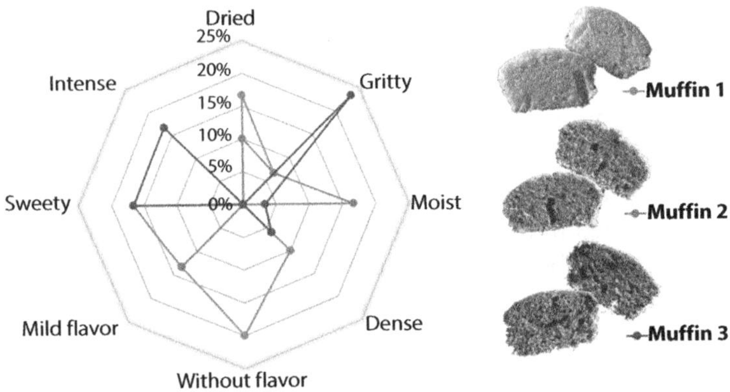

Figure 4.4 Sensory analysis of muffins considering taste and texture.

panelist chose between the following options: without flavor, mild flavor, sweet, and intense; and considering texture, the options were: dried, gritty, moist, and dense. In Figure 4.4, a web graph with a resume of juries' responses is presented.

Considering the taste, most of the panelists indicated that Muffin 1 was without flavor, Muffin 2 was a mild flavor or sweet, and Muffin 3 was sweet or intense. This indicates that adding GPP to muffins results in a product with a sweeter and more intense flavor depending on the percentage of GPP. Bianchi *et al.* [58] proved that the addition of 10% Cabernet Sauvignon marc powder affects positively the taste and the fruitiness of vegan muffins.

On the other hand, the evaluators judged the texture of the muffins: Muffin 1 was dried, Muffin 2 was moist, and Muffin 3 was gritty. It is possible to say that replacing 15% of rice flour for GPP, results in a moister muffin and when the replacement reaches 25% the texture is not acceptable for most consumers, similar behavior was reported by Walker *et al.* [31] in baked goods. Tolve *et al.* [57] proved that the addition of GPP to bread affected the aroma, taste, appearance, and flavor and not influenced their overall acceptability.

Finally, the panelists were asked which of the muffins they would consume, and most of them chose Muffin 2.

4.3.4.3 Physicochemical Analysis of Muffins

In Table 4.2, the physicochemical analysis results for gluten-free muffins are shown. The addition of GPP improved the nutritional and functional

Table 4.2 Physicochemical characteristics of muffins.

Characteristics	Muffin 1	Muffin 2	Muffin 3
Moisture content (%)	12.73±0.34a	16.54±0.38b	17.38±0.28c
Acidity (g tartaric acid/ 100 g)	0.0153±0.0001a	0.0304±0.0001b	0.0610±0.0004c
pH	8.44±0.19a	6.62±0.06b	6.30±0.05c
Ash content (g/100 g dwb)	0.74±0.04a	1.05±0.08b	1.52±0.02c
Lipids (g/100 g dwb)	11.88±0.20a	12.64±0.17b	17.18±0.55c
Proteins (g/100 g dwb)	5.00±0.16a	5.72±0.32b	6.64±0.39c
Crude fiber (g/100 g dwb)	0.05±0.008a	1.47±0.18b	2.19±0.37b

ANOVA. The values given are the average of the analysis carried out in triplicate ($n = 3$). Different letters in the same row indicate significant differences between GPP samples ($p < 0.05$).

properties of gluten-free muffins, which means that it is possible to enrich muffins by incorporating this innovative ingredient.

The three muffin samples showed significant differences ($p < 0.05$) considering moisture content, being Muffin 3 the one with the highest moisture content (17.38 %), it could be due to the sugar content of GPP that acts as a humectant and improve the water adsorption. The same behavior was found by Walker *et al.* [31], Iuga and Mironeasa [60], and Yalcin *et al.* [61]. Acidity has increased its value according to the addition of GPP, and as was expected, pH decreased, the same tendency was reported by Nakov *et al.* [62] in GPP-enriched cakes and it is because of the organic acids in grapes. Lower pH values help to inhibit microbial growth. Ashes increased with the addition of GPP, a similar behavior was found by Nakov *et al.* [62] and Iuga and Mironeasa [60]. When adding 15 and 25 % of GPP to muffins resulted in an increment of lipid percentage of 6 and 44 % respectively, Nakov *et al.* [62], Theagarajan *et al.* [63], and Bender *et al.* [64] observed the same tendency. Protein and crude fiber percentages significantly increased when adding GPP to premixes. Karnopp *et al.* [59] have found the same behavior at preparing fortified muffins made with Bordeaux flour.

It is important to highlight that the addition of GPP has considerably improved the nutritional composition of gluten-free muffins. Many authors have reported that the addition of marc, skins, or seeds improves the nutritional properties of food due to their high content of fiber, carbohydrates, proteins, and polyphenols [64–68].

4.4 Conclusion

Winery waste is a critical problem from an economic and environmental point of view. These bio-waste products have attracted the attention of researchers because of their high content of bioactive compounds and value-added energy products. The circular bio-economy concept proposes a context for the reuse of renewable natural capital to minimize bio-waste. Biorefinery approaches, such as pyro-gasification, have emerged to adequately convert waste into energy, with good results for biowaste from the wine industry. The kinetic analysis of agro-industrial waste from the wine industry was studied by TGA at different heating rates (5, 10, and 15°C/min). It is observed that the heating rate, in the temperature range of the different decomposition stages, does not affect the thermal decomposition behavior and the final value of weight loss is not influenced either. The biochar yields were 16.77, 9.21, and 24.15% at 5, 10, and 15°C/min, respectively. In this respect, the heating rate had a significant influence on the biochar yield.

In addition to thermochemical treatments, the production of flours from marc for the production of gluten-free bakery products is of great interest due to the nutritional properties it provides to foods. Several tests were carried out to determine whether the quality of marc powder changed with drying temperature. According to the results obtained from the proximate analysis of marc flour, significant differences were observed between the lignin content decreased. The variation of drying temperature had little influence on the techno-functional properties of the flours. Moreover, the drying temperature of the marc, in the range considered, did not modify the presence of the different organic groups present in the flour obtained. Based on the results, the marc powder obtained at 75°C was used as an ingredient in premix formulations. The physicochemical analysis of the gluten-free baked goods made with the premixes showed that the addition of marc flour improves the nutritional composition as its content increases, highlighting the increase in protein and crude fiber content. Sensory analysis showed that the incorporation of marc flour had a good level of consumer acceptability. This study showed that there is a potential added value for bio-waste from the wine industry and that it is feasible to use it as an ingredient in gluten-free bakery products.

References

1. Watson, R.R., *Polyphenols in plants: Isolation, purification and extract preparation*, 2nd edition, Academic Press, Cambridge, Massachusetts, USA, 2018.

2. Díez, R.R., *Development of an integrated and green biorefinery from winery waste. Application to wine lees and grape stems*, Universidad de Valladolid, Escuela de Ingenierías Industriales, Valladolid, Spain, 2018.
3. Kyzas, G.Z., Symeonidou, M.P., Matis, K.A., Technologies of winery wastewater treatment: A critical approach. *Desalin. Water Treat.*, 57, 3372, 2014.
4. Basso, D., Weiss-Hortala, E., Patuzzi, F., Baratieri, M., Fiori, L., In deep analysis on the behavior of grape marc constituents during hydrothermal carbonization. *Energies*, 11, 1379, 2018.
5. Troilo, M., Difonzo, G., Paradiso, V.M., Summo, C., Caponio, F., Bioactive compounds from vine shoots, grape stalks, and wine lees: Their potential use in agro-food chains. *Foods*, 10, 342, 2021.
6. Rodriguez, R., Mazza, G., Fernandez, A., Saffe, A., Echegaray, M., Prediction of the lignocellulosic winery wastes behavior during gasification process in fluidized bed: Experimental and theoretical study. *J. Environ. Chem. Eng.*, 6, 5570, 2018.
7. Zalazar-García, D., Torres, E., Rodriguez-Ortiz, L., Deng, Y., Soria, J., Bucalá, V., Rodriguez, R., Mazza, G., Cleaner and sustainable processes for extracting phenolic compounds from bio-waste. *J. Environ. Manage.*, 273, 111154, 2020.
8. Torres-Sciancalepore, R., Fernandez, A., Asensio, D., Riveros, M., Fabani, M.P., Fouga, G., Rodriguez, R., Mazza, G., Kinetic and thermodynamic comparative study of quince bio-waste slow pyrolysis before and after sustainable recovery of pectin compounds. *Energy Convers. Manage.*, 252, 115076, 2022.
9. Zalazar-García, D., Feresin, G.E., Rodriguez, R., Optimal operational variables of phenolic compound extractions from pistachio industry waste (*Pistacia vera* var. Kerman) using the response surface method. *Biomass Convers. Biorefin.*, 12, 3761, 2020.
10. Capossio, J.P., Fabani, M.P., Reyes-Urrutia, A., Torres-Sciancalepore, R., Deng, Y., Baeyens, J., Rodriguez, R., Mazza, G., Sustainable solar drying of brewer's spent grains: A comparison with conventional electric convective drying. *Processes*, 10, 339, 2022.
11. Yuksel, F., Ilyasoglu, H., Baltaci, C., Development of a healthy corn-based snack with sage (*Salvia officinalis* L.) seed. *Bioact. Carbohydr. Dietary Fibre*, 21, 100207, 2020.
12. Ortiz, L.R., Torres, E., Zalazar, D., Zhang, H., Rodriguez, R., Mazza, G., Influence of pyrolysis temperature and bio-waste composition on biochar characteristics. *Renewable Energy*, 155, 837, 2020.
13. Baldán, Y., Riveros, M., Fabani, M.P., Rodriguez, R., Grape pomace powder valorization: a novel ingredient to improve the nutritional quality of gluten-free muffins. *Biomass Convers. Biorefin.*, In press, 2021. https://doi.org/10.1007/s13399-021-01829-8
14. Fernandez, A., Saffe, A., Pereyra, R., Mazza, G., Rodriguez, R., Kinetic study of regional agro-industrial wastes pyrolysis using non-isothermal TGA analysis. *Appl. Therm. Eng.*, 106, 1157, 2016.

15. Baldán, Y., Fernandez, A., Urrutia, A.R., Fabani, M.P., Rodriguez, R., Mazza, G., Non-isothermal drying of bio-wastes: Kinetic analysis and determination of effective moisture diffusivity. *J. Environ. Manage.*, 262, 110348, 2020.
16. Fabani, M.P., Román, M.C., Rodriguez, R., Mazza, G., Minimization of the adverse environmental effects of discarded onions by avoiding disposal through dehydration and food-use. *J. Environ. Manage.*, 271, 110947, 2020.
17. Liu, Q. and Bakker-Arkema, F.W., Stochastic modelling of grain drying: Part 2. Model development. *J. Agric. Eng. Res.*, 66, 275, 1997.
18. Page, G.E., Factors influencing the maximum rates of air drying shelled corn in thin layers, MSc Thesis, Purdue University, West Lafayette, 1949.
19. White, G.M., Ross, I.J., Poneleit, C.G., Fully-exposed drying of popcorn. *Trans. ASAE*, 24, 466, 1981.
20. Henderson, S.M. and Pabis, S., Grain drying theory (I) Temperature effects on drying coefficients. *J. Agric. Eng. Res.*, 6, 169, 1961.
21. Yagcioglu, A., Degirmencioglu, A., Cagatay, F., Drying characteristics of laurel leaves under different drying conditions. *International Congress on Agricultural Mechanization and Energy*, vol. 7, p. 565, 1999.
22. Midilli, A., Kucuk, H., Yapar, Z., A new model for single-layer drying. *Drying Technol.*, 20, 1503, 2002.
23. Saffe, A., Fernandez, A., Mazza, G., Rodriguez, R., Prediction of regional agro-industrial wastes characteristics by thermogravimetric analysis to obtain bioenergy using thermal process. *Energy Explor. Exploit.*, 0, 1, 2018.
24. AOAC, *Official methods of analysis*, 18th edition, The AOAC International, Rockville, Maryland, USA, 2010.
25. Sousa, E.C., Uchôa-Thomaz, A.M.A., Carioca, J.O.B., de Morais, S.M., de Lima, A., Martins, C.G., Alexandrino, C.D., Ferreira, P.A.T., Rodrigues, A.L.M., Rodrigues, S.P., do N. Silva, J., Rodrigues, L.L., Chemical composition and bioactive compounds of grape pomace (*Vitis vinifera* L.), Benitaka variety, grown in the semiarid region of Northeast Brazil. *Food Sci. Technol.*, 34, 135, 2014.
26. Osborne, D.R. and Voogt, P., *The analysis of nutrients in food*, Food Science and Technology, London, England, 1978.
27. Standard test method for acid-insoluble lignin in wood, ASTM D1106-96, 1996.
28. Garau, M.C., Simal, S., Rosselló, C., Femenia, A., Effect of air-drying temperature on physico-chemical properties of dietary fibre and antioxidant capacity of orange (*Citrus aurantium* v. Canoneta) by-products. *Food Chem.*, 104, 1014, 2007.
29. Yusof, Y.A., Mohd Salleh, F.S., Chin, N.L., Talib, R.A., The drying and tabletting of pitaya powder. *J. Food Process Eng.*, 35, 763, 2012.
30. Standard test method for condition monitoring of sulfate by-products in in-service petroleum and hydrocarbonbased lubricants by trend analysis using Fourier transform infrared (FT-IR) Spectrometry, ASTM D7415-18, 2018.

31. Walker, R., Tseng, A., Cavender, G., Ross, A., Zhao, Y., Physicochemical, nutritional, and sensory qualities of wine grape pomace fortified baked goods. *J. Food Sci.*, 79, S1811, 2014.
32. Fernandez, A., Soria, J., Rodriguez, R., Baeyens, J., Mazza, G., Macro-TGA steam-assisted gasification of lignocellulosic wastes. *J. Environ. Manage.*, 233, 626, 2019.
33. Torres, E., Rodriguez-Ortiz, L., Zalazar-García, D., Echegaray, M., Rodriguez, R., Zhang, H., Mazza, G., 4-E (Environmental, economic, energetic and exergetic) analysis of slow pyrolysis of lignocellulosic waste. *Renewable Energy*, 162, 296, 2020.
34. Fernandez, A., Mazza, G., Rodriguez, R., Thermal decomposition under oxidative atmosphere of lignocellulosic wastes: Different kinetic methods application. *J. Environ. Chem. Eng.*, 6, 404, 2018.
35. Fernandez, A., Ortiz, L.R., Asensio, D., Rodriguez, R., Mazza, G., Kinetic analysis and thermodynamics properties of air/steam gasification of agricultural waste. *J. Environ. Chem. Eng.*, 8, 103829, 2020.
36. Völker, S. and Rieckmann, T., Thermokinetic investigation of cellulose pyrolysis - impact of initial and final mass on kinetic results. *J. Anal. Appl. Pyrolysis*, 62, 165, 2002.
37. Di Blasi, C., Modeling chemical and physical processes of wood and biomass pyrolysis. *Prog. Energy Combust. Sci.*, 34, 47, 2008.
38. Yang, H., Yan, R., Chen, H., Lee, D.H., Zheng, C., Characteristics of hemicellulose, cellulose and lignin pyrolysis. *Fuel*, 86, 1781, 2007.
39. Lah, B., Klinar, D., Likozar, B., Pyrolysis of natural, butadiene, styrene–butadiene rubber and tyre components: Modelling kinetics and transport phenomena at different heating rates and formulations. *Chem. Eng. Sci.*, 87, 1, 2013.
40. Abed, I., Paraschiv, M., Loubar, K., Zagrouba, F., Tazerout, M., Thermogravimetric investigation and thermal conversion kinetics of typical North African and Middle Eastern lignocellulosic wastes. *BioResources*, 7, 1200, 2012.
41. Volpe, R., Menendez, J.M.B., Reina, T.R., Messineo, A., Millan, M., Evolution of chars during slow pyrolysis of citrus waste. *Fuel Process. Technol.*, 158, 255, 2017.
42. Sun, J., He, F., Pan, Y., Zhang, Z., Effects of pyrolysis temperature and residence time on physicochemical properties of different biochar types. *Acta Agric. Scand., Section B-Soil Plant Sci.*, 67, 12, 2016.
43. Kaveh, M. and Chayjan, R., Modeling thin-layer drying of turnip slices under semi-industrial continuous band dryer. *J. Food Process. Preserv.*, 41, e12778, 2017.
44. Mohammadi, I., Tabatabaekoloor, R., Motevali, A., Effect of air recirculation and heat pump on mass transfer and energy parameters in drying of kiwifruit slices. *Energy*, 170, 149, 2019.

45. Midilli, A., Kucuk, H., Yapar, Z., A new model for single-layer drying. *Drying Technol.*, 20, 1503, 2002.
46. Sabarez, H., Optimisation of industrial Food Drying Operation. *Int. J. Food Biosyst. Eng.*, 2, 1, 2017.
47. Beres, C., Costa, G.N., Cabezudo, I., da Silva-James, N.K., Teles, A.S., Cruz, A.P., Mellinger-Silvad, C., Tonond, R.V., Cabral, L.M., Freitas, S.P., Towards integral utilization of grape pomace from winemaking process: A review. *Waste Manage.*, 68, 581, 2017.
48. Codex alimentarius, Codex Standard for wheat flour, CXS 152-1985, International Food Standards, 1985, Revised in 1995, Amended in 2016.
49. Sainz, R.L., Szezecinski, A., Fontana, M., Bosenbecker, V.K., Ferri, V.C., Do Nascimento, C.O., Use of grape marc flour in the production of cookies. *BIO Web of Conferences*, vol. 12, p. 04003, 2019.
50. Cilli, L.P., Contini, L.R., Sinnecker, P., Lopes, P.S., Andreo, M.A., Neiva, C.R., Nascimento, M.S., Yoshida, C.M., Venturini, A.C., Effects of grape pomace flour on quality parameters of salmon burger. *J. Food Process. Preserv.*, 44, 1, 2020.
51. Nayak, A., Bhushan, B., Rosales, A., Turienzo, L.R., Cortina, J.L., Valorisation potential of Cabernet grape pomace for the recovery of polyphenols: Process intensification, optimisation and study of kinetics. *Food Bioprod. Process.*, 109, 74, 2018.
52. Zhang, L., Zhu, M., Shi, T., Guo, C., Huang, Y., Chen, Y., Xie, M., Recovery of dietary fiber and polyphenol from grape juice pomace and evaluation of their functional properties and polyphenol compositions. *Food Funct.*, 8, 341, 2017.
53. Praznik, W., Loeppert, R. *et al.*, Dietary fiber and prebiotics, in: *Polysaccharides*, K. Ramawat, and J.M. Mérillon, (Eds.), pp. 891–925, Springer, Cham, 2015.
54. ANMAT, M. @ of P. H. of the A. R., *Guidelines for the authorization of a gluten-free food*, Federal Food Control Program, 2005.
55. Rainero, G., Bianchi, F., Rizzi, C., Cervini, M., Giuberti, G., Simonato, B., Breadstick fortification with red grape pomace: Effect on nutritional, technological and sensory properties. *J. Sci. Food Agric.*, 102, 2545, 2022.
56. Šporin, M., Avbelj, M., Kovač, B., Možina, S.S., Quality characteristics of wheat flour dough and bread containing grape pomace flour. *Food Sci. Technol. Int.*, 24, 251, 2018.
57. Tolve, R., Simonato, B., Rainero, G., Bianchi, F., Rizzi, C., Cervini, M., Giuberti, G., Wheat bread fortification by grape pomace powder: Nutritional, technological, antioxidant, and sensory properties. *Foods*, 10, 75, 2021.
58. Bianchi, F., Cervini, M., Giuberti, G., Rocchetti, G., Lucini, L., Simonato, B., Distilled grape pomace as a functional ingredient in vegan muffins: Effect on physicochemical, nutritional, rheological and sensory aspects. *Int. J. Food Sci. Technol.*, 57, 4847, 2022.

59. Karnopp, A.R., Figueroa, A.M., Los, P.R., Teles, J.C., Simões, D.R., Barana, A.C., Kubiaki, F.T., de Oliveira, J.G., Granato, D., Effects of whole-wheat flour and bordeaux grape pomace (*Vitis labrusca L.*) on the sensory, physicochemical and functional properties of cookies. *Food Sci. Technol.*, 35, 750, 2015.
60. Iuga, M. and Mironeasa, S., Potential of grape by-products as functional ingredients in baked goods and pasta. *Compr. Rev. Food Sci. Food Saf.*, 19, 2473, 2020.
61. Yalcin, E., Ozdal, T., Gok, I., Investigation of textural, functional, and sensory properties of muffins prepared by adding grape seeds to various flours. *J. Food Process. Preserv.*, 46, e15316, 2021.
62. Nakov, G., Brandolini, A., Hidalgo, A., Ivanova, N., Stamatovska, V., Dimov, I., Effect of grape pomace powder addition on chemical, nutritional and technological properties of cakes. *LWT*, 134, 109950, 2020.
63. Theagarajan, R., Malur Narayanaswamy, L., Dutta, S., Moses, J.A., Chinnaswamy, A., Valorisation of grape pomace (cv. *Muscat*) for development of functional cookies. *Int. J. Food Sci. Technol.*, 54, 1299, 2019.
64. Bender, A.B., Speroni, C.S., Salvador, P.R., Loureiro, B.B., Lovatto, N.M., Goulart, F.R., Lovatto, M.T., Miranda, M.Z., Silva, L.P., Penna, N.G., Grape pomace skins and the effects of its inclusion in the technological properties of muffins. *J. Culin. Sci. Technol.*, 15, 143, 2017.
65. Urquiaga, I., Troncoso, D., Mackenna, M.J., Urzúa, C., Pérez, D., Dicenta, S., de la Cerda, P.M., Amigo, L., Carreño, J.C., Echeverría, G., Rigotti, A., The consumption of beef burgers prepared with wine grape pomace flour improves fasting glucose, plasma antioxidant levels, and oxidative damage markers in humans: A controlled trial. *Nutrients*, 10, 1388, 2018.
66. Troilo, M., Difonzo, G., Paradiso, V.M., Pasqualone, A., Caponio, F., Grape pomace as innovative flour for the formulation of functional muffins: How particle size affects the nutritional, textural and sensory properties. *Foods*, 11, 1799, 2022.
67. Marchiani, R., Bertolino, M., Belviso, S., Giordano, M., Ghirardello, D., Torri, L., Piochi, M., Zeppa, G., Yogurt enrichment with grape pomace: Effect of grape cultivar on physicochemical, microbiological and sensory properties. *J. Food Qual.*, 39, 77, 2016.
68. Acun, S. and Gül, H., Effects of grape pomace and grape seed flours on cookie quality. *Qual. Assur. Saf. Crops Foods*, 6, 81, 2014.

5

Tomato By-Products: A Potentially Promising Bioresource for the Recovery of Bioactive Compounds and Nutraceuticals

Toufik Ouatmani[1,2], Hayate Haddadi-Guemghar[1*], Samir Hadjal[2], Lila Boulekbache-Makhlouf[1] and Khodir Madani[1,3]

[1]*Laboratoire de Biomathématiques, Biophysique, Biochimie et Scientométrie (L3BS), Faculté des Sciences de la Nature et de la Vie, Université de Bejaia, Bejaia, Algeria*

[2]*Laboratoire de Développement Analytique, Direction de Recherche et Développement, Cevital Agro-Industrie, Nouveau Quai Port de Bejaia, Bejaia, Algeria*

[3]*Centre de Recherche en Technologies Agro-Alimentaires (CRTAA), Route de Targa-Ouzemour, Bejaia, Algeria*

Abstract

By the recent few years, the substitution of synthetic additives and ingredients with natural ones has gained an industrial focus because of regulatory and other requirements, including consumer concerns for healthy diets. Tomatoes are widely consumed either fresh or in processed form (concentrate, juice, sauce, ketchup) and contribute to a healthier diet, thanks to their benefits. More interestingly, their processing residue could serve as a cheap and natural source of numerous biomolecules (carotenoids, fibers, proteins, oil, and phenolics) with a great potential to be valorized in various fields. Moreover, an efficient management of such waste could constitute a strategic challenge since it is a part of an ecological approach. Also, within the framework of the green chemistry concepts, new technologies (microwaves, ultrasounds, supercritical solvents, enzyme-assisted extraction) and mathematical tools (Designs of Experiments, Desirability function, Artificial Neural Networks) have emerged and have extensively been used to achieve an effective recovery and/or a potential use of the targeted biocompounds. In the present chapter, we will provide an overview about the great potential of tomato

**Corresponding author*: hayate.haddadi@univ-bejaia.dz

Umile Gianfranco Spizzirri (ed.) Nutraceutics from Agri-Food By-Products, (137–172) © 2023 Scrivener Publishing LLC

by-products (mainly peels and seeds) valorization. We will also discuss how some adopted approaches have provided an efficient exploration of the most interesting biomolecules from these by-products.

Keywords: Tomato by-products, bioactive compounds, nutraceuticals, sustainable valorization

5.1 Introduction

Recently, functional and nutraceutical products began to gain more and more importance in business strategy of the food manufacturers, in order to fulfill a growing request for this new type of food. This trend has sparked the search for new sources of natural substances (carotenoids, tocopherols, sterols, terpenes, and polyphenols) apart from fruits and vegetables [1, 2]. Hence, the exploration of agro-industrial wastes has become the object of major interest and a lot of effort has been made to develop strategies in the aim to obtain new phytochemicals useful as drugs, functional food ingredients, or nutraceuticals [3, 4].

On the other hand, the world's population has dramatically increased. Accordingly, the demand for primary agricultural products and crops has been amplified and that has required the entire mobilization of processing facilities in order to ensure the supply of foodstuffs under processed forms [5]. Following this trend, tomato (*Solanum lycopersicon* L) is among the most transformed crops and its processing is highly industrialized; indeed, about 80% of the worldwide production is transformed into various forms (peeled tomatoes, juice, sauce, paste and ketchup) [6]. Meanwhile, side effect of the large-scale industrialization has occurred, including the production of big amounts of waste and their disposal problems. Additionally, regulatory requirements related to environmental protection have required the development of innovative strategy in the aim to smart resolving of this dilemma and thus to well valorize these by-products [7, 8]. In the light of this statement, an awareness is gaining ground through new approaches that are gradually taking root in industrial practices. Hence, residues generated in production lines, once considered of no economic value, are recovered as by-products rich in natural active molecules with the potential to replace synthetic ingredients and additives. Moreover, further benefits could be achieved by reducing cost of their disposal [8, 9].

The present chapter deals with the great opportunities to valorize the by-products of industrial tomato manufacturing. These residues are interesting biosource of natural substances with high value-added for both food, cosmetics, and pharmacological industry (Figure 5.1), and this is

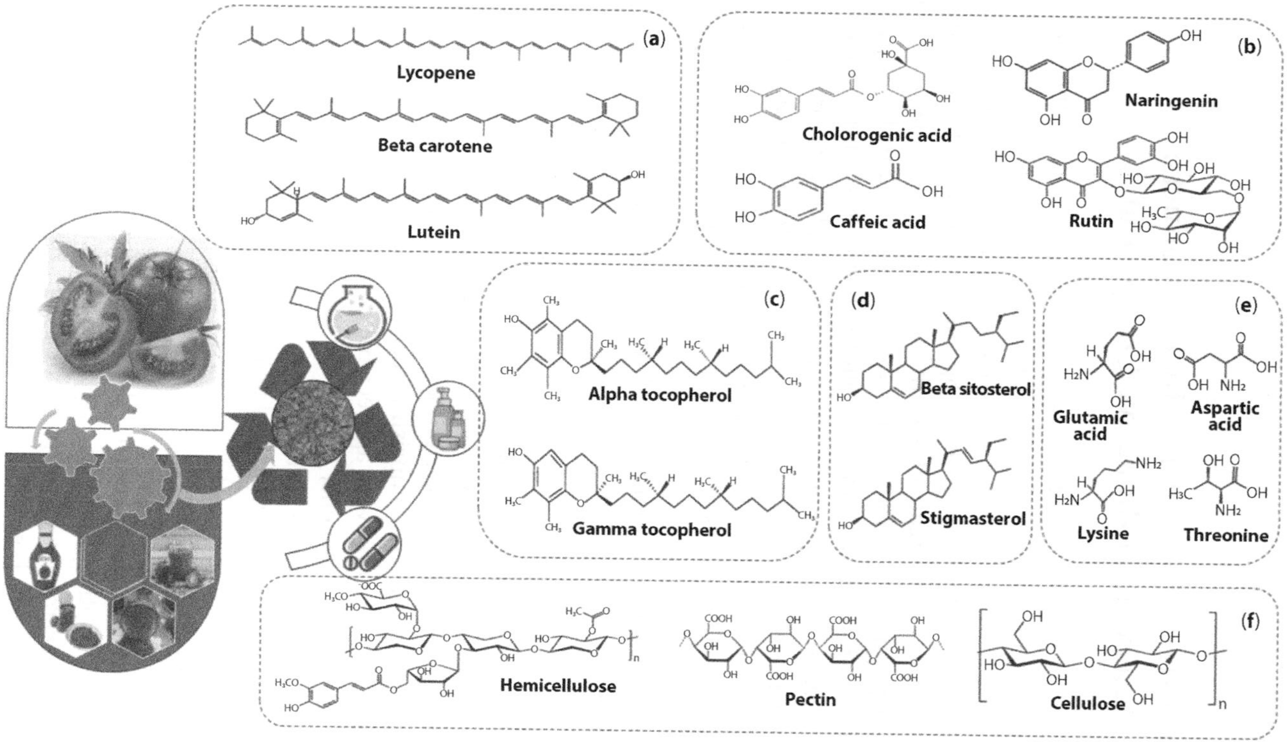

Figure 5.1 Main bioactive compounds from tomato processing by-products (a) carotenoids (lycopene, β-carotene, lutein), (b) phenolic compounds (chlorogenic acid, caffeic acid, rutin, naringenin), (c) tocopherols (α-tocopherol, γ-tocopherol), (d) phytosterols (β-sitosterol, stigmasterol), (e) amino acids (glutamic acid, aspartic acid, lysine, threonine), (f) dietary fibers (cellulose, hemicellulose, pectin).

why they are attracting a great attention through the application of innovative technologies and methods for efficiently extract and explore these biomolecules.

5.2 Industrial Tomato and Its By-Products: A Reservoir of Bioactive Compounds

Tomato (*Solanum lycopersicon* L.) was first domesticated by the indigenous population of present-day Latin America. Indeed, its common name "tomato" derives from the Nahuatl (Aztec language) word "tomatl" [10]. It was then transported to Europe by Spanish colonizers and since then, tomato cultivation has spread all over the world [11]. Nowadays, it is one of the most popular vegetables and is extensively consumed worldwide, either in fresh or processed form. Therefore, thanks to scientific advances in terms of genetics and new farming methods, its cultivation area has significantly increased and reached 5 million hectares for an annual production of 186.6 million tons, in 2020 [10, 12]. About 80% of the harvest is processed into several kinds of manufactured products [6], of which great parts are processed into sauce, paste and ketchup [11] to respond to the upward demand for these tomato based-foods which allows the conservation of this healthier but quickly perishable fruit, and so to ensure its year-round consumption [13, 14].

Tomatoes as well as their based products contain valued compounds like carotenoids, ascorbic acid and phenolics. However, their biomolecules' content depends on cultivar along with processing methods [12, 15, 16]. Furthermore, while tomatoes are processed not negligible amounts (3-7%) of pomace are engendered, comprising peels, seeds, and some wasted tissue of pulp [8, 11]. Therefore, very appreciable amounts of interesting bioactive phytochemicals (lycopene, β-carotene, tocopherols, sterols and polyphenols) might be wasted with the by-products [17] while if they are well exploited they could turn a profit and bring added-value to the industry.

5.2.1 Carotenoids

Carotenoids were discovered during the 18th century by the French chemists Edme Jean Baptiste Bouillon-Lagrange and Nicolas Louis Vauquelin. During their research on carrot pigments, both of them extract what the first described as a "yellow oily material" and the second characterized as a "resinous pigment," referring to carotenes. In 1831, Heinrich Wilhelm Ferdinand Wackenroder succeeded in isolating the actual β-carotene from carrot as a hydrocarbon crystal and named it "carotene". Six years later, Jöns Jacob Berzelius extracted pigments from leaves during their senescence and called the material "xanthophylls". Lutein was discovered in egg yolk by Johann Ludwig Wilhelm Thudichum, in 1869. Four years later, lycopene was isolated by Frederik Anthony Hartsen as red crystals from berries of *Tamus communis* [18, 19]. In 1876, Pierre-Marie-Alexis Millardet succeeded in crystallizing the red piments extracted from the fleshes of ripe tomatoes and named the product "solanorubine". The isolated pigments were highly soluble in organic solvent and hot alcohol but not in water. In the beginning of the 20th century (1903), the British chemist Henry Edward Schunk distinguished the red tomato pigment from the other carotenes and named it "lycopin". He reported it as "a substance that constitutes nearly the whole of the pigment present in the tomato" [20–22]. Montanari was, in 1904, the pioneer who studied the chemical structure of lycopene, describing the molecule as a result of a condensation of two carotene units and giving, according to him, the formula $C_{22}H_{74}$. Six years after Monatanari's findings, Willstätter and Escher discovered the unsaturation of lycopene and determined its currently known formula, $C_{40}H_{56}$. The molecule's microstructure was also examined under a microscope for the first time ever as a pure extracted specimen. The first microscopic images revealed the crystals as clusters of needle-like and long, flat-sided prisms with split or blunt ends, while a deeper red color characterized zones where individual crystals were crossing [20, 23]. It was in 1930 that Paul Karrer and coworkers elucidated the detailed skeleton of lycopene, as a highly unsaturated (13 double bonds) hydrocarbon with an open side chain [20].

All carotenoids share a commune structure of a polymer of 8 isoprenoid units, which constitutes their central hydrocarbon chain with alternance of single and double bonds. Such characteristic skeleton of the long chain polyene confers it a chromophore feature and strongly determines its functional proprieties. Indeed, the number of conjugated double bonds (between 3 and 13) is responsible of the color of the molecule by monitoring the absorbed light energy as well as the transfer of electrons along the conjugated system [24, 25]. Several criteria have been adopted for the classification of carotenoids, considering both structural and functional features [24]. However, the most commonly used and ever-ancient classification is their dividing into (i) Carotenes as highly liposoluble hydrocarbons, including β-carotene and lycopene, and (ii) Xanthophylls which have oxygenated functional groups in their terminal moieties, such as a hydroxyl function in the case of lutein and zeaxanthin. These substituent groups not only increase the molecule's polarity, but also confer other functional roles on it and in its interactions with other molecules like proteins [25, 26].

Carotenoids are key compounds of plants since they act as receptors of light energy during photosynthesis belong to other essential functions like quenching singlet-oxygen (antioxidants), photo-protection, color attraction; some of them are also precursors of plant hormones and vitamin A [27]. In higher plants, biosynthesis of these tetraterpenoid molecules occur from the linkage of two C_5 building isomeric blocs, namely isopentenyl diphosphate and dimethylallyl diphosphate. That can occur through two pathways i.e. mevalonate (in the cytoplasm), and methylerythritol 4-phosphate (in the chloroplast), producing a colorless carotenoid (phytoene), which after enzymatic desaturation and isomerization leads to the formation of lycopene [28]. A variety of carotenoids are synthetized in the aerial parts of tomato plant i.e., leaves, flowers, and fruits. Lutein is the main carotenoid in leaves and ensure the plant photo protection, xanthophylls are abundant in the flowers and give their yellowish color, while lycopene is by far along with β-carotene the most famous carotenoid and specifically linked to tomato since it confers the red color for the rip fruits [29].

On the other side, industrial tomato and its based products constitute major sources of dietary lycopene; the products' content oscillate from 42.3 to 365 ppm, depending on product type, and greatly contribute to the daily intake of this biomolecule [30]. Nevertheless, tomato processing operations induce little loss of lycopene but affect its chemical configuration from the naturally all-*trans* isomer to the *cis* stereoisomers. These latter have better bioavailability but are less stable and can be easily oxidized [31]. The pomace by-produced, during the processing, is interestingly still rich in carotenoids of which lycopene is predominant (1324.84 μg/g DW), but also in β-carotene (235.54 μg/g DW) and lutein (31.16 μg/g DW) [32]. However, when the by-product is separated into peels and seeds, the first fraction (skins) is significantly richer in both lycopene, β-carotene, and lutein with respectively 734.0, 29.3 and 14.5 μg/g DW, while content in the respective carotenoids from seeds was 130.0, 14.4 and 6.5 μg/g DW [33]. Tomato by-products have drawn a lot of attention as potential sources of lycopene and other carotenoids because of how important those molecules are. In fact, a number of techniques have been used to extract them effectively. Indeed, Enzyme-Assisted Extraction (EAE), spectacularly increased both carotenoids and lycopene recovery from a Greek tomato pomace by respectively 6 and 10 folds higher than of those of non-enzyme treated samples. Besides, it was noted in the same paper that high pressure-assisted extraction enhanced the carotenoid yield up to 64% in comparison to conventional extraction while reducing both duration of extraction and volume of solvent [34]. Moreover, Catalkaya and Kahveci [35] used a face-centered central composite design (CCD) and response surface methodology

(RSM) for the optimization of the conditions (enzyme/enzyme ratio = 1, enzyme/substrate ratio = 0.2 ml/g, enzymatic reaction duration of 5 hours, solvent/substrate ratio = 5 ml/g, extraction duration of 1h, at a temperature of 40°C) for enzymatic (using cellulase and pectinase) extraction of lycopene from tomato pomace. Carotenoids, being highly hydrophobic are routinely extracted using conventional non-polar solvents like hexane, petroleum ether, and chloroform which are extremely toxic. Nonetheless, ethanol as a food-grade solvent was effectively used by Li, Pettinato [36] in the Ultrasound-Assisted Extraction of total carotenoids and lycopene. In this study, a CCD and RSM were applied for the optimization of both total carotenoids and lycopene content as well as of the anti-radical power of the ethanolic extracts. The mathematical Desirability function was also successfully used to compute the optimized parameters (sample in 90 mL total volume of extraction with the ratio of 72 mL/g, at 65 °C, during 20 minutes of sonication with a probe setup at 65% amplitude and 33 seconds of pulse duration). The optimized extract was then encapsulated using maltodextrin and inulin as coating agents. Another innovative technology, namely Microwave-Assisted Extraction was also investigated in the work of Ho, Ferruzzi [37] for optimizing the recovery of lycopene from tomato peels. For this, the authors applied a Box-Behnken Design (BBD) and a CCD for the determination of the optimum levels of the operative parameters of extraction. Interestingly, ethyl acetate (which is also a food grade solvent) was found to be more effective than hexane in the MAE of all-*trans* lycopene, leading to the extraction of up to 13 mg/100g after the sample was irradiated for 60 seconds under 400 W microwave's power and an energy of 24 kJ equivalents. MAE under the optimized setup was found to give a similar yield in *cis* lycopene as that of conventional extraction but, in contrast, engender significant higher yields in both total and all-*trans* lycopene. Nour, Corbu [38] performed a comparative study between MAE for 5 minutes, UAE for 50 minutes, and maceration for 7 days of tomato pomace in numerous vegetable oils for their carotenoid enrichment, and then evaluated the impact of such a strategy on the physico-chemical and the stability of the enriched oils. Maceration for a week presented the best enrichment result with up to 38.6 mg/kg enrichment of extra virgin sunflower oil, sonication of the different oils at 5% w/v of the pomace allowed the increase of their carotenoids content by 29.2 to 38.0 mg/kg, extra virgin olive and rice oil being the most enriched. Irradiation for only 5 minutes allowed the enrichment of unrefined corn, Extra virgin sunflower, and refined rapeseed oil with respectively 30.6, 29.1 and 32.3 mg/kg, making of the MAE the most efficient method when considering the ratio time to the solubilized quantity. However, the findings of stability tests of the enriched oils were not correlated; they were explained by the eventual pro-oxidant behavior of carotenoids and/or other bioactive molecules like the phenolic compounds and tocopherols under certain conditions.

Therefore, it must be noticed that carotenoids are very vulnerable to heat, light, and reactive chemicals like peroxides and metals, which is due to their specific chemical skeleton with many conjugated double bonds. Hence, precautions must be taken during the material preparation, as well as the procedure of extraction, in order to prevent the carotenoids from losing [39].

Regarding their industrial application, carotenoids principally β-carotene and lycopene are increasingly attracting the manufacturers' interest, particularly in Europe where their use as natural or nature-identical dyes has been authorized by the European Commission [40, 41]. Another interesting function of carotenoids is their potent antioxidant capacity. Indeed, their tetraterpene structure characterized by a multitude of conjugated double bonds provides them the ability to scavenge singlet oxygen and peroxyl radicals and to absorb excess light energy [40, 42]. Among all carotenoids, lycopene majorly extracted from tomatoes constitutes the most powerful antioxidant, i.e., its antioxidant capacity is 2.2 higher than β-carotene and 100 folds better than that of alpha tocopherol [42]. Its regular intake is inversely associated to risk of cancers like prostate tumors [43–45]. Likewise, precursors of vitamin A such as β-carotene and other none-provitamin ones like lycopene show anti-obesity and anti-inflammatory activities [46] but also anti-aging properties of skin protection from harmful radiation and free reactive radicals [47]. Therefore, to get benefit from their nutritional, cosmetic, and health-promoting virtues, carotenoids like lycopene and β-carotene could be provided in food supplement [41, 47].

5.2.2 Phenolic Compounds

Polyphenols constitute wide variety of biomolecules. They are synthesized during the secondary metabolism of plants for their own protection needs against predators and UV radiation. Also, some of them serve to attract pollinating agents or to provide structural functions and sensory attributes of plants [48–51]. Phenolic compounds are grouped into five principal classes i.e., phenolic acids, flavonoids, stilbenes, lignans and tannins [9, 51, 52]. Their commune point is a chemical skeleton with at least one aromatic ring having a bind with a hydroxyl functional group [51]. The classification of polyphenols is generally based on the number of benzene rings in addition to the type, the position, and the number of the attached functional substituents [53, 54]. Their chemical structure varies from the simplest monomeric phenolic acid (protocatechuic acid) to the complex polymerized condensed tannins (proanthocyanidins) [55, 56]. Actually, their wide diversity in term of chemical structure is conferring to them versatile industrial uses including pharmaceutical, cosmetics, textile as well as in food and packaging [9].

In a review about anticancer substances from tomato, Martí, Roselló [29] reported that tomato rip fruit contains essentially flavonoids and phenolic acids; naringenin chalcone, being the main polyphenol of tomato with 182 µg/g FW, rutin and quercetin were also found in tomato in quit equal contents of 45 and 44 µg/g FW respectively. Chlorogenic acid constitute the main phenolic acid and its content was up to 33 µg/g FW. However, phenolic content could be influenced by the genotype [29, 57, 58], by site of cultivation [59] as well as by irrigation treatment [58]. Even interestingly, processing of tomato into their based products e.g., type a sauce, highly increase some polyphenols content like naringenin; indeed, the amount of this flavonoid is boosted by heating under acidic medium, during which naringenin chalcone is dramatically converted to naringenin, increasing its content 20 folds in the processed tomato versus to fresh fruit [60].

As discussed above, molecules of industrial interest like carotenoids might be wasted with the residue of tomato industrial processing. Similarly, considerable amounts of phenolic compounds could be discarded with tomato pomace if it they are not recovered. According to the study of Kalogeropoulos, Chiou [17], a diversity of polyphenols content were reported in tomato pomace including flavonoids (naringenin, chrysin, quercetin, kaempferol, catechin, epicatechin, luteolin), phenolic acids (chlorogenic, caffeic, *p*-coumaric, ferulic, sinapic, phloretic, vannilic, syringic, cinnamic, *p*-OH-benzoic), and a stilbene (resveratrol); by far, naringenin was the predominant phenol of tomato pomace with 328.6 µg/g DW equally to 63.5% of the sum of quantified polyphenols (517.5 µg/g DW). Chrysin was the second most abundant flavonoid with 29.5 µg/g DW, chlorogenic and caffeic acids were the most quantified phenolic acids with respectively 51.7 µg/g DW and 33.5 µg/g DW, while resveratrol was the unique detected and quantified stilbene in the tomato processing pomace with 7.7 µg/g DW. in an appealing work of Perea-Domínguez, Hernández-Gastelum [61], a sequential extraction of polyphenols from a Mexican tomato pomace was performed; the procedure consisted in optimizing the quantity of recovered phenolics by qualitative cascade extractions of targeted groups of polyphenols namely soluble free phenols, acid-hydrolysable, alkaline-hydrolysable and bound phenolics. Actually, very interesting amounts of both phenolic acids and flavonoids with respectively 52.22 and 73.72 mg by g of a previously defatted and dried extract (DE); bound form of naringenin was the predominant polyphenol with 25.22

mg/g DE, other phenolic compounds were also identified and quantified, that are flavonoids (catechin, kaempferol, quercetin and rutin) and five phenolic acids whose caffeic and *p*-coumaric acids were the main phenolic acids (11.79 and 9.57 mg/g DE respectively) although ferulic, gallic and chlorogenic acids were quantified in small amounts. Moreover, some green emerging technologies like ohmic heating extraction allow the enhancement of bound phenolic recovery by breaking the ester bonds as an example, as well as the preservation of thermolabile molecules, since it permits a fast heating of the reactional medium. Coelho, Pereira [62] used the ohmic heating for the extraction of polyphenol from tomato pomace, and they observed an increase of 77 and 61% in rutin and naringenin recovery in a reduced duration of 16% less in comparison to the control.

Phenolic compounds of separate fractions of tomato pomace have also been the subject of interest; the skin fraction have particularly concerned the research community [63], since it is well known that bioactive molecules with antioxidant capacity, UV light protective and defense functions such as polyphenols are mainly concentrated in the exocarpe of fruits and vegetables [64]. However, tomato seeds could also contain non negligible content in polyphenolics [64–66]. Moreover, eight flavonoids (quercetin-3-O-βglucoside, rutin, isorhamnetin, kaempferol, naringenin, quercetin, apigenin, and myricetin) and eight phenolic acids (caffeic, vanillic, ferulic, sinapic, chlorogenic, p-coumaric, trans-cinnamic, and gallic acids) have successfully been identified by Valdez-Morales, Espinosa-Alonso [66] in the peels and have also been detected (except apigenin) in the seeds; the latter fraction showed appreciable contents of flavonoids (67-99 μg/g DW) and phenolic acids (89-126 μg/g DW) while the peels were slightly richer in phenolic acids (67-433 μg/g DW) and significantly in flavonoids (122-774 μg/g DW). Indeed, during the tomato fruit is ripening, flavonoids are accumulated in the skin simultaneously to the decrease of chlorophyll content [54]. More recently, Tamasi, Pardini [67] have identified two flavonoids (rutin and kaempferol-3-O-rutinoside) and three hydroxycinnamic acids (chlorogenic, caffeic and *p*-coumaric) in the "comone" variety skins from Italy. Rutin (170.1 μg/g FW) and chlorogenic acid (40 μg/g FW) were the most concentrated phenolics. High Voltage Atmospheric Cold Plasma (HVACP) is a relatively underknown emerging technology compared to EAE, UAE and MAE. This non thermal innovative technique consists of treatment of the sample by an ionized inert gas flow which leads to the production of oxygen and nitrogen species [68]. These reactive species alter the chemical features of cell walls, enhancing the penetration of solvent into the cells and subsequent release of their contents [69]. HVACP was applied by Bao, Reddivari [70] to extract phenolic compounds from tomato pomace, and they successfully enhanced the yield of extraction as well as the antioxidant capacity of the phenolic extract. More specifically, pretreatment with air plasma enhanced caffeic acid content by 25.9%, and chlorogenic content recovered was also increased by 18% with nitrogen plasma; That resulted thanks to the disturbance and the increased hydrophilicity of well walls, as well to the breaking of bound phenolic linkage, leading to a better release of those biomolecules. Tomato seeds are the other important fraction of tomato pomace, but unlike to the peels, are under-considered regarding their polyphenols content. Therefore, in the recent work of Ouatmani, Haddadi-Guemghar [8], we went ahead and sequentially extracted the oil content from tomato seeds before optimizing the MAE of polyphenol from the defatted extract. The oil's fatty acid profile was very interesting. In addition, CCD and RSM were used for the building and modeling of the experimental design of MAE of polyphenols as well the antioxidant capacity of the defatted extract, and the Desirability function for the computing of the optimum levels of the factors giving the highest polyphenol content with the highest antioxidant capacity. MAE applied with the optimized parameters (700 W, 70 seconds, and water as solvent) allowed the increase of the yield by 24% in a very short time of irradiation, in comparison to the conventional method of extraction.

Flavonoids are an important group of polyphenols with a compelling capacity to scavenge free radical and to chelate metals. Certainly, their highly reactive hydroxyl substituents are conferring to them relevant biological proprieties including antiviral, antiallergic, anti-inflammatory and anticancer [71]. As reported in the afore-discussed studies of Valdez-Morales, Espinosa-Alonso [66] and of Tamasi, Pardini [67], the glycosylated forms of quercetin namely quercetin-3-O-β-D-glucoside (isoquercetrin) and quercetin 3-O-rutinoside (rutin) were the principal flavonoids quantified in tomato by-products. These latter, being more soluble are more bioavailable than non-glycosylated quercetin and could provide a multitude of biological actions i.e., antimicrobial, antidiabetic, anti-inflammatory, anti-Alzheimer, anticancer, antioxidant, and iron chelation [72]. They have also been reported to prevent alcohol induced cytotoxicity of liver as well as to enhance its detoxification [73]. Specifically, besides to its pharmacological uses related to the class of flavonoids, rutin (also called vitamin P) could be used as a preservative, colorant, antioxidant and as ingredient is poly-vitaminic formulations [74, 75]. Likewise, naringenin has been associated to tomato and its based products [60] but also with its industrial by-product [61]. This flavanone is a potent bioactive metabolite which has focused much interest and has extensively been explored in drug formulations, thanks to its therapeutic proprieties [76, 77].

Phenolic acids constitute another valuable class of polyphenols. From a chemical point of view, they represent the least complex form of phenols, consisting of a benzene ring with a carboxylic functional group [9]. These bio substances are produced through the shikimate pathway during the secondary metabolism of plants; they are enzymatically synthetized via the aromatic amino-acids tyrosine and phenylalanine to subsequently giving rise to two important subgroups of polyphenols i.e., the C6-C1 hydroxybenzoic and C6-C3 hydroxycinnamic acids [78]. The latter are ubiquitous in plants ever in free or in various conjugated forms and are considered the most abundant phenolic compounds, accounting for up to one third of the phenolics in food. More even interestingly, hydroxycinnamic acids comprising predominantly caffeic acid and its conjugated form 5-O-caffeoylquinic acid (chlorogenic acid) are the most abundant phenolic acids in all parts of the rip fruits (especially in the skins) of the Solanaceae family (whose tomatoes are a part) [79]. Regarding to tomato case, this is closely in accordance with the findings seen above, viz. those of Tamasi, Pardini [67] and of Valdez-Morales, Espinosa-Alonso [66]. On the other side, hydroxycinnamic acids have been linked to various bioactivities including antioxidant, anticancer and antimicrobial [80] and were consequently investigated in several areas like cosmeceutical formulations [78, 81], and as antioxidants [79]. More explicitly, caffeic and chlorogenic acids have been of paramount importance and have found an effective exploration in advanced food and medical research, i.e., as ingredients in smart food packaging and in nanocomposite sensors to detect molecules responsible for the progression of certain diseases [81]. Also, numerous clinical trials have demonstrated physiologic actions of chlorogenic acids in the protection of the body's systems, including neurological, gastrointestinal, cardiovascular, hepatic and renal, as well as to other important roles in the prevention and treatment of metabolic syndromes, in addition to anti-carcinogenesis, anti-inflammatory and antimicrobial activities [82].

5.2.3 Vitamin E (Tocopherols and Tocotrienols)

In the 1920s, Vitamin E was isolated and characterized for the first time by Evans and Bishop from wheat germ oil [83]. Since this time, its several isoforms were discovered and extracted from various origins [84], and their health virtues have been investigated [85].

Chemically, vitamin E is an aggregate of eight isomers belonging to two groups, viz. tocopherols and tocotrienols (cited as tocochromanols or tocols) whose components in each group are conventionally referred to as α, β, γ, and δ [86–88]. Their commune structural skeleton is designed with a chromanol ring bound to a hydrophobic side chain at its C2 position; tocopherols are distinguished from tocotrienols by their saturated isoprenoid side chain whereas the latter are characterized by unsaturated bonds in the R3 phytyl chain, comprising three carbon-carbon double bonds (at C3', C7', and C11'). The forms α, β, γ, and δ of both tocopherols and tocotrienols are similarly determined by the number and the position of CH3 groups on the phenolic ring. In other words, α forms have three methyl substituents, β and γ each one with two methyl groups but at different positions, while δ forms are mono-methylated [87–89]. Tocopherols can naturally be found in eight stereoisomers because of their three stereogenic chiral centers (at C2, C4', and C8' positions) while tocotrienols have just one chiral Carbone in the phenolic ring (at C2 position), giving only two isomers [87, 90]. The just-cited structural differences could strongly impact both sensitivity and activity of each of molecule [90].

Tocols are biosynthesized by photosynthetic plants through the simultaneous synthesis and the further condensation of their two moieties, namely the chroman ring (as a polar head) derived from *p*-hydroxyphenyl pyruvic acid via shikimic acid pathway, and the isoprenoid side chain (as a highly lipophile tail) derivative from geranylgeranyl di-phosphate via the non-mevalonate deoxyxylulose phosphate pathway [91]. Tocols are principally concentrated in the seeds of plants even if lesser amounts are found in their photosynthetic parts. Alpha tocopherol is predominant in chloroplasts while seeds are generally much richer in γ analogue [92]. Conventional edible oils are the most important source of vitamin E in human diet. Nonetheless, there is a proportionality in terms of concentrations of the different counterparts in each type of oil, for example, some of them (sunflower, olive, cottonseed oil) are mainly rich in α tocopherol, others (soybean, canola, corn, rapeseed oil) in γ tocopherol [90, 92], while palm oil which is the most worldwide consumed oil is particularly rich in γ tocotrienol [90, 93]. Furthermore, vitamin E rich-oil could also be extracted from non-conventional sources such as plant by-products including kernels (apricot), germ (wheat) and seeds (grape and tomato) [94]. Tomato fruit was reported to contain mainly α and γ tocopherol [95, 96] and after tomatoes are processed, interesting amounts of α tocopherol are preserved in the tomato-based products [96, 97]. Even more interestingly, very attractive quantities of both α (313.6 μg/g) and γ tocopherol (231.8 μg/g DW) were found in the pomace [96]. More recently, δ isoform was also successfully quantified in both skin (0.5) and seeds (0.4 μg/g FW) by Figueira, Pereira [98] who also confirmed the predominancy of α analogue in the peels (12.3) and the γ one in the seeds (11.1 μg/g FW). Likewise, tomato seed oil has been reasonably reported to contain primarily γ tocopherol but appreciable amounts of α analogue were also quantified in this non-conventional oil [94, 99–104]. Supercritical Fluid Extraction (SFE) is among the greenest technologies and widely used to extract lipophilic compounds. A chemical element like CO_2 is submitted to specific temperature and pressure over its critical point to give it certain physical proprieties, and it attains a state between liquid and gaz. the applied conditions are specifically set up to provide it the attributes (density and solvating power) of those of an equivalent solvent [105]. SFE was used by Romano, Aiello [106] to extract oleoresin from both the whole and a separate fraction of peels and compared to dichlomethane and liquid CO_2 as solvent of extraction. After characterization of the extract, they concluded that CO_2 supercritical extraction gives the best amounts of both γ and α tocopherols from the peel fraction, with respectively 5.75 and 0.80 g/kg, as compared to conventional extraction and by liquid CO_2 extraction. Moreover, according to Vági, Simándi [107] the optimal SFE of total tocols content from tomato pomace (5.95 g/kg) were under a pressure of 300 bar and a temperature of 60° C. SFE was also investigated by Eller, Moser [101], using CO_2 as solvent for the recovery of US tomato seed oil in parallel with accelerated solvent extraction by using ethanol and hexane as solvent, in order to compare both quantity and quality of the final product. The authors noticed that SFE gave the highest tocopherol content (1.11 mg/g), followed by ethanol and hexane with, respectively 1.08 and 0.98 mg/g. On another side, enzymatic-assisted aqueous extraction (EAAE) of Turkish tomato seed oil gave, surprisingly, a greatly exceptional composition in tocopherols. δ and β analogues were cited as predominant with respectively 209.88 and 128.5 mg/L of oil, followed by α (79.24) and lastly γ tocopherol with 33.16 mg/L oil, while a different order was reported in terms of tocopherol homologues content in the cold-pressed oil as γ>β>α>δ, and the total tocopherol content (256 mg/L) was less than in that of the EAAE with 450.8 mg/L of oil [108]. Moreover, Yilmaz, Aydeniz [102] found that roasting as a pretreatment before the cold-press extraction of tomato seed oil not only increased the yield of extraction but also enhanced both γ and α tocopherols content from, respectively, 182.74 and 0.76 mg/kg to 280.51 and 1.29 mg/kg.

Vitamin E (comprising all its constituents) is a liposoluble molecule with a high potency to prevent and/or to delay lipids autoxidation by quenching peroxyl free radical and thus it acts as a breaking-chain antioxidant [109]. Besides to its potent antioxidant capacity [110], other biological functions and health benefits have been attributed to this biomolecule like anti-inflammatory and antidiabetic [90, 111], prevention of cancer and cardiovascular diseases [90] and regulation of immune functions [112].

The lipophilicity of both tocopherols and tocotrienols (tocols) is mainly due to their hydrocarbon tail, but a relative difference in lipo-solubility as affected by the degree of methylation of the phenolic ring should be observed between α, β, γ, and δ forms [113]. On the other side, the hydroxyl group in the position 6 of the chromanol head of tocols provides them their antioxidant capacity to quench free radicals by hydrogen donation [109, 113]. Among the tocols, α tocopherol is the most studied and conventionally considered as the most biologically active form of vitamin E [89, 114]. Indeed, the high *in vivo* antioxidant capacity of α tocopherol is related to its best bioavailability compared to other tocols; in other words, the tocols' biological activity depends strongly on their affinity with the α-Tocopherol Transfer Protein (α-TTP) which is responsible for vitamin E transport into nascent very low density lipoproteins of liver tissues and also for its subsequent secretion into blood [114–116]. In contrast, an absolute consensus is not yet established about the supremacy of α homologue in term of *in vitro* antioxidant capacity; Seppanen, Song [117] reviewed the effect of incorporation of tocols on the oxidative stability of several food systems including fats, refined and fried oils and other food products and reported a widespread differences in behavior of the presumed antioxidants. Indeed, although trimethylation of the phenolic ring ensures a better stability for the tocoperoxyl radical following the neutralization of peroxyl, α tocoperoxyl appears to be unable to react with other reactive species [118] while carbons without methylation in the functional ring still electrophile and could trap more reactive radicals such as reactive nitrogen species. In this area, γ tocopherol was reported to be more effective than its α homologue since the former has the ability to reduce NO2 and afterward to scavenge its product NO, avoiding the free radical reaction with unsaturated fatty acids [89, 119, 120]. Furthermore, among to their structural pattern, tocols *in vitro* reactivity could be affected by their concentration [121] as well as the reactional medium including solvent, temperature, light, and the presence of either synergistic or antagonistic substances [113].

5.2.4 Sterols

Sterols are very important biomolecules for animals and plants because of the crucial functions they perform, including a structural role (they are active component of cell walls and regulate their permeability) and also functional roles since they constitute precursors of the biosynthesis of steroids and even possess hormonal activity [122]. In plants, sterols are playing a key role in their response to abiotic stress relative to the environment as well as a regulatory action in their growth and development [123, 124].

Sterols have a wide diversity in terms of chemical structure but are characterized by a common basic skeleton consisting of a tetracyclic steroid ring with two angular methyl groups C18 and C19 and one hydroxyl group in the position C3, the cyclopenoperhydrophenanthrene ring is linked in the C17 position with a side chain comprising 8 to 10 carbons. Both isoprenoid ring and the lateral chain might comprise unsaturated double bonds and also the C3 hydroxyl group could be either esterified with fatty or phenolic acids or glycosylated by linking to a carbohydrate [123, 125, 126]. In both animals and plants, sterols are biosynthesized through the mevalonate pathway with squalene as a precursor. However, unlike animal tissues that contain cholesterol as the predominant sterol, β-sitosterol, stigmasterol and campesterol constitute the most plentiful phytosterols, although cholesterol is found in minor amounts in some plants from the Solanaceae family [123]. Free phytosterols can specifically be attached in the C24 of the side chain with a methyl group to give 24-methyl sterols (campestrol) or with an ethyl group to give 24-ethyl sterols (β-sitosterol and stigmasterol), constituting the two major groups of sterols in higher plants [127].

Due to their structural and functional roles, phytosterols are present in all plant tissues, but their corresponding amounts may differ depending on the specific role of each kind of these biomolecules, e.g. β-sitosterol and stigmasterol are related to structure and regulation of cell membranes, while campesterol is a precursor of brassinosteroids and hence plays a key hormonal role [123, 125]. In fact, the ratio campesterol/sitosterol is so important in the physiological system of higher plants and strongly affects their growth and development as well as their response to biotic and abiotic stress that composition of sterols is sensitively regulated [124, 127]. Furthermore, as aforesaid sterols can be found in plants in free or conjugated form. Thus, phytosterols are stored as steryl esters (as a pool of sterols) in the seeds to be mobilized during the germination stage and the glycosylated form has been reported to exert cell integrity protection in the solanum genus [125].

Sterols are abundant in all organs of the tomato plant, including roots, stem, leaves, and seeds. Their content is comprised between 0.1 to 0.2% DW with a characteristic predominance of the glycosylated form in the vegetative parts while seeds contain mainly free and esterified sterols but their glycosylated form content rises during germination [128]. Kalogeropoulos, Chiou [17] conducted a sterol analysis of industrial tomatoes (Heinz hybride genotype) and their by-products, reporting a total sterol content of 208 μg/g total sterols in DW of whole tomatoes and 833 μg/g in DW of the by-products (a mix of skins and seeds), which equates to 4 folds the fruit's content. β-sitosterol and stigmasterol were the predominant phytosterols in both tomato fruit (respectively 151.7 and 67.3 μg/g) and its industrial pomace (378.8 and 91.5 μg/g of each respective phytosterol). Vági, Simándi [107] have applied a full factorial design 3^2 for the optimization of SFE of bioactive compounds from Hungarian tomato pomace. Under the optimized conditions of temperature (80° C) and pressure (380 bar), they obtained not less than 62.1 μg/g DW of β-sitosterol which was much higher than that obtained by hexane-Soxhlet conventional extraction (6.8 μg/g DW). Additionally, oil extracted from industrial tomato seed was reported to show high content in total sterols, ranging from 1.54 to 2.07 mg/g [129], and can be considerably enhanced when using innovative methods of extraction like enzyme-assisted extraction, with 3.98 mg/g total phytosterol content in the oil [103], accelerated solvent extraction with hexane (10.81 mg/g) and ethanol (10.55 mg/g), and can reach 12.3 mg/g in the extracted oil by using CO_2 SCE method [101]. These findings are extremely intriguing compared to the sterol content reported by Fernandes and Cabral [130] in conventional edible oil I.e., corn, rapeseed, sunflower and soybean with 7.3, 7.7, 3.5 and 2.6 mg/g. By far, β-sitosterol is considered the major sterol in tomato seed oil, with quantities oscillating from 51.3 to 59.02%, seconded by stigmasterol (10.84-14.6%) and cholesterol (7.5-17.1%) which is conventionally known as a zoosterol but is characteristic of Solanaceae plants, other phytosterols like Δ5-avenasterol (6.7-11.23%) and campesterol (4.83–6.7%) were found with quite relevant amounts in tomato seed oil [103, 104, 129].

Phytosterols are analogues of cholesterol and differ only in the chemical structure of their side chain [130, 131]. Contrary to the latter, which is frequently linked to cardiovascular diseases, plant sterols have a variety of health-promoting proprieties such as the ability to prevent coronary thrombosis and other cardiovascular diseases brought on by high plasma levels of cholesterol and low-density protein cholesterol. [131, 132]. Functionally, they act as competitors of cholesterol in the intestine and lessen its absorption [133]. Indeed, high authorities in Europe, the United states, Canada, Australia, and New-Zealand have recognized phytosterols' ability to lower LDL-cholesterol in blood, and daily intake recommendations of 1.5 to 2.4 g have been established [132]. Moreover, numerous *in vitro* and animal model experiments have supported their antitumor activity against the common cancers (colon, prostate, lung, liver, breast, ovary, stomach, etc.), by means of diver mechanisms like anti-tumor occurrence or, in the event that they do, to permit their perception while delaying their spread [131, 134–136]. β-sitosterol, the most plentiful of phytosterols have garnered a lot of attention. Babu and Jayaraman [137] reviewed the most relevant *in vitro* and *in vivo* studies that looked at the biological proprieties of this plant's bioactive compound and numbered many activities, including anxiolytic and sedative, anti-inflammatory and analgesic, immunomodulatory, antimicrobial, hepatoprotective, lipid lowering, antioxidant, and antidiabetic.

Dietary sources of phytosterols include vegetable oils, nuts, seeds, cereal grains, beans, along with industrial preparations like fat-based (spreads and margarine) and fiber-based products which are essentially rich in β-sitosterol, campesterol, and stigmasterol. Alternatively, functional enriched-sterol foods or supplements could be offered to meet the daily recommended dose [131, 138]. However, such formulations are hampered by free sterols' low solubility in lipids and their insolubility in water-based matrices. Therefore, chemical and/or enzymatic modifications might be carried out in order to enhance their bioavailability as well as their functional uses. These modifications aim to improve the hydrophobicity of free phytosterols by their esterification and transesterification in one hand, and to higher their solubility in water by synthesis of new derivatives, on the other hand [132]. In addition, other techniques like nano and microencapsulation have been reported to improve phytosterols stability and bioavailability, when they were applied in the formulation of novel functional rich-sterol foods [139, 140].

5.2.5 Amino Acids

Following a lengthy correspondence between Jöns Jakob Berzelius and Gerrit Jan Mulder regarding the results of Mulder's experiments, Mulder first used, in 1838, the term "protein," which was derived from the Greek word "Primarius." He performed experimental trials on products derived from acid and alkali treatment of silk, blood serum, egg albumin, gelatin, fibrin, and wheat glutenin, and he noted a similar structural behavior between the matters he obtained, concluding that these substances share the same fundamental structure of a basic oxidized radical with the ability to react with sulfur and phosphorus, giving the organic matter the so-named protein produced by plant and transferred into herbivorous tissues and, later, to carnivores through feeding [141]. Nevertheless, still basing himself on the letter content exchanged between Mulder and Berzelius, Hartley [142] claimed the Berzelius' authorship of the word "protein" because it was he who recommended it to his confrere. Amino acids (AA) are the product of acid or/and hydrolysis of proteins. Till now, twenty AA are regarded as constituting the human proteins, of which nine (valine, leucine, isoleucine, threonine, methionine, lysine, histidine, phenylalanine, and tryptophan) are considered as essential since the body cannot synthesized them. The history of discovery of each of AA has been well illustrated in the pertinent and comprehensive publication of Vickery and Schmidt [143], who sketched the chronology and the circumstance of history of each one.

Proteins, along with carbohydrates and fats are the major components of food and directly contribute to the physicochemical of foodstuffs [144–146]. Cereal grains and legume seeds are the richest plant sources of proteins; the first are deficient in lysine and threonine, and the last ones are poor in methionine. These three AA along with isoleucine are essential for human and synthetized by plants via aspartate pathway [147]. Tyrosine, phenylamine, and tryptophan are aromatic AA, biosynthesized via shikimate pathway; the two latter being also essential for humans [148, 149]. The several pathways besides the mechanisms of amino acid biosynthesis regulation were extensively studied [150]. Also, the synthesis of essential AA, their regulation, and their accumulation in plants were deeply discussed in more details, by Galili, Amir [151]. Amino acids, as fully known to serve as the building blocks for the biosynthesis of proteins, are precursors for other molecules such as phenols, polyamines, and glycosinolates which play a critical role in signaling and stress adaptation in plants [152], and also contribute in fruit fresh flavors like those developed during the ripening of tomatoes [153].

Tomato free AA content was reported to be about 2% in the dry matter, with glutamic acid as the most predominant, whose concentration in total free AA content increases with ripening stage, from 12.7 in green fruit to 55% in a fully ripened crop [153, 154]. The predominancy of glutamate in eleven Italian varieties of "San Marzano" was confirmed in the study of Loiudice, Impembo [155], who reported a mean value of 2.4 g/kg, followed by γ amino butyric acid (GABA) and glutamine, with respectively 1.56 and 1.29 g/kg. These three AA were reported to represent 65% of whole amino acids content. Tomato-based products including canned goods, juices, and ketchup were also studied for their AA content; juices (1.64-2.53 g/kg) and ketchup (1.39-2.12 g/kg) showed lower content as compared with canned products (4.46-7.85 g/kg), in which very significant amounts of AA were noticed [156]. Even so, the solid waste generated after processing of tomatoes still retaining significant amounts of high-quality protein. Indeed, the sonication (37 kHz; 140/340 W) of by-product samples from ten industrial tomato varieties for one hour at 30°C allowed the recovery of not less than 4.21 g/kg glutamine, 6.59 g/kg aspartic acid, and up to 45.1 g/kg glutamate [157]. The peel fraction of tomato pomace have been reported to contain about 10.5 g protein per 100g DW sample, including glutamic acid as the most dominant AA with 14.56% and nine essential AA, representing 47.57% of overall protein content [158]. Even more remarkably, tomato seeds showed a high content of protein of 28.66 g/100g DW [159], with a balanced composition in AA, of which all the essential ones for adults could be found. Glutamic and aspartic acids (known as umami taste amino acids) were the most present in the defatted tomato seed meal with 6.45 and 4.06%, in a basis DW. Threonine and lysine were also significantly quantified in tomato seed protein with respectively 7.8 and 6.6 g/100g, representing higher amounts than of soy flour and egg, which offers opportunities to use tomato protein as supplements in cereal-based diets, known as poor in these AA [160]. Moreover, Sarkar and Kaul [159] estimated the protein efficiency ratio (PER) score of tomato seed isolate as equivalent to 2.66, which is superior to that of the standard value (2.5). According to them, the isolate's detailed AA composition showed better values versus those of the WHU/FAO/UNU references and have a negligible antinutritional factors effect due the insignificant content of phytate and lower trypsin inhibitory activity.

Therefore, exploration of tomato proteins may offer unique advantages over other unconventional sources, making this protein of particular interest. Mechmeche, Kachouri [161] used a central composite rotatable design (CCRD) with response surface methodology (RSM) for the optimization of protein isolate from a defatted tomato seed meal. The optimum conditions determined by the postulated model (water to sample ratio of 82.81; extraction duration of 49.76 h and mixing time of 24.56 minutes) were experimentally checked, under which they allowed the isolation of 80.37% protein from the meal, with a high nutritional quality and functional capacity. The same experimental design CCRD-RSM was also investigated by Zhang, Pan [162] who successfully optimized a process for the papain enzymatic hydrolysis of defatted tomato seed meal proteins to efficiently extract both glutamic and aspartic acid. CCRD-RSM was used to get the optimal conditions (pH=3, papain activity of 18.22%, temperature and time of hydrolysis of 40°C and 6 hours) for the optimized concentrations in glutamic (727.6 mg/L) and aspartic acid (149.9 mg/L).

In a chemical point of view, AA have the particularity to have both simultaneously a carboxylic and an amino group (amphoteric), permitting their versatile use as anions or cations in functional ionic preparation [163]. Furthermore, AA are actively involved in the thermal and enzymatic changes in food during storage and processing, which result on new flavors and aromatic substances [146]. Indeed, their amino group does condense with the carbonyl group of reducing sugar, initiating the cascade Maillard reaction and producing the Amadory Rearrangement Products; these latter are responsible for the requested flavors of roasted food like coffee, bread, and meat [164, 165]. Also, some amino acids like umami taste AA, strongly contribute to taste of plants and can be explored as natural additives for food taste improvement [162]. On the medical side, amino acids, especially glutamine, have been investigated as potential targeted agents for transport of anti-tumor therapy [166]. They also play crucial roles in immune system; as examples, glutamate and aspartate are involved in diverse key functions and the metabolism of leucocytes, lysine is directly implicated in the regulation of cytokines production and the lymphocytes proliferation, while threonine enhances antibody synthesis in lymphocytes [167]. Indeed, early aged broilers' diet supplementation with 1 to 3 g threonine per kg, not only promotes their growth but also improves their immunity and intestine health and reduces malonaldehydes in their serum [168]. Likewise, the supplementation of overweight individuals diet, for four weeks, with essential AA and phytosterols could lower their blood content in both triglycerides as well as total and LDL- cholesterols [169].

5.2.6 Dietary Fibers

Although the laxative-associated effect of some cereal brans has been known for a long time, the notion of dietary fiber is relatively recent [170]. Indeed, the term "Fibre" appeared for the first time in the paper of Hipsley [171], referring to the poly-carbohydrates of diet. Dietary fibers (DF), conventionally referred to as a wide range of polysaccharides, with the exception of non-resistant starch, that escape gastrointestinal enzymatic action [172, 173]. DF, including cellulose, hemicellulose, lignin, and pectin, constitute the principal components of plant cell walls and contribute, along with other sugars, proteins, and phenolics to their complexity, thus defining their structure-function features [174]. They can be classified, depending their solubility into soluble dietary fibers (SDF) like pectin, galactomannan, and inulin, and insoluble dietary fibers (IDF) such as cellulose, hemicellulose and lignin [170]. Cereals, vegetables, and fruits are the principal source of DF in human diet. Thus, a balanced diet should provide both SDF and IDF, but also their conjugated forms together with the bound bioactive compounds like phenolics and proteins [175].

Well-ripe tomatoes were reported to contain 1.34 g total DF in 100 g FW, of which 0.15 are SDF and 1.19 g/100 g are IDF [176]. Lee, Lee [177] analyzed some tomato-based products for their DF content and reported, respectively, 0.76 and 0.79 g total DF/100 g FW in tomato juice and ketchup. On another side, tomato processing by-product really represents a very rich source of DF. Indeed, Del Valle, Cámara [178] declared that dried tomato pomace contained up to 59% neutral DF, as well as not less than 7.55% in pectin. In another work, Alvarado, Pacheco-Delahaye [179]

[179] also reported more than 40.5% and 9% of, respectively, IDF and SDF by DW of tomato pomace, representing a total DF of 49.5%. Furthermore, dried tomato seed fraction was found to contain 34.65% in total DF, of which 30.54% was an insoluble fraction, and the rest was the SDF content [180], while the skin fraction is much richer in DF since its content could reach, after a designed treatment, until about 86%, with 14.33% in SDF content. Moreover, the extract showed very good physico-chemical properties including the capacity to swell and/or to retain water, to adsorb fat, and to retard glucose diffusion. Its osmotic pressure, phenolic content and antioxidant activity were also determined. According to the findings, tomato peels fiber was highly suggested as a supplement in functional foods [13].

On the other side, pectin constitutes a very needed molecule in the industry thanks to the characteristic physicochemical properties, allowing its versatile uses in both food-based products, pharmaceutics, nutraceuticals, and in packaging. This polysaccharide with a unique colloidal feature could be used as a gelling, emulsifier, thickener, and coating agent among other uses, like as a prebiotic or in edible films [181]. Therefore, non-conventional sources of pectin like agro-industrial waste and by-products, as cheap and readily available natural origins, were explored for pectin provision by adopting various strategies. In the light of this trend, Ninčević Grassino, Ostojić [182] used high hydrostatic pressure (HHPE) to extract pectin from tomato peels, and enhanced the yield of extraction by 15%, compared to conventional extraction. In addition, the HHPE was achieved in only 45 minutes while conventional extraction needed 180 minutes. Also, Sengar, Rawson [183] evaluated the effect of different methods of extraction, namely UAE, MAE, combined UAE and MAE (UMAE), ohmic-assisted extraction (OhAE), as well as a combined UAE and OhAE on both yield and quality of the recovered tomato peel pectin. MAE, set up at a power of 900 W and a temperature of 88.7 °C, gave the best yield of 25.42% pectin in DW of tomato peels in only 3.34 minutes of irradiation. However, regarding the quality of the extract it was the UMAE which resulted in high degree of esterification of 73.33% and galacturonic acid content of 913.3 g/kg of extracted pectin. In addition, this later extract showed the best lightness among all others, with 67.22%. In fact, some criteria like degree of esterification or methoxy substituents are crucial for the gelling capacity of pectin. Moreover, the color strongly influences the sensorial acceptance of the added food.

Besides the specifically technical uses of dietary fibers (more precisely pectin) in numerous fields, these latter have been also closely associated with health benefits. Indeed, Hipsley [171] have related occurrence of pregnancy toxemia to fiber-poor diets. Also, the study of Brauchla, Juan [184] established an inverse correlation between DF intake and childhood obesity risk, leading to diabetes occurring. The anti-diabetes action is postulated to be due a better metabolism of glucose. Other studies have also linked SDF and IDF intake with the prevention of cardiovascular diseases [185, 186], by the regulation of fat metabolism and lowering of LDL cholesterol in the blood [187]. DF consumption has been also related to the prevention of various cancers like of colorectum [188], stomach [189], esophagus [190], and breast [191]. The modulation of gut microbiota, which leads to a healthier background in the intestine, and the reduction of both proteolytic fermentation duration and the contact of the induced harmful species with the mucosal wells by fecal viscosity and bulking are the mechanisms attributed to the cancer-preventing effect of dietary fibers [192].

5.3 Conclusion

Daily intake of fresh tomatoes and tomato-based products is contributing to the promotion of a healthier diet with a proven impact on the prevention of degenerative and aging-related diseases. Even more interestingly, the high quantities of waste generated after the processing of tomatoes could also represent a valuable niche of molecules with high nutritional and/or therapeutic value. Hence the growing interest that the by-products of the tomato industry are arousing through innovative approaches and strategies to better exploit these cheap bioresources. This is also part of a global circular economy approach that adopts the principles of resource optimization by depleting waste through green processes and re-injecting it into the production operations loop as additives or ingredients, for example, or as to use it as medicines. This also contributes to the creation of added value throughout the whole food processing chain.

References

1. Manzoor, M. *et al.*, Food hydrocolloids: Functional, nutraceutical and novel applications for delivery of bioactive compounds. *Int. J. Biol. Macromol.*, 165, 554–567, 2020.
2. Kumar, K. *et al.*, Food waste: A potential bioresource for extraction of nutraceuticals and bioactive compounds. *Bioresour. Bioprocess.*, 4, 1, 18, 2017.

3. Cano-Lamadrid, M. and Artés-Hernández, F., By-products revalorization with non-thermal treatments to enhance phytochemical compounds of fruit and vegetables derived products: A review. *Foods*, 11, 1, 2022.
4. Augustin, M.A. *et al.*, Recovery of wasted fruit and vegetables for improving sustainable diets. *Trends Food Sci. Technol.*, 95, 75–85, 2020.
5. Benítez, J.J. *et al.*, Valorization of tomato processing by-products: Fatty acid extraction and production of bio-based materials. *Materials*, 11, 11, 2018.
6. Pataro, G. *et al.*, Recovery of lycopene from industrially derived tomato processing by-products by pulsed electric fields-assisted extraction. *Innovative Food Sci. Emerg. Technol.*, 63, 102369, 2020.
7. Boccia, F. *et al.*, Food waste and bio-economy: A scenario for the Italian tomato market. *J. Cleaner Prod.*, 227, 424–433, 2019.
8. Ouatmani, T. *et al.*, A sustainable valorization of industrial tomato seeds (cv Rio Grande): Sequential recovery of a valuable oil and optimized extraction of antioxidants by microwaves. *J. Food Process. Preserv.*, 46, 1, e16123, 2021.
9. Albuquerque, B.R. *et al.*, Phenolic compounds: Current industrial applications, limitations and future challenges. *Food Funct.*, 12, 1, 14–29, 2021.
10. Sauvage, C., Arnoux, S., Causse, M., Population genomics of tomato, in: *Population Genomics*, pp. 1–39, Springer International Publishing, Cham, 2020.
11. Laranjeira, T. *et al.*, Sustainable valorization of tomato by-products to obtain bioactive compounds: Their potential in inflammation and cancer management. *Molecules*, 27, 5, 2022.
12. Rivero, A.G., Keutgen, A.J., Pawelzik, E., Antioxidant properties of tomato fruit (Lycopersicon esculentum Mill.) as affected by cultivar and processing method. *Horticulturae*, 8, 6, 2022.
13. Navarro-González, I. *et al.*, Chemical profile, functional and antioxidant properties of tomato peel fiber. *Food Res. Int.*, 44, 5, 1528–1535, 2011.
14. Chanforan, C. *et al.*, The impact of industrial processing on health-beneficial tomato microconstituents. *Food Chem.*, 134, 4, 1786–1795, 2012.
15. Pérez-Conesa, D. *et al.*, Changes in bioactive compounds and antioxidant activity during homogenization and thermal processing of tomato puree. *Innovative Food Sci. Emerg. Technol.*, 10, 2, 179–188, 2009.
16. Georgé, S. *et al.*, Changes in the contents of carotenoids, phenolic compounds and vitamin C during technical processing and lyophilisation of red and yellow tomatoes. *Food Chem.*, 124, 4, 1603–1611, 2011.
17. Kalogeropoulos, N. *et al.*, Bioactive phytochemicals in industrial tomatoes and their processing by-products. *LWT - Food Sci. Technol.*, 49, 2, 213–216, 2012.
18. Zia-Ul-Haq, M., Historical and introductory aspects of carotenoids, in: *Carotenoids: Structure and Function in the Human Body*, M. Zia-Ul-Haq, S. Dewanjee, M. Riaz (Eds.), pp. 1–42, Springer International Publishing, Cham, 2021.

19. Hartsen, F.A., Note à la substance colorante des baies rouges. *C.R. Acad. Sci. (Paris)*, 76, 385, 1873.
20. Matlack, M.B. and Sando, C.E., A contribution to the chemistry of tomato pigments: The coloring matter in American red and purple tomatoes (Lycopersicum esculentum). *J. Biol. Chem.*, 104, 2, 407–414, 1934.
21. Millardet, P.M.A., *Note sur une substance colorante nouvelle (solanorubine) decouverte dans la tomate*, Berger-Levrault, Nancy, 1876.
22. Schunck, C.A. and Brown, H.T., The xanthophyll group of yellow colouring matters. *Proc. R. Soc. Lond.*, 72, 477–486, 165–176, 1904.
23. Willstätter, R. and Escher, H.H., Über den Farbstoff der Tomate. *Zeit. Physiol. Chem.*, 64, 1, 47–61, 1910.
24. Riaz, M., Zia-Ul-Haq, M., Dou, D., Chemistry of carotenoids, in: *Carotenoids: Structure and Function in the Human Body*, M. Zia-Ul-Haq, S. Dewanjee, M. Riaz (Eds.), pp. 43–76, Springer International Publishing, Cham, 2021.
25. Britton, G., Structure and properties of carotenoids in relation to function. *FASEB J.*, 9, 15, 1551–1558, 1995.
26. Fernandes, A.S. *et al.*, Carotenoids: A brief overview on its structure, biosynthesis, synthesis, and applications, in: *Progress in Carotenoid Research*, vol. 1, pp. 1–17, 2018.
27. Maoka, T., Carotenoids as natural functional pigments. *J. Nat. Med.*, 74, 1, 1–16, 2020.
28. Majer, E. *et al.*, Rewiring carotenoid biosynthesis in plants using a viral vector. *Sci. Rep.*, 7, 1, 41645, 2017.
29. Martí, R., Roselló, S., Cebolla-Cornejo, J., Tomato as a source of carotenoids and polyphenols targeted to cancer prevention. *Cancers*, 8, 6, 2016.
30. Rao, A.V., Waseem, Z., Agarwal, S., Lycopene content of tomatoes and tomato products and their contribution to dietary lycopene. *Food Res. Int.*, 31, 10, 737–741, 1998.
31. Shi, J. and Maguer, M.L., Lycopene in tomatoes: Chemical and physical properties affected by food processing. *Crit. Rev. Food Sci. Nutr.*, 40, 1, 1–42, 2000.
32. Szabo, K. *et al.*, Carotenoid recovery from tomato processing by-products through green chemistry. *Molecules*, 27, 12, 2022.
33. Knoblich, M., Anderson, B., Latshaw, D., Analyses of tomato peel and seed by-products and their use as a source of carotenoids. *J. Sci. Food Agric.*, 85, 7, 1166–1170, 2005.
34. Strati, I.F., Gogou, E., Oreopoulou, V., Enzyme and high pressure assisted extraction of carotenoids from tomato waste. *Food Bioprod. Process.*, 94, 668–674, 2015.
35. Catalkaya, G. and Kahveci, D., Optimization of enzyme assisted extraction of lycopene from industrial tomato waste. *Sep. Purif. Technol.*, 219, 55–63, 2019.
36. Li, J. *et al.*, A comprehensive optimization of ultrasound-assisted extraction for lycopene recovery from tomato waste and encapsulation by spray drying. *Processes*, 10, 2, 308, 2022.

37. Ho, K.K.H.Y. *et al.*, Microwave-assisted extraction of lycopene in tomato peels: Effect of extraction conditions on all-trans and cis-isomer yields. *LWT - Food Sci. Technol.*, 62, 1, Part 1, 160–168, 2015.
38. Nour, V. *et al.*, Effect of carotenoids, extracted from dry tomato waste, on the stability and characteristics of various vegetable oils. *Grasas Aceites*, 69, 1, e238, 2018.
39. Trombino, S. *et al.*, Valorization of tomato waste as a source of carotenoids. *Molecules*, 26, 16, 2021.
40. Farkas, Á., Bencsik, T., Deli, J., Carotenoids as food additives, in: *Pigments from Microalgae Handbook*, E. Jacob-Lopes, M.I. Queiroz, L.Q. Zepka (Eds.), pp. 421–447, Springer International Publishing, Cham, 2020.
41. Meléndez-Martínez, A.J., An overview of carotenoids, apocarotenoids, and vitamin A in agro-food, nutrition, health, and disease. *Mol. Nutr. Food Res.*, 63, 15, 1801045, 2019.
42. Honda, M., Nutraceutical and pharmaceutical applications of carotenoids, in: *Pigments from Microalgae Handbook*, E. Jacob-Lopes, M.I. Queiroz, L.Q. Zepka (Eds.), pp. 449–469, Springer International Publishing, Cham, 2020.
43. Mazidi, M., Ferns, G.A., Banach, M., A high consumption of tomato and lycopene is associated with a lower risk of cancer mortality: Results from a multi-ethnic cohort. *Public Health Nutr.*, 23, 9, 1569–1575, 2020.
44. Li, N. *et al.*, Tomato and lycopene and multiple health outcomes: Umbrella review. *Food Chem.*, 343, 128396, 2021.
45. Mirahmadi, M. *et al.*, Potential inhibitory effect of lycopene on prostate cancer. *Biomed. Pharmacother.*, 129, 110459, 2020.
46. Tourniaire, F., Mounien, L., Landrier, J.-F., Carotenoids as anti-obesity supplements, in: *Pigments from Microalgae Handbook*, E. Jacob-Lopes, M.I. Queiroz, L.Q. Zepka (Eds.), pp. 541–557, Springer International Publishing, Cham, 2020.
47. Bin-Jumah, M. *et al.*, Application of carotenoids in cosmetics, in: *Carotenoids: Structure and Function in the Human Body*, M. Zia-Ul-Haq, S. Dewanjee, M. Riaz (Eds.), pp. 747–756, Springer International Publishing, Cham, 2021.
48. Bertelli, A. *et al.*, Polyphenols: From theory to practice. *Foods*, 10, 11, 2021.
49. Singh, S., Kaur, I., Kariyat, R., The multifunctional roles of polyphenols in plant-herbivore interactions. *Int. J. Mol. Sci.*, 22, 3, 2021.
50. Tomás-Barberán, F.A. and García-Villalba, R., Structural diversity of polyphenols and distribution in foods, in: *Dietary Polyphenols*, pp. 1–29, 2020.
51. Tijjani, H. *et al.*, Polyphenols: Classifications, biosynthesis and bioactivities, in: *Functional Foods and Nutraceuticals: Bioactive Components, Formulations and Innovations*, C. Egbuna and G. Dable Tupas, (Eds.), pp. 389–414, Springer International Publishing, Cham, 2020.
52. Mitra, S. *et al.*, Polyphenols: A first evidence in the synergism and bioactivities. *Food Rev. Int.*, 39, 4, 1–23, articles updated to 17 June 2021.
53. Vuolo, M.M., Lima, V.S., Maróstica Junior, M.R., Chapter 2 - Phenolic compounds: Structure, classification, and antioxidant power, in: *Bioactive*

Compounds, M.R.S. Campos (Ed.), pp. 33–50, Woodhead Publishing, Cambridge, 2019.
54. Chaudhary, P. *et al.*, Bioactivities of phytochemicals present in tomato. *J. Food Sci. Technol.*, 55, 8, 2833–2849, 2018.
55. Zhang, Y. *et al.*, A brief review of phenolic compounds identified from plants: Their extraction, analysis, and biological activity. *Nat. Prod. Commun.*, 17, 1, 1–14, 2022.
56. Qi, Q. *et al.*, Anthocyanins and proanthocyanidins: Chemical structures, food sources, bioactivities, and product development. *Food Rev. Int.*, 39, 4, 1–29, articles updated to 17 June 2021.
57. Martínez-Valverde, I. *et al.*, Phenolic compounds, lycopene and antioxidant activity in commercial varieties of tomato (Lycopersicum esculentum). *J. Sci. Food Agric.*, 82, 3, 323–330, 2002.
58. Patanè, C. *et al.*, Fruit yield, polyphenols, and carotenoids in long shelf-life tomatoes in response to drought stress and rewatering. *Agronomy*, 11, 10, 2021.
59. Martí, R. *et al.*, Polyphenol and l-ascorbic acid content in tomato as influenced by high lycopene genotypes and organic farming at different environments. *Food Chem.*, 239, 148–156, 2018.
60. Tomas, M. *et al.*, Industrial processing versus home processing of tomato sauce: Effects on phenolics, flavonoids and in *vitro* bioaccessibility of antioxidants. *Food Chem.*, 220, 51–58, 2017.
61. Perea-Domínguez, X.P. *et al.*, Phenolic composition of tomato varieties and an industrial tomato by-product: Free, conjugated and bound phenolics and antioxidant activity. *J. Food Sci. Technol.*, 55, 9, 3453–3461, 2018.
62. Coelho, M. *et al.*, Extraction of tomato by-products' bioactive compounds using ohmic technology. *Food Bioprod. Process.*, 117, 329–339, 2019.
63. Lu, Z. *et al.*, Sustainable valorisation of tomato pomace: A comprehensive review. *Trends Food Sci. Technol.*, 86, 172–187, 2019.
64. Gharbi, S. *et al.*, Tunisian tomato by-products, as a potential source of natural bioactive compounds. *Nat. Prod. Res.*, 31, 6, 626–631, 2017.
65. Toor, R.K. and Savage, G.P., Antioxidant activity in different fractions of tomatoes. *Food Res. Int.*, 38, 5, 487–494, 2005.
66. Valdez-Morales, M. *et al.*, Phenolic content and antioxidant and antimutagenic activities in tomato peel, seeds, and by-products. *J. Agric. Food Chem.*, 62, 23, 5281–5289, 2014.
67. Tamasi, G. *et al.*, Characterization of nutraceutical components in tomato pulp, skin and locular gel. *Eur. Food Res. Technol.*, 245, 4, 907–918, 2019.
68. Keener, K.M. and Misra, N.N., Chapter 14 - Future of cold plasma in food processing, in: *Cold Plasma in Food and Agriculture*, N.N. Misra, O. Schlüter, P.J. Cullen (Eds.), pp. 343–360, Academic Press, San Diego, 2016.
69. Domonkos, M. *et al.*, Applications of cold atmospheric pressure plasma technology in medicine, agriculture and food industry. *Appl. Sci.*, 11, 11, 4809, 2021.

70. Bao, Y., Reddivari, L., Huang, J.-Y., Development of cold plasma pretreatment for improving phenolics extractability from tomato pomace. *Innovative Food Sci. Emerg. Technol.*, 65, 102445, 2020.
71. Korkina, L.G. and Afanas'Ev, I.B., Antioxidant and chelating properties of flavonoids, in: *Advances in Pharmacology*, H. Sies (Ed.), pp. 151–163, Academic Press, Cambridge, Massachusetts, 1996.
72. Alizadeh, S.R. and Ebrahimzadeh, M.A., O-Glycoside quercetin derivatives: Biological activities, mechanisms of action, and structure–activity relationship for drug design, a review. *Phytother. Res.*, 36, 2, 778–807, 2022.
73. Lee, S. *et al.*, Relative protective activities of quercetin, quercetin-3-glucoside, and rutin in alcohol-induced liver injury. *J. Food Biochem.*, 43, 11, e13002, 2019.
74. Patel, K. and Patel, D.K., Chapter 26 - The beneficial role of rutin, a naturally occurring flavonoid in health promotion and disease prevention: A systematic review and update, in: *Bioactive Food as Dietary Interventions for Arthritis and Related Inflammatory Diseases*, Second Edition, R.R. Watson and V.R. Preedy (Eds.), pp. 457–479, Academic Press, Cambridge, Massachusetts, 2019.
75. Naseer, B. *et al.*, Evaluation of antioxidant and antimicrobial potential of rutin in combination with butylated hydroxytoluene in cheddar cheese. *J. Food Process. Preserv.*, 45, 1, e15046, 2021.
76. Salehi, B. *et al.*, The therapeutic potential of naringenin: A review of clinical trials. *Pharmaceuticals*, 12, 1, 2019.
77. Zeng, W. *et al.*, Naringenin as a potential immunomodulator in therapeutics. *Pharmacol. Res.*, 135, 122–126, 2018.
78. Taofiq, O. *et al.*, Hydroxycinnamic acids and their derivatives: Cosmeceutical significance, challenges and future perspectives, a review. *Molecules*, 22, 2, 2017.
79. Teixeira, J. *et al.*, Hydroxycinnamic acid antioxidants: An electrochemical overview. *Biomed. Res. Int.*, 2013, 251754, 2013.
80. Sova, M. and Saso, L., Natural sources, pharmacokinetics, biological activities and health benefits of hydroxycinnamic acids and their metabolites. *Nutrients*, 12, 8, 2020.
81. Contardi, M. *et al.*, Hydroxycinnamic acids and derivatives formulations for skin damages and disorders: A review. *Pharmaceutics*, 13, 7, 2021.
82. Lu, H. *et al.*, Chlorogenic acid: A comprehensive review of the dietary sources, processing effects, bioavailability, beneficial properties, mechanisms of action, and future directions. *Compr. Rev. Food Sci. Food Saf.*, 19, 6, 3130–3158, 2020.
83. Evans, H.M., The pioneer history of vitamin E, in: *Vitamins & Hormones*, R.S. Harris *et al.* (Eds.), pp. 379–387, Academic Press, New York, 1962.
84. Kundu, S. and Sarkar, D., A year away to 100th year of vitamin E synthesis. *J. Heterocycl. Chem.*, 58, 9, 1741–1748, 2021.

85. Birringer, M. *et al.*, History of vitamin E research, in: *Vitamin E in Human Health*, P. Weber *et al.* (Eds.), pp. 7–18, Springer International Publishing, Cham, 2019.
86. Ju, J. *et al.*, Cancer-preventive activities of tocopherols and tocotrienols. *Carcinogenesis*, 31, 4, 533–542, 2009.
87. Niki, E. and Abe, K., Chapter 1 Vitamin E: Structure, properties and functions, in: *Vitamin E: Chemistry and Nutritional Benefits*, pp. 1–11, The Royal Society of Chemistry, London, 1745, 2019.
88. Hussain, N. *et al.*, Biosynthesis, structural, and functional attributes of tocopherols in planta; past, present, and future perspectives. *J. Agric. Food Chem.*, 61, 26, 6137–6149, 2013.
89. Yang, C.S. *et al.*, Vitamin E and cancer prevention: Studies with different forms of tocopherols and tocotrienols. *Mol. Carcinog.*, 59, 4, 365–389, 2020.
90. Shahidi, F. and De Camargo, A.C., Tocopherols and tocotrienols in common and emerging dietary sources: Occurrence, applications, and health benefits. *Int. J. Mol. Sci.*, 17, 10, 2016.
91. Munné-Bosch, S. and Alegre, L., The function of tocopherols and tocotrienols in plants. *Crit. Rev. Plant Sci.*, 21, 1, 31–57, 2002.
92. Saini, R.K. and Keum, Y.-S., Tocopherols and tocotrienols in plants and their products: A review on methods of extraction, chromatographic separation, and detection. *Food Res. Int.*, 82, 59–70, 2016.
93. Ahsan, H., Ahad, A., Siddiqui, W.A., A review of characterization of tocotrienols from plant oils and foods. *J. Chem. Biol.*, 8, 2, 45–59, 2015.
94. Hassanien, M.M.M. *et al.*, Phytochemical contents and oxidative stability of oils from non-traditional sources. *Eur. J. Lipid Sci. Technol.*, 116, 11, 1563–1571, 2014.
95. Abushita, A.A. *et al.*, Determination of antioxidant vitamins in tomatoes. *Food Chem.*, 60, 2, 207–212, 1997.
96. Capanoglu, E. *et al.*, Changes in antioxidant and metabolite profiles during production of tomato paste. *J. Agric. Food Chem.*, 56, 3, 964–973, 2008.
97. Chun, J. *et al.*, Tocopherol and tocotrienol contents of raw and processed fruits and vegetables in the United States diet. *J. Food Compos. Anal.*, 19, 2, 196–204, 2006.
98. Figueira, J.A., Pereira, J.A.M., Câmara, J.S., Quantification of δ-, γ- and α-tocopherol in tomatoes using an improved liquid-dispersive solid-phase extraction combined with ultrahigh pressure liquid chromatography. *Food Anal. Methods*, 10, 7, 2507–2517, 2017.
99. Solaberrieta, I. *et al.*, Valorization of tomato seed by-products as a source of fatty acids and bioactive compounds by using advanced extraction techniques. *Foods*, 11, 16, 2022.
100. Müller, L. *et al.*, Antioxidant capacity of tomato seed oil in solution and its redox properties in cultured macrophages. *J. Agric. Food Chem.*, 61, 2, 346–354, 2013.

101. Eller, F.J. *et al.*, Extraction and analysis of tomato seed oil. *J. Am. Oil Chem. Soc.*, 87, 7, 755–762, 2010.
102. Yilmaz, E. *et al.*, Sensory and physico-chemical properties of cold press-produced tomato (*Lycopersicon esculentum L.*) seed oils. *J. Am. Oil Chem. Soc.*, 92, 6, 833–842, 2015.
103. Zuorro, A. *et al.*, Enzyme-assisted production of tomato seed oil enriched with lycopene from tomato pomace. *Food Bioprocess Technol.*, 6, 12, 3499–3509, 2013.
104. Lazos, E.S., Tsaknis, J., Lalas, S., Characteristics and composition of tomato seed oil. *Grasas Aceites*, 49, 5-6, 440–445, 1998.
105. Phelps, C.L., Smart, N.G., Wai, C.M., Past, present, and possible future applications of supercritical fluid extraction technology. *J. Chem. Educ.*, 73, 12, 1163, 1996.
106. Romano, R. *et al.*, Characterisation of oleoresins extracted from tomato waste by liquid and supercritical carbon dioxide. *Int. J. Food Sci. Technol.*, 55, 10, 3334–3342, 2020.
107. Vági, E. *et al.*, Supercritical carbon dioxide extraction of carotenoids, tocopherols and sitosterols from industrial tomato by-products. *J. Supercrit. Fluids*, 40, 2, 218–226, 2007.
108. Ozyurt, V.H., Çakaloğlu, B., Otles, S., Optimization of cold press and enzymatic-assisted aqueous oil extraction from tomato seed by response surface methodology: Effect on quality characteristics. *J. Food Process. Preserv.*, 45, 5, e15471, 2021.
109. Barouh, N. *et al.*, Tocopherols as antioxidants in lipid-based systems: The combination of chemical and physicochemical interactions determines their efficiency. *Compr. Rev. Food Sci. Food Saf.*, 21, 1, 642–688, 2022.
110. Miyazawa, T. *et al.*, Vitamin E: Regulatory redox interactions. *IUBMB Life*, 71, 4, 430–441, 2019.
111. Di Vincenzo, A. *et al.*, Antioxidant, anti-inflammatory, and metabolic properties of tocopherols and tocotrienols: Clinical implications for vitamin E supplementation in diabetic kidney disease. *Int. J. Mol. Sci.*, 20, 20, 2019.
112. Lewis, E.D., Meydani, S.N., Wu, D., Regulatory role of vitamin E in the immune system and inflammation. *IUBMB Life*, 71, 4, 487–494, 2019.
113. Kamal-Eldin, A. and Appelqvist, L.-Å., The chemistry and antioxidant properties of tocopherols and tocotrienols. *Lipids*, 31, 7, 671–701, 1996.
114. Arai, H. and Kono, N., α-Tocopherol transfer protein (α-TTP). *Free Radic. Biol. Med.*, 176, 162–175, 2021.
115. Arita, M. *et al.*, Human α-tocopherol transfer protein: cDNA cloning, expression and chromosomal localization. *Biochem. J.*, 306, 2, 437–443, 1995.
116. Hosomi, A. *et al.*, Affinity for α-tocopherol transfer protein as a determinant of the biological activities of vitamin E analogs. *FEBS Lett.*, 409, 1, 105–108, 1997.

117. Seppanen, C.M., Song, Q., Saari Csallany, A., The antioxidant functions of tocopherol and tocotrienol homologues in oils, fats, and food systems. *J. Am. Oil Chem. Soc.*, 87, 5, 469–481, 2010.
118. Burton, G.W. and Ingold, K.U., Vitamin E as an *in vitro* and *in vivo* antioxidanta. *Ann. N. Y. Acad. Sci.*, 570, 1, 7–22, 1989.
119. Cooney, R.V. *et al.*, Gamma-tocopherol detoxification of nitrogen dioxide: Superiority to alpha-tocopherol. *Proc. Natl. Acad. Sci.*, 90, 5, 1771–1775, 1993.
120. Jiang, Q. *et al.*, Gamma-tocopherol, a major form of vitamin E in diets: Insights into antioxidant and anti-inflammatory effects, mechanisms, and roles in disease management. *Free Radical Biol. Med.*, 178, 347–359, 2022.
121. Lampi, A.-M. *et al.*, Antioxidant activities of α- and γ-tocopherols in the oxidation of rapeseed oil triacylglycerols. *J. Am. Oil Chem. Soc.*, 76, 6, 749–755, 1999.
122. Heftmann, E., Functions of sterols in plants. *Lipids*, 6, 2, 128–133, 1971.
123. Rogowska, A. and Szakiel, A., The role of sterols in plant response to abiotic stress. *Phytochem. Rev.*, 19, 6, 1525–1538, 2020.
124. Schaller, H., The role of sterols in plant growth and development. *Prog. Lipid Res.*, 42, 3, 163–175, 2003.
125. Ferrer, A. *et al.*, Emerging roles for conjugated sterols in plants. *Prog. Lipid Res.*, 67, 27–37, 2017.
126. Piironen, V. *et al.*, Plant sterols: Biosynthesis, biological function and their importance to human nutrition. *J. Sci. Food Agric.*, 80, 7, 939–966, 2000.
127. Valitova, J.N., Sulkarnayeva, A.G., Minibayeva, F.V., Plant sterols: Diversity, biosynthesis, and physiological functions. *Biochem. (Moscow)*, 81, 8, 819–834, 2016.
128. Duperon, R., Thiersault, M., Duperon, P., High level of glycosylated sterols in species of solanum and sterol changes during the development of the tomato. *Phytochemistry*, 23, 4, 743–746, 1984.
129. Giuffrè, A.M., Capocasale, M., Zappia, C., Tomato seed oil for edible use: Cold break, hot break, and harvest year effects. *J. Food Process. Preserv.*, 41, 6, e13309, 2017.
130. Fernandes, P. and Cabral, J.M.S., Phytosterols: Applications and recovery methods. *Bioresour. Technol.*, 98, 12, 2335–2350, 2007.
131. Shahzad, N. *et al.*, Phytosterols as a natural anticancer agent: Current status and future perspective. *Biomed. Pharmacother.*, 88, 786–794, 2017.
132. He, W.-S., Zhu, H., Chen, Z.-Y., Plant sterols: Chemical and enzymatic structural modifications and effects on their cholesterol-lowering activity. *J. Agric. Food Chem.*, 66, 12, 3047–3062, 2018.
133. Li, X. *et al.*, The bioavailability and biological activities of phytosterols as modulators of cholesterol metabolism. *Molecules*, 27, 2, 2022.
134. Ramprasath, V.R. and Awad, A.B., Role of phytosterols in cancer prevention and treatment. *J. AOAC Int.*, 98, 3, 735–738, 2015.

135. Bradford, P.G. and Awad, A.B., Phytosterols as anticancer compounds. *Mol. Nutr. Food Res.*, 51, 2, 161–170, 2007.
136. Woyengo, T.A., Ramprasath, V.R., Jones, P.J.H., Anticancer effects of phytosterols. *Eur. J. Clin. Nutr.*, 63, 7, 813–820, 2009.
137. Babu, S. and Jayaraman, S., An update on β-sitosterol: A potential herbal nutraceutical for diabetic management. *Biomed. Pharmacother.*, 131, 110702, 2020.
138. Srigley, C.T. and Haile, E.A., Quantification of plant sterols/stanols in foods and dietary supplements containing added phytosterols. *J. Food Compos. Anal.*, 40, 163–176, 2015.
139. Tolve, R. *et al.*, Microencapsulation as a tool for the formulation of functional foods: The phytosterols' case study. *Foods*, 9, 4, 2020.
140. Pavani, M. *et al.*, Novel encapsulation approaches for phytosterols and their importance in food products: A review. *J. Food Process Eng.*, 45, 8, e14041, 2022.
141. Vickery, H.B., The origin of the word protein. *Yale J. Biol. Med.*, 22, 5, 387–93, 1950.
142. Hartley, H., Origin of the word 'protein'. *Nature*, 168, 4267, 244–244, 1951.
143. Vickery, H.B. and Schmidt, C.L.A., The history of the discovery of the amino acids. *Chem. Rev.*, 9, 2, 169–318, 1931.
144. Tandang-Silvas, M.R.G. *et al.*, Molecular design of seed storage proteins for enhanced food physicochemical properties. *Annu. Rev. Food Sci. Technol.*, 2, 1, 59–73, 2011.
145. de Castro, R.J.S. *et al.*, Whey protein as a key component in food systems: Physicochemical properties, production technologies and applications. *Food Struct.*, 14, 17–29, 2017.
146. Belitz, H.D., Grosch, W., Schieberle, P., Amino acids, peptides, proteins, in: *Food Chemistry*, H.D. Belitz, W. Grosch, P. Schieberle (Eds.), pp. 8–91, Springer Berlin Heidelberg, Berlin, Heidelberg, 2004.
147. Azevedo, R.A. *et al.*, The biosynthesis and metabolism of the aspartate derived amino acids in higher plants. *Phytochemistry*, 46, 3, 395–419, 1997.
148. Tzin, V. and Galili, G., New insights into the shikimate and aromatic amino acids biosynthesis pathways in plants. *Mol. Plant*, 3, 6, 956–972, 2010.
149. Maeda, H. and Dudareva, N., The shikimate pathway and aromatic amino acid biosynthesis in plants. *Annu. Rev. Plant Biol.*, 63, 1, 73–105, 2012.
150. Umbarger, H.E., Amino acid biosynthesis and its regulation. *Annu. Rev. Biochem.*, 47, 1, 533–606, 1978.
151. Galili, G., Amir, R., Fernie, A.R., The regulation of essential amino acid synthesis and accumulation in plants. *Annu. Rev. Plant Biol.*, 67, 1, 153–178, 2016.
152. Heinemann, B. and Hildebrandt, T.M., The role of amino acid metabolism in signaling and metabolic adaptation to stress-induced energy deficiency in plants. *J. Exp. Bot.*, 72, 13, 4634–4645, 2021.

153. Petro-Turza, M., Flavor of tomato and tomato products. *Food Rev. Int.*, 2, 3, 309–351, 1986.
154. Sorrequieta, A. *et al.*, Free amino acid production during tomato fruit ripening: A focus on l-glutamate. *Amino Acids*, 38, 5, 1523–1532, 2010.
155. Loiudice, R. *et al.*, Composition of San Marzano tomato varieties. *Food Chem.*, 53, 1, 81–89, 1995.
156. Choi, S.-H. *et al.*, Free amino acid and phenolic contents and antioxidative and cancer cell-inhibiting activities of extracts of 11 greenhouse-grown tomato varieties and 13 tomato-based foods. *J. Agric. Food Chem.*, 59, 24, 12801–12814, 2011.
157. Szabo, K. *et al.*, Antimicrobial and antioxidant properties of tomato processing by-products and their correlation with the biochemical composition. *LWT*, 116, 108558, 2019.
158. Elbadrawy, E. and Sello, A., Evaluation of nutritional value and antioxidant activity of tomato peel extracts. *Arabian J. Chem.*, 9, S1010–S1018, 2016.
159. Sarkar, A. and Kaul, P., Evaluation of tomato processing by-products: A comparative study in a pilot scale setup. *J. Food Process Eng.*, 37, 3, 299–307, 2014.
160. Brodowski, D. and Geisman, J.R., Protein content and amino acid composition of protein of seeds from tomatoes at various stages of ripeness. *J. Food Sci.*, 45, 2, 228–229, 1980.
161. Mechmeche, M. *et al.*, Optimization of extraction parameters of protein isolate from tomato seed using response surface methodology. *Food Anal. Methods*, 10, 3, 809–819, 2017.
162. Zhang, Y. *et al.*, Umami taste amino acids produced by hydrolyzing extracted protein from tomato seed meal. *LWT - Food Sci. Technol.*, 62, 2, 1154–1161, 2015.
163. Ohno, H. and Fukumoto, K., Amino acid ionic liquids. *Acc. Chem. Res.*, 40, 11, 1122–1129, 2007.
164. Mottram, D., Meat flavour, in: *Understanding Natural Flavors*, J.R. Piggott and A. Paterson (Eds.), pp. 140–163, Springer US, Boston, MA, 1994.
165. Luo, Y., Li, S., Ho, C.-T., Key aspects of amadori rearrangement products as future food additives. *Molecules*, 26, 14, 2021.
166. Lieu, E.L. *et al.*, Amino acids in cancer. *Exp. Mol. Med.*, 52, 1, 15–30, 2020.
167. Li, P. *et al.*, Amino acids and immune function. *Br. J. Nutr.*, 98, 2, 237–252, 2007.
168. Chen, Y.P. *et al.*, Effects of threonine supplementation on the growth performance, immunity, oxidative status, intestinal integrity, and barrier function of broilers at the early age. *Poult. Sci.*, 96, 2, 405–413, 2017.
169. Coker, R.H. *et al.*, Nutritional supplementation with essential amino acids and phytosterols may reduce risk for metabolic syndrome and cardiovascular disease in overweight individuals with mild hyperlipidemia. *J. Endocrinol. Diabetes Obes.*, 3, 2, 2015.

170. Mudgil, D., Chapter 3 - The interaction between insoluble and soluble fiber, in: *Dietary Fiber for the Prevention of Cardiovascular Disease*, R.A. Samaan (Ed.), pp. 35–59, Academic Press, Cambridge, Massachusetts, 2017.
171. Hipsley, E.H., Dietary "fibre" and pregnancy toxaemia. *Br. Med. J.*, 2, 4833, 420–2, 1953.
172. Kay, R.M., Dietary fiber. *J. Lipid Res.*, 23, 2, 221–242, 1982.
173. Guardiola-Márquez, C.E., Santana-Gálvez, J., Jacobo-Velázquez, D.A., Association of dietary fiber to food components, in: *Science and Technology of Fibers in Food Systems*, J. Welti-Chanes *et al.*, (Eds.), pp. 45–70, Springer International Publishing, Cham, 2020.
174. Ayala Soto, F.E. and Serna Saldívar, S.O., Architecture, structure and chemistry of plant cell walls and their constituents, in: *Science and Technology of Fibers in Food Systems*, J. Welti-Chanes *et al.*, (Eds.), pp. 3–14, Springer International Publishing, Cham, 2020.
175. la Peña, M.M.-d. *et al.*, Dietary fiber in fruits and vegetables, in: *Science and Technology of Fibers in Food Systems*, J. Welti-Chanes *et al.*, (Eds.), pp. 123–152, Springer International Publishing, Cham, 2020.
176. Li, B.W., Andrews, K.W., Pehrsson, P.R., Individual sugars, soluble, and insoluble dietary fiber contents of 70 high consumption foods. *J. Food Compos. Anal.*, 15, 6, 715–723, 2002.
177. Lee, Y. *et al.*, Analytical dietary fiber database for the National Health and Nutrition Survey in Korea. *J. Food Compos. Anal.*, 21, S35–S42, 2008.
178. Del Valle, M., Cámara, M., Torija, M.-E., Chemical characterization of tomato pomace. *J. Sci. Food Agric.*, 86, 8, 1232–1236, 2006.
179. Alvarado, A., Pacheco-Delahaye, E., Hevia, P., Value of a tomato by-product as a source of dietary fiber in rats. *Plant Foods Hum. Nutr.*, 56, 4, 335–348, 2001.
180. Isik, F. and Yapar, A., Effect of tomato seed supplementation on chemical and nutritional properties of tarhana. *J. Food Meas. Charact.*, 11, 2, 667–674, 2017.
181. Chandel, V. *et al.*, Current advancements in pectin: Extraction, properties and multifunctional applications. *Foods*, 11, 17, 2022.
182. Ninčević Grassino, A. *et al.*, Application of high hydrostatic pressure and ultrasound-assisted extractions as a novel approach for pectin and polyphenols recovery from tomato peel waste. *Innovative Food Sci. Emerg. Technol.*, 64, 102424, 2020.
183. Sengar, A.S. *et al.*, Comparison of different ultrasound assisted extraction techniques for pectin from tomato processing waste. *Ultrason. Sonochem.*, 61, 104812, 2020.
184. Brauchla, M. *et al.*, Sources of dietary fiber and the association of fiber intake with childhood obesity risk (in 2–18 year olds) and diabetes risk of adolescents 12–18 year olds: NHANES 2003–2006. *J. Nutr. Metab.*, 2012, 736258, 2012.

185. Liu, S. *et al.*, A prospective study of dietary fiber intake and risk of cardiovascular disease among women. *J. Am. Coll. Cardiol.*, 39, 1, 49–56, 2002.
186. Kokubo, Y. *et al.*, Dietary fiber intake and risk of cardiovascular disease in the Japanese population: The Japan Public Health Center-based study cohort. *Eur. J. Clin. Nutr.*, 65, 11, 1233–1241, 2011.
187. Soliman, G.A., Dietary fiber, atherosclerosis, and cardiovascular disease. *Nutrients*, 11, 5, 2019.
188. Nomura, A.M.Y. *et al.*, Dietary fiber and colorectal cancer risk: The multiethnic cohort study. *Cancer Causes Control*, 18, 7, 753–764, 2007.
189. Zhang, Z. *et al.*, Dietary fiber intake reduces risk for gastric cancer: A meta-analysis. *Gastroenterology*, 145, 1, 113–120.e3, 2013.
190. Coleman, H.G. *et al.*, Dietary fiber and the risk of precancerous lesions and cancer of the esophagus: A systematic review and meta-analysis. *Nutr. Rev.*, 71, 7, 474–482, 2013.
191. Aune, D. *et al.*, Dietary fiber and breast cancer risk: A systematic review and meta-analysis of prospective studies. *Ann. Oncol.*, 23, 6, 1394–1402, 2012.
192. Zeng, H., Lazarova, D.L., Bordonaro, M., Mechanisms linking dietary fiber, gut microbiota and colon cancer prevention. *World J. Gastrointest. Oncol.*, 6, 2, 41–51, 2014.

6

Valorization and Functionalization of Cereal-Based Industry By-Products for Nutraceuticals

Akin, M.[1]**, Jukic, M.**[2]**, Lukinac, J.**[3]**, Yilmaz, B.**[4]**, Özogul, F.**[5] **and Rocha, J.M.**[6,7,8*]

[1]*Department of Horticulture, Faculty of Agriculture, Iğdır University, Iğdır, Turkey*
[2]*Josip Juraj Strossmayer University of Osijek, Faculty of Food Technology, Osijek, Croatia*
[3]*Department of Process Engineering, Josip Juraj Strossmayer University of Osijek, Faculty of Food Technology, Osijek, Croatia*
[4]*Department of Nutrition and Dietetics, Cukurova University, Sarıcam, Adana, Turkey*
[5]*Department of Seafood Processing Technology, Faculty of Fisheries, Cukurova University, Balcali, Adana, Turkey*
[6]*Universidade Católica Portuguesa, CBQF - Centro de Biotecnologia e Química Fina – Laboratório Associado, Escola Superior de Biotecnologia, Rua Diogo Botelho Porto, Portugal*
[7]*LEPABE—Laboratory for Process Engineering, Environment, Biotechnology and Energy, Faculty of Engineering, University of Porto, Rua Dr. Roberto Frias, s/n, Porto, Portugal*
[8]*ALiCE—Associate Laboratory in Chemical Engineering, Faculty of Engineering, University of Porto, Rua Dr. Roberto Frias, s/n, Porto, Portugal*

Abstract

Cereal and cereal-based products represent an important resource for food and feed purposes. From the cultivation of cereals to its processing and consumption of the end-goods, substantial amounts of by-products and food wastes are generated.

**Corresponding author*: jmfrocha@fe.up.pt; ORCID: http://orcid.org/0000-0002-0936-2003
Meleksen Akin: ORCID: http://orcid.org/0000-0002-9513-8365
Marko Jukić: ORCID: https://orcid.org/0000-0001-8790-5256
Jasmina Lukinac: ORCID: https://orcid.org/0000-0002-0998-0258
Birsen Yilmaz: ORCID: https://orcid.org/0000-0002-4866-2818
Fatih Özogul: ORCID: https://orcid.org/0000-0002-0655-0105

Umile Gianfranco Spizzirri (ed.) Nutraceutics from Agri-Food By-Products, (173–222) © 2023 Scrivener Publishing LLC

Such by-products encompass the residual materials from cereal cultivation, cereal fractions from milling, as well as from the agri-food industries. Dry and wet milling products and co-products, and baked, pasta and breakfast products, breakfast cereals and baby food, snack products, malt, alcoholic and non-alcoholic beverages are examples of products from the brewing, starch, and baking industries. Cereal by-products are known to be composed by numerous valuable compounds with technological and commercial interest, for instance, dietary fibers, carbohydrates and sugars, minerals, proteins, antioxidants (such as polyphenols and vitamins), among many others. Nutraceuticals can be classified into three groups as herbal (β-glucan, ascorbic acid, quercetin, luteolin, cellulose, gallic acid, pectin, lycopene, lignin, capsaicin, α-tocopherol, α-carotene, selenium, zeaxanthin etc.), animal (conjugated linolenic acid, eicosapentaenoic acid, docosahexaenoic acid, sphingolipids, choline, lecithin, calcium, selenium, zinc, creatine etc.), as well as microbial (*Saccharomyces boulardii, Bifidobacterium bifidum, B. longum, B. infantis, Lactobacillus acidophilus, Streptococcus salvarius*, etc.). This chapter focuses on the current state-of-the-art on the valorisation and functionalization of cereal-based by-products and food wastes from agriculture and cereal-based industries, resorting to innovative biotechnological processes, and specifically with application in nutraceuticals, viz. in dietary supplements and food additives.

***Keywords*:** Cereal and pseudocereal by-products, food wastes, starch and baking industries, dietary fibers, carbohydrates and sugar, minerals, proteins, antioxidants

***Extended Keywords*:** Cereal and pseudocereal by-products, arabinoxylans, Arabinoxylan-oligosaccharides (AXOS); α-L-arabinofuranosyl (Araf); Arabinose; β-glucans, dietary fibers, biopolymers; Exopolysaccharides (EPS); fermentable oligosaccharides, disaccharides, monosaccharides and polyols (FODMAPs);

bacteriocin-like inhibitory substances (BLIS); bio-based polyethylene; bio-based polyethylene terephthalate; bio-based polyurethane; mono- (MAG), Di- (DAG) and triacylglycerols (TAG); free fatty acids (FFA); sterol esters (SE); ferulic acid (FA); γ-aminobutyric acid (GABA); polyhydroxyalkanoates (PHA); polylactic acid (PLA);

single cell oils (SCO); single cell proteins (SCP); solid-state fermentation (SSF); Lactic acid bacteria (LAB); subcritical water extraction (SWE); supercritical carbon dioxide extraction (SC-CO_2); ultrasonic(/Ultrasound)-assisted extraction (UAE); pulsed electric field (PEF);

β-glucan, ascorbic acid, quercetin, luteolin, cellulose, gallic acid, pectin, lycopene, hemicellulose, lignin, capsaicin, α-tocopherol, α-carotene, selenium, zeaxanthin, monounsaturated fatty acids, conjugated linolenic acid, eicosapentaenoic acid, docosahexaenoic acid, sphingolipids, choline, lecithin, calcium, selenium, zinc, creatine, *Saccharomyces boulardii, Bifidobacterium bifidum, B. longum, Lactobacillus acidophilus, Streptococcus salvarius.*

6.1 Introduction

Cereals and cereal derivatives have been essential foods in the human's diet for many years, providing a substantial percentage of dietary energy and nutrients in both developed and developing countries [1]. Cereals provide 50-70% of the dietary energy needs of the majority of the world's population [2]. The examination of dental calculus from various sites in Europe, the Near East, and Africa suggests that late Neanderthals consumed wild *Triticeae* grains in a cooked form about 40 and 50 thousand years ago [3]. The first neolithic crops, including wheat, rice, maize, barley, and sorghum, contributed to the development of the earliest agricultural societies. Since that time, they have continued to be the main source of energy for the vast majority of the population [4]. Cereals and cereal products are composed of around 75% carbohydrates (primarily starches), and 6% to 15% protein, contributing to more than 50% of the global energy supply [1]. According to the latest data (July 2021) provided by Food and Agriculture Organization (FAO), the production of cereals in the world has reached 2817 million tons while the world area harvested for cereals nes is 4,559,417. The three main kinds of cereal produced globally in 2018 were maize (38.7%), rice (26.4%), and wheat (24.8%), together accounting for 89.9% of global cereal production [5]. However, FAO foresees that in 2022, there will be a decline in global cereal production (16 million tonnes from the record output estimated for 2021) and the largest decrease will be in the production of maize, followed by wheat and rice, with a possible increase in barley and sorghum production [5].

Food waste and loss have recently become a major public concern and a priority on the global and state political agendas (Target 12.3 in the new United Nations Sustainable Development Goals: *By 2030, halve per capita global food waste at the retail and consumer levels and reduce food losses along production and supply chains, including post-harvest losses.*) [6]. Globally, approximately 14% of food produced is lost between harvest and sale, while an estimated 17% of total world food output is squandered (11% in households, 5% in the food service and 2% in retail) [7]. Global food waste is estimated to be 1.6 billion tonnes of "primary product equivalents." It has been reported that the total food waste for the edible portion is 1.3 billion tonnes. The carbon footprint of food waste is estimated to be 3.3 billion tonnes of CO_2-equivalent greenhouse gases released into the atmosphere per year [8]. According to World Counts Statistics, 101,965,025 tons of food are lost or wasted only in a month, July 2022 [9]. Food waste and losses can happen in any step(s) of the food supply chain from production steps to home processing [10]. Interestingly, cereals have the largest per-capita food waste and loss of any food category, even more than perishable foods, such as vegetables and fruits. Consumer cereal waste rises in accordance with a country's gross domestic product [11]. Food waste and loss cause not only high socioeconomic costs but also environmental and moral issues since hunger affects nearly 12% of the world's population [12].

Bran and germ are two main by-products obtained during the traditional milling procedures in the cereals processing step. During milling, cereal grains are generally processed to remove the fibrous bran; nonetheless, bran that is separated from the starchy endosperm of the grain is a significant by-product of milling. The outer layer of the grain is usually higher in many nutrients (vitamins and minerals) and dietary fiber, however, it is frequently devalued and used as animal feed [13]. Not only the functional components of cereal processing by-products have been reported but also their potential health benefits have been shown by many researchers [14–16]. Cereal-based industry by-products (bran, germ, aleurone, husk, hull, proteins and brewers' spent grain) have various bioactive components including nutraceutical ingredients which can be used in different industries [17]. This chapter focuses on a detailed insight into the valorisation and functionalization of cereal-based industry by-products for nutraceuticals.

6.2 Definition and Regulatory Framework of Nutraceuticals

Nutraceuticals have been introduced by Dr Stephen DeFelice in 1989, defining them as "food or part of a food that provides medical or health benefits, including the prevention and/or treatment of a disease" [18]. However, nutraceuticals still do not have an internationally recognized definition (Table 6.1) that distinguishes them from other food-derived categories such as food supplements, herbal products, prebiotics, probiotics, functional foods, and fortified foods. Although the definitions of nutraceuticals are often confused with food supplements, it should be noted that nutraceuticals differ from them in terms of clinical evidence, efficacy and safety requirements. Different countries and researchers suggest some definitions that are somewhat similar to each other but not the same [19, 20]. Similarly, the regulatory frameworks for nutraceuticals also differ according to the countries (Figure 6.1) [19, 21].

Table 6.1 Different definitions of nutraceuticals.

Definition	Reference
Food or part of food that provides medical or health benefits, including the prevention and/or treatment of a disease.	[18]
A diet supplement that delivers a concentrated form of a biologically active component of food in a non-food matrix to enhance health.	[22]
Any substance that is a food or a part of a food and is able to induce medical and health benefits, including the prevention and treatment of disease.	[23]
A nutraceutical is a product isolated or purified from foods that are generally sold in medicinal forms not usually associated with food. A nutraceutical is demonstrated to have a physiological benefit or provide protection against chronic disease.	[24]
A foodstuff (as a fortified food or dietary supplement) that provides health benefits in addition to its basic nutritional.	[25]
Nutritional products that provide health and medical benefits, including the prevention and treatment of disease.	[26]
Nutraceuticals differ from dietary supplements in the following aspects: →Nutraceuticals must not only supplement the diet but should also aid in the prevention and/or treatment of disease and/or disorder. →Nutraceuticals are represented for use as a conventional food or as the sole item of meal or diet.	[27]
Nutraceutical, a dietary supplement, has a potential to deliver a concentrated form of a presumed bioactive agent from a food, presented in a non-food matrix and used to enhance health in dosages that exceed those that could be obtained from normal foods. These are sold in presentations similar to drugs: pills, extracts, and tablets.	[28]
Numerous classes of compounds found in natural and processed foods are claimed to have beneficial effects on human health and wellness, e.g., vitamins, carotenoids, flavonoids, curcuminoids, polyunsaturated fatty acids, proteins, peptides, dietary fibers, oligosaccharides, and minerals.	[29]

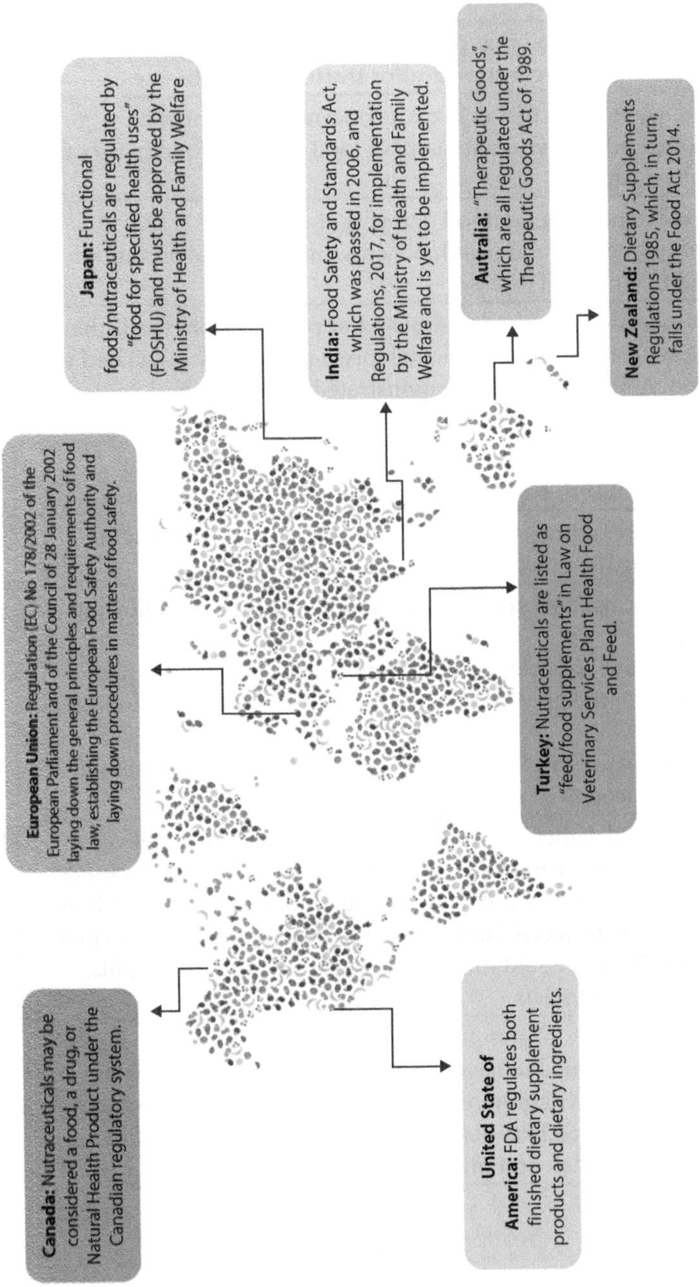

Figure 6.1 Regulatory framework of nutraceuticals in different countries.

According to the definition of the European Council Regulation (EC) No. 178/2002 "food or foodstuff is any substance or product, whether processed, partially processed or unprocessed, intended to be, or reasonably expected to be ingested by humans" [30]. Moreover, one of the aims of this Regulation is "to provide the basis for the assurance of a high level of protection of human health and consumers' interest in relation to food, taking into account, in particular, the diversity in the supply of food including traditional products, whilst ensuring the effective functioning of the internal market. It establishes common principles and responsibilities, the means to provide a strong science base, efficient organisational arrangements and procedures to underpin decision-making in matters of food and feed safety". According to this information, it is possible to say that nutraceuticals are covered by this regulation [30]. Nutraceuticals can be taken into the human body through regular foods as naturally high-content ingredients and/or as components enriched in a food matrix for the purpose of inhibiting a disease [31]. Due to the similarity and overlaps between medicinal products which also have properties for treating or preventing disease in human beings, food supplements and nutraceuticals, proper definitions are needed to draw a clear framework for nutraceuticals.

6.3 By-Products From Cereal Cultivation

Global food production is foreseen to increase around 12% by 2025 [32]. Cereals represent the most prominent contribution to this increment according to Food and Agriculture Organization (FAO), with wheat accounting for 10%, rice (14%) and maize (13%). Cereals are considered as staple food in human and animal nutrition, and myriad kinds of cereals are grown globally including wheat, rice, corn, barley, and oat [33]. High amounts of residues as corn stover and wheat straw, as well as by-products including endosperm, germ, bran and husk are generated during processing [34]. This growing cereal production trend may lead to excessive amount of processing bio-waste, which can be utilized as animal feed or in bioenergy conversion. In the meantime, studies conducted over the years discovered that these organic wastes are rich in valuable components important in nutrition such as dietary fiber. Dietary fibers are categorized as water-soluble (generally non-starch polysaccharides) and water-insoluble (cellulose, hemicelluloses, non-starch polysaccharides and lignin) [33]. Hemicellulose is the second most abundant cereal waste (20–35%) which has a promising potential for high-value implementations. Hemicelluloses has an amorphous nature with a lower degree of polymerization, thus utilized in polymeric materials, medicine, food, and energy [35].

Straw and stover are the two main fibrous residues from cereal cultivation. Straw is the residue from the slender-stemmed fine grain crops such as wheat, barley and rice, while stovers are the thick-stemmed residues from coarse grain crops like maize, sorghum and millet [36]. Lignocellulosic bio-waste, which is composed of cellulose, hemicelluloses and lignin, is the most abundant renewable resource to be considered for bioethanol transformation [37]. Bioethanol can be produced from cheap lignocellulosic biomass like corn stover, rice and wheat straws. The rice straw is potentially the most favorable biowaste with respect to available biomass for biofuel production, followed by wheat straw, and corn stover. However, the use of lignocellulosic biomass in ethanol production is still under development [38].

6.4 By-Products From Cereal Processing

6.4.1 Cereal Grain Composition

There are myriad kinds of cereals belonging to the grass family Poaceae family, also known as Gramineae. Wheat, barley, rye, rice, corn, oat, triticale (hybrid of wheat and rye) and sorghum are the cereals most important on a global scale [39]. Cereal grains have a very rich nutritional composition with endosperm, germ and bran as main structural components. These structural components composition change according to species and growing conditions [40]. For example, brown rice has the highest endosperm content followed by barley, corn and wheat [41]. Wheat has the richest bran composition, and corn has the highest germ composition [40]. The endosperm has notable amounts of starch and protein, as well as small contents of minerals, fatty acids and fibers [41]. The germ is composed from proteins, lipids, and vitamins. The bran cell wall consists of hemicellulose rich in cellulose, arabinoxylans and β-glucan, vitamins, and minerals. Bioactive compounds like polyphenols, phytosterols, carotenoids, tocopherols and trace elements are primarily located in the germ and bran [40]. Cereal health benefits are mainly associated with their dietary fiber content and bioactive compounds composition. Therefore, cereal grain structure composition is important in food industry to conserve certain nutrients following processing which depends largely on the applied method.

6.4.2 Cereal Grain Processing Techniques

Cereals are subjected to particular processing techniques to generate various food components with better flavor, appearance, texture, and shelf life. The most popular cereal grain processing applications are milling, pearling, fermentation and malting. Heat processing, puffing, and extrusion are the other less utilized procedures. The choice of grain processing methods depends largely on the nutrient composition of the structural parts [40].

6.4.2.1 Milling

Milling is the most traditional grain processing technique based on the principle of removing the germ and bran from the endosperm to obtain flour. Together with moving nutrients as dietary fiber and phytate; aleurone, testa, and pericarp parts are achieved with this process. The milling method is categorized as dry and wet and generates a high amount of biowaste and by-products [42].

Dry milling which is mainly utilized in grains like wheat, barley, rice and corn depends on the principle of removing the seed coat, germ, aleurone, and subaleurnoe fractions from the endosperm. This method is widely used in maize food production to obtain refined endosperm compounds with various particle dimensions including flour, meals and grits [42]. In the case of barley and rice, prior to dry milling process pretreatments as cleaning and shelling are implemented to remove the husk and bran, then the endosperm is used for flour production. The bran and broken grains are the major by-products of rice-milling [40]. Tempering–degerming systems are utilized to help fractioning. First the grain is cleaned with tempering and tuned to certain moisture content based on the cereal type, after than the germ and pericarp are separated from the endosperm in a determinator by crude abrasion. As a final step, these grain elements are takeoff utilizing gravity tables and aspirators [42].

The wet milling method is mostly used for wheat and corn, but this process is also appropriate for rice, barley, oat and sorghum. Physical, chemical and biochemical events are substantiated to separate germ, bran, starch and gluten. The aim is to obtain the highest starch concentration, eventually germ, bran and gluten are generated as by-products [14]. The highest germ and bran amounts are generated during corn wet-milling. Other notable by-products of wet-corn milling are gluten and steep liquor. Besides, starch and syrups are generated as food components [40]. The bio-waste obtained from wet milling is also utilized in bioethanol production. Starch and gluten are the major by-products of wheat wet-milling [43]. Soluble protein fibers are generated by enzymatic starch digestion. Fiber and proteins of cereal grains are-water soluble, and widely utilized to improve nutritional and functional properties of myriad food products such as breakfast cereals, soups, food bars, and beverages [44].

6.4.2.2 *Pearling*

Pearling is popular in rice milling with applications in durum wheat as well. The germ, testa, pericarp, aleurone, and subaleurone layers are separated from the grain using abrasion, after than the grain can be converted to flour implementing roller milling [34]. Pearling also decrease free fatty acids and lipase activity in bran which improves storage [45]. Pearled barley is a good source of β-glucan and is largely utilized in bakery and other food products [40]. Further studies are needed to optimize pearling in wheat and barley for uniform grain composition which also affects the flour particle size.

6.4.2.3 *Fermentation*

The fermentation applications date back to antique ages to conserve food and improve nutritional properties. The process is carried out through various microorganisms including yeasts and lactic acid bacteria. The fermentation process degrades grain structure and alter the texture and flavor leading to improved nutritional and functional quality of cereal by-products [39]. The grain composition of phenolics, vitamins, and minerals, as well as protein digestibility and phytic acid degradation increases during the fermentation process. The end products of fermentation are bread, porridge, beverages, etc [39, 40]. Bioactive compounds, which have antimicrobial and antioxidant properties, are synthesized through fermentation from cereal by-products [39].

6.4.2.4 *Malting*

The malting process consists of soaking/steeping, germination, and heating/kilning steps to produce alcoholic beverages as beer. Barley is the main cereal utilized in malting, especially in the brewing sector [46]. Although not common, the malting procedures is also applied in rice, wheat, buckwheat, and sorghum [40]. Cell wall, starch granules, and proteins go through enzymatic breakdown during malting to synthesize amino acids and peptides. During brewing, barley is soaked to germinate and degrade endosperm. In this process, the germination event is incomplete to assure enzymatic degradation of the endosperm and starch granules formation. The last step of malting is kilning for product consistency [46].

6.4.2.5 *Other Procedures*

Husking is applied to remove rice awn from the grain. The husking success depends on the husk tightness and grain humidity. Polishing is another application to separate rice bran by mild abrasion (generally in the form of gentle brushing) for better grain translucency. Grading is performed to separate rice components through sieving and aspiration. Oat grains are subjected to thermal treatment and dehulling prior to milling to obtain groats [34]. Flaking is a procedure to flake oat groats. Heat treatment, dehulling, cutting, and flaking steps deactivate enzymes to prevent oat groat destruction during rolling. Heat application deactivates lipase activity in oat flour production, which also affects the starch and protein pattern. Dry-grind method is applied to corn grains to ground them into coarse powder for ethanol production through fermentation. The applied process depends on the cereal specie, grain structure and composition to obtain germ, bran, hull, etc [40]. Another important aspect which needs to be considered is the biomass generated during processing which could lead to environmental issues. Therefore, it is important to consider the recovery of these by-products to profitable products and sustainable environment.

6.4.3 Cereal Grain By-Products

Following cereal processing various by-products with functional and nutritional properties are obtained. The most well-known cereal by-products are germ, bran, protein, fiber, hull, etc. These by-products are rich in dietary fibers, oil and phytochemicals. Bran consists around 25% of the cereal grain. Bran comprises the starchy endosperm and is mainly used in animal feeding. The bran is rich in vitamins, minerals, lipids, dietary fibers including cellulose, lignin, β-glucan, arabinoxylans, etc [40]. Rice bran is a good source of fiber, protein, lipid, and polyphenols [47]. Especially maize bran is rich in cellulosic fractions and is widely utilized in biofuel production, which is an excellent way of energy recovery contributing to sustainable environment [48]. The germ is also rich in bioactive compounds. Rice germ contains fatty acids, protein, dietary fibers, minerals and vitamins like vitamin E, B1, B2, and B6. The rice germ is utilized as animal feedstock and in brewing. Wheat germ has a very high nutritional profile with high amount of vitamins (especially vitamin E and B group), dietary fibers, and minerals [40]. The wheat germ is a very good source of unsaturated fatty acids (primarily oleic and linoleic acids), and phytochemicals as flavanoids and sterols. Wheat germ oil has wide application in cosmetic and food industry, as well as in insect biological control [49]. Brewers' spent grain (BSG) is obtained from the barley mash. BSG represents 85% of the total biowaste from the brewing and it is estimated that yearly 30,000 tons of BSG are wasted in a global scale. Although BSG is rich in cellulose, non-cellulosic polysaccharides, protein, fiber and phenolic acids (such as ferulic acid, caffeic acid, *p*-coumaric acid, hydroxycinnamic acid, and tocols), it is generally utilized as cattle feed or discarded. The research performed so far on BSG implementation in food industry suggests improving pre-treatments which will reduce BSG negative effects on food quality [39, 50]. Milling and bioprocessing methods as fermentation are potential tools to incorporate BSG in food products.

Although cereal by-products are used as animal feedstock, it is crucial to find new industries where these biomasses could be utilized. These by-products are a good source of dietary fibers, proteins, starch, fatty acids, vitamins, minerals and various phytochemical compounds with antioxidant properties. Therefore, they have a good potential in the food and pharmaceutical industry. Consequently, it is vital to explore the nutraceutical and functional components of these by-products.

6.4.4 Nutraceutical Compounds of Cereal Grain By-Products

Cereal grain by-products are a rich source of various phytonutrients and bioactive compounds with potential health benefits. Eventually, these by-products have good potential in the food and pharmaceutical industry as nutraceuticals. Therefore, categorizing by-products nutraceuticals based on cereal grain structure is pretty important.

6.4.4.1 Categorization of Cereal By-Products' Nutraceuticals

6.4.4.1.1 Polyphenols

Polyphenols are bioactive compounds with potential health improving properties, which is attributed to their antioxidant and anti-inflammatory activities. Phenolic acids (benzoic and cinnamic acids) and flavonoids (anthocyanidins, quinones, flavonols, flavones, flavanones), as well as tannins are the major polyphenols found in cereals [51]. The bran is the richest by-product structure in polyphenols [40].

Cereals contain high amounts of *penolics* found mainly bound to the cell wall compounds in bran. Phenolics existing in cereals grain bran, aleurone, and husk are benzoic acids (such as *p*-hydroxybenzoic, vanillic, and protocatechuic acids) and cinnamic acids (such as ferulic, *p*-coumaric, sinapic, and caffeic acids) derivatives. Ferulic acid is the most abundant in cereal grain with 75% located in the husk bound to the cell wall polysaccharides, 15% in the endosperm, and the rest in the aleurone part [52]. Vanillic and *p*-coumaric acids are also mainly located in the husk. Wheat bran has 2- to 3-folds higher concentrations of phenolics compared to the respective flours [40].

Flavonoids, also referred as anthosyanins, are categorized as secondary plant metabolites with phenolic nature responsible. Flavonoids have a positive impact on human and animal welfare thanks to their antioxidant and immunostimulating features. Anthocyanins are responsible in plant pigmentation and the major anthocyanins found in cereals are cyanidin-3-glucoside, cyanidin-3-rutinoside (reddish-purple color), pelargonidin-3-glucoside (pink color), delphinidin-3-rutinoside, and delphinidin 3-glucoside (blue color) [53]. Anthocyanins accumulate generally in the outer layers of the grain. For example, flavonoids are found in the pericarp and aleurone layer of purple and blue wheat grain varieties. In the case of maize, the pericarp is also rich in anthocyanins, whereas the aleurone layer has a small amount. The aleurone layer of barley and rye accumulates blue anthocyanins, which is not very usual in cereal crops [40, 53].

Tannins, also known as proanthocyanidins or procyanidins, are formed of polymerized flavanols and are responsible for the astringent flavor in foods. Tannins are bound to proteins, carbohydrates and minerals, which decrease their digestibility. However, tannins have a high antioxidant capacity with many potential health benefits. Tannins are found in the testa layer of sorghum [54]. The tannin level of brown rice and corn decrease with milling degree, which indicates that these compounds are concentrated in the bran. Tannins are present also rice, wheat and corn germ, as well as rice husk. Wild and pigmented cereal genotypes represent higher tannin amounts. For example, proanthocyanidins not found in cultivated rice are present in black and red rice. Similarly, naked barley and two-row barley have lower tannin rates than six-row barley. Barley tannin–protein interaction is very important in brewing; high tannin barley varieties often cause undesirable haze in beer [40].

6.4.4.1.2 Carotenoids

Carotenoids are lipid-soluble pigments with bioactive properties. These compounds are classified as carotenes (e.g. lycopene and β-carotene) and xanthophylls (e.g. oxygen by-products of carotenes, including astaxanthin, lutein, and zeaxanthin). Carotenoid content depends on the cereal specie and variety, growing conditions, kernel tissue origin (e.g. endosperm, germ, bran, and aleurone fractions). The total carotenoid composition is usually 9.55–62.96 μg/g in maize, 3.82–19.50 μg/g in sorghum, 1.178–3.345 μg/g in wheat. Xanthophylls, especially lutein, are the major carotenoids found in rice, wheat, maize and sorghum, which is followed by β-cryptoxanthin and zeaxanthin, as well as minor amounts of carotenes, like β- and α-carotene [55]. Carotenoids are mainly found in the germ of wheat barley and oats, as well as in the endosperm of yellow corn. Lutein is primarily concentrated in the endosperm and aleurone in maize, as well as in the aleurone and germ of wheat, barley, and oats. Zeaxanthin is mostly distributed in the wheat germ and corn endosperm, together with dominating the aleurone layers [56, 57].

6.4.4.1.3 Dietary Fibers

The dietary fibers include polysaccharides, resistant starch and non-digestible oligosaccharides. Dietary fibers are classified as soluble and insoluble fibers according to their water solubility. The insoluble dietary fibers includes cellulose, hemicellulose, lignins, while the soluble dietary fiber is composed of pectin, gums, mucilage, and hemicellulose [40]. Cell wall non-starch polysaccharides are the primary components of dietary fibers and significantly contribute to myriad health effects of the cereals. They are primarily distributed in the grain bran. The endosperm also contains minor amounts of fibers. Fiber rich parts can be generated through different milling methods [58]. Rice husk is rich in dietary fiber [40]. Buckwheat brans is significantly richer in fibers than refined flour, as well as buckwheat hull contains more fiber than the groats and brans [59, 60].

Cellulose is the dominant constituent of the cell wall. Cellulose can be converted into soluble oligosaccharides as a source of prebiotics [40]. The bran is a rich source of cellulose. The conversion of cellulose into value-added products is a popular research topic among academics and industry players. The cellulose is a promising alternative material for synthetic products thanks to its biodegradability, renewability, biocompatibility, and nontoxicity [61].

Arabinoxylans are among the most important group of hemicelluloses, which are found in the endosperm cell walls and outer layers of cereal grains [62]. Arabinoxylans are the primary structural constituents of the brain cell wall and are extracted from deproteinized and destarched bran. These polysaccharides are categorized as water extractable and water-unextractable. Feruloylated arabinoxylans are the primary non-starch polysaccharides from the cereal cell walls [63]. Arabinoxylans vary depending on molecular mass, xylose to arabinose ratio, and ferric acid content. Arabinose to xylem ratio may vary between 0.3-1.1 higher in bran than in endosperm [64]. Arabinoxylan is composed of a linear β-D-(1-4)-linked xylopyranose backbone, which can be substituted at C(O)-2, C(O)-3, or both in different positions. The main substituent is single α-l-arabinofuranose bounded to the xylan at C(O)-1 [65]. Arabinoxylans with high substitution are generally soluble in water and do not aggregate. Arabinoxylans with low substitution tend to aggregate. Besides, arabinoxylans can have arabinofuranosyl residues esterified at C-5 with ferulic acid, which influences the extraction of arabinoxylans and antioxidant characteristics in wheat. The insoluble arabinoxylans forms within the cell wall have higher amounts of bound phenolic acids which prevents their extraction [40, 66]. Arabinoxylans are of interest to scientists and industry due to their effect on bread quality. These compounds have an excellent viscosity and water-holding features, thus used as food thickening and stabilizing agents. Arabinoxylans show also antioxidant properties and have influence on health like increasing immunity and reducing the risk of coronary heart disease [62].

β-Glucans are main constituents of starchy endosperm and aleurone layer of cereal grains. The localization of *β*-glucans within the grain impacts the isolation and purification processes. In oat and barley, *β*-glucans are found in the starchy endosperm whereas in wheat the highest amount is located in the subaleurone layer with minor levels in the starchy endosperm [67]. *β*-Glucans are water soluble and their structure is composed of cellotriose and cellotetraose fractions bound by mixture of *β*-1–3 linkages. Cereal *β*-glucans shows high diversity with respect to tri- to tetramer ratio, the level of longer cellulosic oligomers and the mixture ratio of *β*-1–3 linkages [67]. These structural characteristics are significant factors affecting physical features like water solubility, viscosity, and gelation properties, and thereby impact the functionality of *β*-glucans when added to food products. These polysaccharides are of nutritional and commercial importance due to their health benefits as well including blood glucose level regulation and blood serum cholesterol level reduction [68].

6.4.4.1.4 Lipids and Fatty Acids

Cereal lipids can be categorized into three main groups based on their location in the grain, as well as the extraction technique. These groups are starch lipids, non-starch lipids and starch surface lipids. The aleurone cells also contains lipids and minerals in addition to proteins [69]. Cereal grain lipids comprise more than 50% polyunsaturated fatty acids [70]. Lipids and fatty acids are constituents of intercellular membranes, spherosomes, protein, and starch fractions of cereal grains by-products. The major lipids of the brewers' spent grains are triglyceride followed by free fatty acids. Lipid concentration of wheat brewers' spent grains is higher than that of malt [40]. Both total lipid and fatty acid composition of malt are significant factors affecting fermentation process [69]. The lipid content of rice germ is around 25%, and corn germ lipid concentration is between 9–50% [71]. The most prominent fatty acids in rice husk are linoleic, oleic, stearic, and palmitic acids [40]. Lipids have favorable health benefits as well, such as enhancing fat-soluble vitamins absorption and maintaining skin elasticity [70].

6.4.4.1.5 Proteins

Cereal grain proteins are grouped in four categories: glutelins, albumins, globulins, and prolamins, and their solubility are in alkali, water, salt, and alcohol, respectively [40]. Wheat bran protein distribution is as follow: 16.64% albumin, 13.62% globulin, 31.01% alcohol-soluble protein [72]. The ratio of albumin, globulin, glutelin and prolamin has been reported to be 5–10, 7–17, 75–81, and 3–6% in brown rice, 4–6, 6–13, 79–83, and 2–7% in milled rice, and 24–43, 13–36, 22–45 and 1–5% in rice bran [73].

6.4.4.1.6 Starch

Starch is a primary source of energy, which is synthesized in the endosperm, and helps in grain germination and plant growth [74]. Starch fractions are generally distributed in the starchy endosperm. The non-structural polysaccharides mainly consist of fructans and starch in barley grain [75]. Starch is one of the major constituents of bran depending on milling degree. The bran starch can be converted to glucose and utilized as sugar feed in the fermentation process of lactic acid, ethanol, and butanol production [65].

6.4.4.1.7 Prebiotics

Prebiotics are described as non-digestible food constituents that improve human health by promoting probiotic bacteria's growth and/or activity in the intestinal tract. These are nondigestible carbohydrates, such as polysaccharides, fructans, resistant starch, and oligosaccharides. Cereal grains contain polysaccharides and oligosaccharides which show probiotic features. The dietary fibers including β-glucan, α-glucan, arabinoxylans, and cellulose, as well as resistant starch, which are available in wheat bran are probiotic sources [76]. Regular intake of prebiotics could reduce the risk of diseases as irritable bowel syndrome, inflammatory bowel diseases, type 2 diabetes, colorectal cancer, obesity, osteoporosis, and infectious diarrhea. These beneficial effects are related to prebiotics' capability to motivate beneficial microbiota growth, hamper pathogen microorganisms' increase, stimulate immune functions, lower stomach emptying rate, enhance viscosity, and decrease polysaccharide concentrations [77].

6.5 Biotechnological Processes for the Valorization and Functionalization of Cereal-Based By-Products and Wastes

Over the years, a number of techniques have been developed to isolate specific compounds from certain plant materials, including cereal by-products, to obtain new products that have higher functional and nutritional value than the starting material. Roughly, these techniques can be divided into physical, chemical, enzymatical and microbiological, but often they must be combined to ultimately achieve the best possible result and make the added value of new products as high as possible. The choice of technique depends on not only the nature of the starting material, but also on the target product to be obtained. In addition, the recovery of pure substances from by-products should be considered from an economic point of view. Often, it is not cost-effective to isolate individual components, but the added value of new products can be achieved by using various techniques to concentrate the mass of by-product material in certain intermediate processing steps, and in this way, various concentrated mixtures of value-added substances are obtained. To date, most of the techniques for recovering nutraceuticals from cereal by-products have been applied on a laboratory and pilot scale, and there are very few practical industrial examples of obtaining nutraceuticals from cereal by-products. The reason is that it is currently more profitable to use these techniques to increase the efficiency of processes in bio-industrial production (biofuel, biogas, bioethanol) than in the production of pure substances that can be used as nutraceuticals. However, by improving existing techniques and developing new ones, as well as reducing process costs, there are more and more practical applications for the production of various food supplements and food additives from cereal by-products.

6.5.1 Techniques for Recovery of the Valuable Compounds From the Biomass of Cereal By-Products

Valuable compounds of interest (hemicellulose compounds, phytochemicals, proteins, lipids) can be found in the biomass of cereal by-products in free or bound state and with different solubility, both in water and in different solvents. Therefore, extraction/isolation procedures of these substances must be adapted to their properties. The extraction yield depends largely on their solubility and the nature of the bonds by which they are linked to other substances in the cell walls. Of all the valuable compounds, cereal by-products contain hemicellulose substances most frequently and to a great extent. In addition to hemicellulose compounds, the biomass of cereal by-products contains a whole range of other valuable substances such as cellulose, starch, proteins, lipids, lignin etc., which, whether in bound or free form, make the recovery of target compounds a demanding task. Compounds of interest are usually isolated by solid-liquid extraction, the principle of which is based on the separation of the target product or a group of products from the matrix of the starting material and its transfer to the solvent phase [78, 79]. The complete process of recovery of target compounds from the by-product biomass can be divided in four main steps: preprocessing, extraction, purification (separation), and product formation. Although some authors consider the separation of micro- and macromolecular compounds as the second step of the overall process to recover target substances from biomass, this step can be considered as part of the extraction procedure [80].

6.5.1.1 Pre-Processing

Since the biomass of cereal by-products can vary widely (by-products of dry and wet milling, grain polishing, brewer's spent grain (BSG), distillers' dried grains with solubles (DDGS), etc.), the biomass should be processed in certain ways. In the case of by-products with a high moisture content (e.g. BSG), the first step in the overall process of their utilization is drying, which ensures their preservation until further processing [81, 82], although these materials can also be used in their original form in the utilization processes of the target compounds if all the necessary conditions are ensured so that they can be processed in a short time after their production.

One of the most common pre-process techniques, sometimes referred to as macroscopic pre-treatment, for the dry cereal by-product biomass is milling or grinding to obtain the smallest and most uniform particles possible, which increases the total active surface area and facilitates the extraction process [33]. Pre-treatment may include wet milling of biomass with high water content. In the case of liquid by-product streams, vacuum or thermal concentration, or freeze-drying may also be used [83]. The choice of pre-treatment technology depends on not only the type of biomass and target product, but also to a large extent on the economic viability of the process used.

6.5.1.2 *Extraction*

The most important step in the entire recovery process of the target compounds from the cereal by-products is extraction. Among the several different types of extraction, the most usual method is solid-liquid extraction with the use of different solvents. The extraction process itself can be divided into different phases: solvent penetration into the cell wall matrix, dissolution of solutes in the solvents, diffusion of solutes from the matrix, and collection of extracted compounds [84].

Conventionally, solvent extraction can be performed with water or various chemical solvents using different techniques (maceration, percolation, reflux extraction). Due to the small amount of water-soluble target compounds in the biomass of cereal by-products, conventional water extraction is rarely used because the extraction yield is very low. To avoid the problems associated with water extraction, extraction with chemical solvents is often used in practice. The use of various chemical solvents has been shown to be much more effective, but one of the major problems with the use of chemicals is the risk to the environment, as well as safety issues and higher process costs. To overcome the shortcomings of conventional extraction techniques (low yield, long extraction times and a large volume of solvents), they are usually complemented by other techniques that can be used as a pre-treatment step or as a concomitant extraction procedure to improve the extraction process and increase the yield of the target compound. Some of these assisting extraction techniques are ultrasound-assisted extraction, microwave-assisted extraction, enzyme-assisted extraction, pulsed electric field extraction, pressurized liquid extraction, as well as supercritical fluid extraction [84].

One of the most commonly used techniques to assist solvent extraction is treatment with ultrasound. When ultrasound of various power and frequencies is applied to a mixture of solvent and cell wall matrix, acoustic cavitation occurs, which breaks up the tissue and facilitates diffusion of the solvent into the matrix, greatly increasing the efficiency of extraction. One of the advantages of this technique is that it can be easily combined with other assisting techniques [85]. In microwave-assisted extraction, electromagnetic radiation leads to rapid heating and pressure build-up that enhances the mass transfer of bioactive compounds into the surrounding solvent, thus increasing extraction efficiency, shortening extraction time, and reducing the amount of solvent [86, 87].

Considering the selectivity of the method, enzyme-assisted extraction has proven to be the best so far. Although the yield is somewhat lower compared to chemical extraction, the selectivity of the method and no negative impact on the environment are the main reasons for using enzymatic extraction. By selecting and combining enzymes, certain compounds can be specifically broken down so that they do not interfere with the extraction of the desired product. The solubility of certain compounds can also be increased by controlled enzymatic action. In this way, selectivity can be significantly improved, and in combination with other extraction processes and supporting techniques, the yield of the desired product can also be increased [33]. Another of the newer assisting techniques is a pulsed electric field. When this electric field is applied, an electric potential is generated and destroys membrane structures, which leads to increased permeability and facilitated mass transfer [80].

Pressurized liquid extraction, also known as accelerated solvent extraction or pressurized solvent extraction, is solvent extraction performed at elevated temperatures (above the boiling point at atmospheric pressure) and high pressure, where the solvent remains in liquid form. These conditions reduce the cohesive and adhesive interactions between molecules in the cell wall, break down the matrix, and improve the diffusion of the solvent into the matrix by reducing the surface tension of the solvent. So-called "green" solvents such as various alcohols (ethanol, methanol, heptane and hexane), alcohol-water mixtures or pure water are usually used in this extraction technique. Ethanol-water mixtures and pure water are considered the most "environmentally friendly" solutions. When water with a temperature between 100°C and 374°C (the temperature of 374°C is the critical point of water at 22.1 MPa) is used as a solvent, this technique is called subcritical water extraction [88, 89]. The principle of supercritical fluid extraction is based on phenomena that occur when solvents are near or above their critical points. In this state, the solvents have a solubility similar to that of a liquid and a diffusivity similar to that of a gas, which greatly increases the efficiency of the extraction. Near the critical point, small changes in pressure result in large changes in the density of the supercritical fluid, allowing precise adjustment of extraction conditions. Therefore, supercritical fluid extraction not only serves as an extraction technique but can also be considered as a very effective method for the separation and purification of the target compound, which can lead to a reduction of the necessary steps in the recovery of valuable compounds. The most commonly used solvent in supercritical fluid extraction of bioactive compounds is CO_2, but supercritical water extraction is also widely used [84, 90].

6.5.1.3 *Purification (Separation)*

Usually, the extract after extraction is a mixture of compounds with similar solubility in the extraction solvent, and if the pure target compound is to be obtained, it must be separated from other substances in the extract in a certain way. Besides, the solvent must also be separated. Considering the wide range of compounds that can be found in the extract, as well as taking into account the properties of the target compound and its final purity, there are quite a number of different separation and purification techniques that can be roughly divided into separation techniques based on size, shape, density, adsorption properties, etc. One of the most common separation techniques in the industry is precipitation. It is generally used to separate and concentrate macromolecules such as polysaccharides and proteins. Alcohol precipitation can be used both before and after the main extraction step. Ethanol washing can be used to remove lipids prior to extraction. The gradual precipitation method with different concentrations of water-ethanol mixtures can be used to fractionate extracts containing a mixture of polysaccharides with different molecular weights [90]. Other common precipitation methods used primarily to separate and fractionate proteins include isoelectric precipitation (adjusting the pH to a value close to the protein pI) and precipitation by changing the ionic strength, usually by adding neutral salts (e.g., ammonium sulphate). After precipitation, the precipitates can be recovered by centrifugation or filtration. Pressure-driven membrane separation processes are also commonly used to separate the target compounds. Depending on the pore size and pressure applied, these techniques can be divided into microfiltration, ultrafiltration, nano-filtration, and reverse osmosis. Of all separation techniques, chromatographic methods are considered the most effective when high purity of the target compound is required [85]. Considering bioactive compounds, adsorption column chromatography is the most commonly used chromatographic method, and the most commonly used adsorbent is silica gel. Other chromatographic separation methods include ion exchange, gel filtration, and simulated moving bed chromatography [80, 84].

6.5.1.4 *Product Formation*

The final step in obtaining target compounds from by-product biomass is product formation. Usually, powder preparations are targeted, but there are other examples where the final product is in liquid form. Various techniques can be used to dry the target substance: drum, pneumatic, freeze, vacuum and spray drying. One of the most efficient drying methods is spray drying, which can also be used for encapsulation purposes. The advantages of this technique are the production of a stable solid product from a liquid in only one continuous step, the possibility of drying heat-sensitive compounds, and the fact that it is usually inexpensive [91].

6.5.2 Recovery of Hemicellulose Compounds

Hemicellulose substances (arabinoxylans (AXs) and mixed-linkage (1→3)(1→4)-βD-glucans (MLG) represent a group of substances found in abundant amounts in cereal by-products in most cases. Their amount, as well as their ratio to each other, depends on the type of cereal from which they are obtained and on the origin of the by-product, considering the anatomical part of the cereal grain. While AXs is present in considerable amounts in both the husk and the bran of all cereals, the proportion of MLG varies considerably depending not only on the type of cereal, but also on whether the biomass of the by-product contains husk or bran. Thus, there is almost no MLG in the husk of cereals, and the amount in the bran depends greatly on the type of cereal from which the by-product is derived. Their content is highest in barley and oat bran, they are found in moderate amounts in wheat and rye bran, while the lowest amount is found in rice and corn bran [33]. Most of the phenolic acids in the matrix of cereal by-products, especially in cereal bran fractions, are present in insoluble bound form, bound to polysaccharides via an ester linkage and to lignin via an ether linkage [40, 92]. This is the reason why extracted arabinoxylans often contain a significant amount of phenolic acids at the same time. Therefore, if pure components of hemicellulose and phenolic compounds are to be obtained, great attention must be paid to the purification step [93].

Moreover, most AXs are present in bound (insoluble) form in cereal bran and only small amounts are present as free (soluble) AXs. Considering this, ordinary water extraction yields only very small amounts of AXs, since the water un-extractable AXs (WUAX) dominate over the water extractable AXs (WEAX) [62]. The water extraction can be carried out at different temperatures, on which the yield of extracted hemicellulose compounds as well as their form (degree of polymerization) depends. At lower temperatures (<100°C), the yield is relatively low because most hemicellulose compounds remain bounded in the remaining lignocellulosic material although extracted hemicellulose substances remain more or less in their native form. The yield can be significantly increased by using the so-called hydrothermal process at temperatures above 100°C and elevated pressure. This facilitates the penetration of water into the lignocellulose-hemicellulose matrix, which leads to the breaking of cross-linkages and acceleration of the extraction process. As the temperature is increased, the yield of extracted hemicellulose compounds also increases, while their degradation also occurs [33, 62]. Extraction of hemicellulose compounds is usually carried out under subcritical conditions at 160°C to 200°C and a pressure of 4-15 MPa [33]. Under these conditions, the polarity, viscosity, density, and surface tension of water decrease, resulting in increased diffusivity of water and enhanced penetration into the matrix. In this way, the solubilisation of hemicellulose compounds is increased and an autohydrolysis process occurs as the ionization constant of water also increases, leading to the depolymerisation of hemicellulose and the cleavage of acetyl groups by hydronium ions [33, 89, 94]. In addition, these acetyl groups support the formation of new hydronium ions, which further enhances the hydrolysis of hemicellulose [95]. Another type of hydrothermal process is water extraction by steam explosion, in which saturated steam is introduced into the reactor under high pressure for a short time. The temperature rises rapidly to 160-260°C and the autohydrolysis process begins as soon as the steam enters the matrix. After that, the pressure is quickly released and explosive decompression occurs, leading to the degradation of the polymer structure [33, 96]. From the above it can be concluded that in the hydrothermal extraction processes the yield of hemicellulose compounds is directly proportional to the water temperature and inversely proportional to the product purity and molecular weight of the extracted hemicellulose compounds. The size of the biomass particles also plays an important role. If the particles are small, extraction is faster and there is less time for autolytic processes, thus compounds with higher molecular weight are extracted. With larger particles, autohydrolysis takes longer and is therefore more pronounced, and hemicellulose compounds with lower molecular mass are obtained after extraction [97]. Moreover, higher temperatures during hydrothermal extraction reduce the purity of the extracted hemicellulose compounds, as some cellulose oligomers and various lignin fractions are also released from the lignocellulosic matrix [93, 98].

As for chemical extraction of hemicellulose compounds, although there are some examples of extraction with acids, alkaline solutions have been shown to be much more effective, as acidic solutions have been shown to largely degrade hemicellulose substances. The most commonly used alkaline for extraction of hemicellulose compounds are sodium (NaOH) or calcium hydroxide solutions ($Ca(OH)_2$), although other alkaline solutions can be used ($Ba(OH)_2$, NH_4OH, and KOH). Alkaline solutions destroy the cell wall matrix, releasing the polysaccharides and improving their reactivity. The solubility of hemicellulose is increased in this way, as hydroxyl ions (OH^-) break hydrogen and ester cross-linkages between cellulose, lignin and hemicellulose compounds. The extraction capacity of hemicellulose is further enhanced by the conversion of uronic acid residues into a negatively charged form under alkaline conditions, which increases the mutual repulsion of different polysaccharide molecules [33, 62, 86]. When barium hydroxide solution is used as the extraction medium, Ba^{2+} ions have the ability to separate AXs and glucans due to their interaction with pentoses, which improves the selectivity of the extraction product [62]. Alkaline hydrogen peroxide (H_2O_2) solution can also be used to effectively release the hemicellulose compounds from the biomass originated from cereal by-products, because under alkaline conditions, an active hydroperoxide anion (HOO^-) is formed, which promotes the dissolution of hemicellulose compounds and lignin [95]. However, chemical extraction has some limiting factors such as environmental and safety problems and high process costs.

To overcome the problems encountered in solvent extraction, various assisting techniques have been used to increase the efficiency of the extraction process. Usually, ultrasonic pre-treatment and various enzymes are used to increase the extraction yield and better control the process. A mixture of different enzymes can also be used to remove other polymers that may interfere with the hemicellulose compounds. Destarching of the by-product biomass is usually done with α-amylase prior to extraction, and proteins can be degraded with protease. By controlling the type and conditions of hemicellulase addition, water-insoluble hemicelluloses become more soluble and extractable. The endo-β-(1,4)-xylanases are widely used when arabinoxylan and β-glucan are to be extracted to increase the yield and purity of the target compound [33, 62]. It is believed that satisfactory results in terms of yield and purity can only be achieved by combining different techniques of pre-treatment, extraction, separation and purification. Sometimes a multi-step extraction process is required. For example, soluble arabinoxylans can be extracted first by hydrothermal extraction or by alkali solution, and then the remaining arabinoxylans can be further extracted by enzyme-assisted extraction using xylanase [99]. However, the limiting factor in the extraction of hemicellulose substances is that the use of various methods to isolate and purify these compounds can reduce their molecular mass, which may affect their functional properties and potential for health [33].

The extraction procedures for obtaining arabinoxylans and β-glucan are similar. In fact, the final product is often a mixture of these compounds, as the isolation of pure β-glucan from cereals for food purposes is still not economically justified. If a purer β-glucan product is to be produced, extensive hydrolysis of the arabinoxylans by xylanase and multiple precipitation of the extract with ammonium sulphate should be performed [100, 101]. Another valuable end product can be obtained by extracting the hemicellulose compounds from the cereal by-products. By adjusting the extraction conditions, the degree of hydrolysis of the hemicellulose compounds, especially arabinoxylan, can be controlled to some extent. Thus, the final product after the extraction process is often a mixture of arabinoxylan oligosaccharides with different chain lengths, which have a prebiotic effect and thus increase the nutritional value [95].

The so-called dry fractionation is another promising technique for the extraction of hemicellulose compounds from the by-product matrix of cereal bran. This technique uses electrostatic separation to separate fractions from the aleurone layer and endosperm from the pericarp fraction, which is rich in hemicellulose compounds. These two fractions acquire opposite charges when milled bran hits the surface of the charge tube. When an electric field is applied to these fractions, they are separated in an electric separator. The fraction rich in hemicellulose can be further sieved to remove starch and protein particles. In this way, the yield and purity of the final product can be increased [33, 102, 103].

6.5.3 Recovery of Phenolic Compounds

Since phytochemicals can occur in various states in the biomass of cereal by-products (free soluble or bound soluble and insoluble form), there are a number of different approaches to their recovery. Of the large number of valuable phytochemicals found in cereal by-products, the most attention has received for the isolation of phenolic compounds. Due to the wide range of phenolic compounds in terms of the number of hydroxyl groups and aromatic rings, concentration, solubility, acidity, as well as the type of biomass, there is no universal procedure for their isolation. Complexes with carbohydrates, proteins or other compounds also complicate the extraction process. Therefore, it is necessary to choose the optimal pre-treatment method and the type of extraction depending on the properties of the phenolic compounds. Less polar phenolic compounds can be extracted with hexane, while more polar compounds are mainly extracted with polar solvents such as water-methanol or water-ethanol mixtures. As with hemicellulose substances, new extraction techniques are increasingly being used to increase not only the extraction yield of phenolic compounds but also their bioavailability. In addition to conventional solvent extraction techniques, the most recent procedures for the recovery of the phenolic compounds include: ultrasound-assisted extraction from wheat bran [104], rice bran [105], and BSG [106]; supercritical CO_2 extraction from wheat bran [107], rice bran [108], and BSG [109]; microwave-assisted extraction from rice bran [110] and BSG [109]; pressurized aqueous ethanol solution extraction from wheat bran [111] and corn bran [112]; steam explosion from wheat bran [113]. Another source of valuable phenolic compounds is wastewater from the nixtamalization process of corn processing. Polyphenols can be recovered from wastewater by gradual application of membrane and ultra-filtration [114].

6.5.4 Recovery of Proteins

Although proteins are present in almost all cereal by-products, protein extraction from cereal by-products has not been widely commercialized. Mostly, protein-rich by-product fractions from wet milling of grains are used as animal feed, but more and more efforts are being made to improve extraction processes to make them viable for nutraceutical production. Today, although rare, there are some examples of rice protein preparations from rice bran being commercialized in the marketplace [115]. Rice bran proteins are considered hypoallergenic. Usually, the proteins are obtained by alkaline extraction combined with isoelectric precipitation [10]. The important step in the recovery process of proteins from rice bran is the removal of lipids to prevent oxidation and ensure better extractability and quality of the final product. As elevated temperatures are generated during lipid removal, proteins may be denatured and their extractability reduced [80, 116]. For this reason, special attention must be paid to the removal of lipids, and for this purpose, hexane can be used at lower temperatures [117].

Although this conventional approach is widely used, the yield is low, so more and more researches are focused on various modifications and improvements of the procedures to improve the yield and quality of the protein preparations obtained and to shorten the extraction time. Enzyme extraction is the most commonly used among the assisting techniques for the extraction of protein from rice bran. A mixture of different carbohydrates is usually used to dissolve the cell wall matrix and increase the extractability of proteins. Proteases can also be used to hydrolyse rice bran proteins to improve their solubility and extractability [118]. Ultrasound-assisted [119] and microwave-assisted extraction [120] can also be used to further improve extraction efficiency. Although they have not yet been used commercially, some promising techniques can be used to extract proteins from other grain by-products. For example, there are examples of protein extraction from wheat bran using alkaline extraction and isoelectric precipitation [121], and the combined use of ultrasound-assisted extraction and ultrafiltration for protein extraction from BSG [122].

6.5.5 Cereal By-Products in Solid-State Fermentation

Solid-state fermentation can be used for the production of various enzymes, antibiotics, bioactive compounds, and organic acids. Due to its simplicity, this process is often enforced as the first choice for the disposal of solid organic wastes from agriculture or the food industry, as solid-state fermentation allows their use without complex pre-treatment procedures. Almost all cereal by-products can serve as a carrier for conducting solid-state fermentation, i.e. as a substrate for the growth of various microorganisms during fermentation [123].

The most common examples of the use of cereal by-products in solid-state fermentation are the use of bran fractions of various kinds of cereal. For example, wheat bran is used as a substrate for the growth of *Aspergillus niger* to produce α-amylase [124] and polygalacturonase [125], and for the growth of *Bacillus subtilis* for protease production [126] and *Escherichia coli* for phytase production [127]. Rice bran is used for the production of hemicellulase and cellulase from *Phanerochaete chrysosporium* [128]. In addition, wheat and rice bran are used in the production of various antibiotics: tetracycline by *Streptomyces viridifaciens* [129], neomycin by *Streptomyces marinensis* [130], meroparamycin by *Streptomyces* sp. MAR01 [131] and lovastatin by *Penicillium funiculosum* NCIM 1174 [132]. Moreover, wheat bran is used for the production of lactic acid from *Lactobacillus amylophilus* GV6 [123, 133].

6.6 Market of High-Added Value Nutraceuticals From Cereal-Based By-Products and Wastes

A nutraceutical is a bioactive substance that has a beneficial effect on human health. They are natural compounds found in food or in some non-food plant species, from which they are extracted and then added to food. They are also available on the market in concentrated form as powders, capsules or tablets. Foods that contain nutraceuticals are called functional foods. DeFelice (1995) defined the term "nutraceutical" as any substance that is a food or part of a food and provides medicinal or health benefits, including the prevention or treatment of disease. These are products that come in various forms, either as purified and concentrated nutrients, dietary supplements, herbal products, or natural components of plants and processed foods such as cereals, soups, and beverages. Depending on their origin, nutraceuticals can be classified into three groups, which are herbal, animal, and microbial [134]:

- *Herbal nutraceuticals:* β-glucan, ascorbic acid, quercetin, luteolin, cellulose, lutein, gallic acid, pectin, lycopene, hemicellulose, lignin, capsaicin, α-tocopherol, α-carotene, selenium, zeaxanthin, monounsaturated fatty acids;
- *Animal nutraceuticals*: conjugated linolenic acid, eicosapentaenoic acid, docosahexaenoic acid, sphingolipids, choline, lecithin, calcium, coenzyme Q10, selenium, zinc, creatine;
- *Microbial nutraceuticals*: *Saccharomyces boulardii*, *Bifidobacterium bifidum*, B. *longum*, B. *infantis*, *Lactobacillus acidophilus*, *Streptococcus salvarius* (subs. *Thermophilus*).

Within the EU, the legal classification of nutraceuticals is based on their effect on the human body. If a substance only contributes to maintaining the health of the organism, it can be classified as food. However, if it has a proven modifying effect on one or more physiological functions of the body, i.e., it is involved in prevention and treatment, it is classified as a drug [135]. Among nutraceuticals, several categories stand out. Thus, a group of dietary supplements (for example, antioxidants, enzymes, probiotics and prebiotics), a group of herbal nutraceuticals and phytopreparations and a group of nutrients (for example, minerals, vitamins, fatty acids, amino acids) has been distinguished. Today, there is an interest in using the by-products of the cereal industry to extract the starch, proteins, lipids, dietary fiber, and bioactive compounds from these by-products to produce nutraceuticals used in the production of functional foods. Depending on the jurisdiction, the products may be claimed to prevent chronic disease, improve health, delay aging, prolong life, or support body structure or function. Antioxidant phenolic compounds, for example, have been extracted from agricultural wastes and used as food additives and in the manufacture of nutraceuticals and cosmetics.

6.6.1 Cereal Processing By-Products

Cereals like wheat (*Triticum* spp.), rye (*Secale* spp.), barley (*Hordeum* spp.), oat (*Avena* spp.), rice (*Oryza* spp.), millet (*Pennisetum* spp.), corn (*Zea* spp.), sorghum (Sorghum spp.) and triticale (x Triticosecale Wittmack) are one of the most important sources of food for human consumption. Global cereal production exceeds 2.7 million tons [136]. With the increase in the number of inhabitants on earth, the challenge of feeding all inhabitants adequately at the right time also increases. On the other side, food waste is a major environmental pollution problem, and it is estimated that about 1.3 trillion tons are thrown away each year. Food waste is any food or inedible parts of food that have been removed from the food supply chain to be recycled or disposed of in compost, incineration, energy production, or the ocean [137]. To reduce food loss and waste, it is necessary to find an alternative use for waste and prevent the overproduction of food. One of the proposed solutions is to recycle industrial by-products with the aim of isolating bioactive compounds such as polyphenols, fibers, vitamins and minerals with multiple health-promoting effects that can be extracted from waste from the cereal industry. This would contribute to less waste, the high economic potential for the production of value-added ingredients and functional products, more sustainable development as well as protection of the environment [138].

Various plant secondary metabolites and by-products from agricultural and industrial processing of crops are underutilized in terms of their nutritional value (rich in proteins, essential amino acids, micronutrients, dietary fiber, and phenolic compounds) and functional ingredients. The food and processing industries contribute to the generation of large amounts of waste, which has a negative impact on the environment and the economy. Since by-products generated during processing of raw materials and food production are valuable sources of useful bioactive compounds (phenolic compounds, antioxidants, fibers) that can be used as nutraceuticals and pharmaceuticals, it is proposed to use by-products as a sustainability tool to combat hunger, climate change and food waste. Cereals and their ingredients are considered functional foods and nutraceuticals because they contain important fiber, energy, minerals, vitamins, and antioxidants necessary for the functioning of the human body. From the group of dietary fiber, beta-glucans, and arabinoxylans are particularly important, and from the group of carbohydrates resistant starch and galactooligosaccharides and fructo-oligosaccharides.

Cereal processing produces various nutrient-rich by-products (e.g., germ and outer layers (bran) and brewer's spent grains) which are used as animal feed, fuel, substrate for bio-refineries, or waste. The above applications only partially utilize the cereal industry by-products. The highest proportion of endosperm is found in brown rice, the highest proportion of bran in wheat, and the highest proportion of germ in corn [139, 140].

Cereal industry by-products are potentially a good source of essential compounds for human nutrition but have limited applications in food production due to their challenging technological properties. The development of innovative biotechnologies will enable the broader application of cereal industry by-products in the form of new functional foods.

Cereal by-products are formed during dry milling (to produce flour), wet milling (to produce starch and glucose) pearl milling, and malting [34]. Considering the two different milling processes, dry and wet milling, cereal fractions with different nutritional compositions are obtained for different uses (food and non-food products or as animal feed). In addition to the brewing process, cereal by-products are obtained through processes such as malting, brewing and grain distillation. Each of the above techniques is specific to a particular grain. For example, dry milling is used in processing wheat and rye, wet milling is used in processing corn and sorghum, pearl milling is used in processing oats and barley, and malting is used in processing barley. The main cereal by-product generated during these techniques is bran, germ, hull (husk) and wastewater from polishing. The goal of dry milling is to separate bran, germ, and husk from the starchy endosperm, while wet milling aims to recover the maximum amount of undamaged starch granules from by-products such as germ, bran, and gluten. During the malting process, polished grains and by-products such as the seed coat (testa and pericarp), aleurone and subaleurone layers, and germ are produced. Throughout the production of malt, the enzymes consume sugars and starches that can be fermented in the grain, while the spent grain remains as a by-product [141].

The cereal fractions obtained during milling are a rich source of biofunctional molecules, dietary fiber, minerals, vitamins, lignans, phytoestrogens, and phenolic compounds. Since grain is one of the most widely grown crops in the world, its processing produces a large amount of biomass that is disposed of as waste. Following the principle of sustainable and circular bio-economy, new strategies for the use of biomass in the production of value-added products are prescribed. Some by-products of the cereal industry contain constituents that are of particular research interest because of their potential health-promoting properties, as they have been proposed as a promising antioxidant, anti-inflammatory, and anticarcinogenic compounds.

6.6.2 Nutraceutical From Cereal-Based By-Products

According to the research of the FAO, the most economically important cereals are wheat, rice, maize, barley, oats, sorghum, rye and millet as the second most important cereal of global economic importance [136]. Wheat (*Triticum spp.*) by-products are bran (processed by dry milling) and germ. Wheat bran is the outer layer of wheat, consisting of the outer and inner pericarp, the seed coat, and the epidermis of the grain [142]. Wheat bran is a good source of dietary fiber (β-glucans, arabinoxylan); phytochemicals (alkylresorcinols, ferulic acid, flavonoids), micronutrients (minerals, vitamins), and phytosterols. Another by-product of wheat grain processing is wheat germ, which is produced during milling. This by-product is a rich source of vitamin E, group B vitamins, protein, fiber and minerals, essential amino acids, unsaturated fatty acids (oleic acid, linoleic acid and a-linolenic acid) and functional phytochemicals (flavonoids, sterols, octacosanols and glutathione) [143, 144]. Rye (*Secale spp.*) by-product of conventional milling is bran, rich in fibers and bioactive compounds (phenolic acids, phytosterols, tocopherols) [145–147]. Dry milling of corn (*Zea spp.*) produces by-products such as corn flour, corn bran, and feed. In contrast to dry milling of corn, wet milling produces the following by-products: liquefied corn product, starch molasses, germ meal, gluten meal, hydrolysed corn protein, and condensed fermented corn extracts [148, 149]. Barley (*Hordeum spp.*) is the main raw material for malting and brewing, which can be milled or malted. Each of the above barley processing methods produces different by-products. Malting produces the barley malt sprouts, while brewing leaves the brewers' spent grains [150, 151]. In cereals such as barley and oat, the cell walls are rich in the non-starch polysaccharide beta-glucan, the main component of soluble dietary fiber. Beta-glucan has a beneficial effect on the human body, lowering plasma cholesterol levels and the glycemic index, and reducing the risk of colon cancer [152]. The processing of oats (*Avena spp.*) produces the by-product oatmeal [153]. The main post-production process of rice (*Oryza* spp.) is milling, leaving by-products such as rice meal (rich in thiamine) and rice bran (contains B-complex vitamins, proteins, and amino acids) [47].

Various nutraceuticals can be obtained from cereal by-products, of which three main groups stand out: dietary fiber (cellulose, beta-glucan and arabinoxylans from the cereal bran), phytochemicals like polyphenols (phenolic acids, flavonoids, tannins) carotenoids, phytosterols and micronutrients), and prebiotic carbohydrates. Cereal by-products are a source of compounds such as ferulic acid, β-glucan, arabinoxylan, sterols, vitamins, polyphenol oryzanol, which have a positive effect on human health, preventing diseases such as diabetes, cardiovascular diseases, diseases derived from inflammatory processes, etc. When it is mentioned about dietary fiber in cereal by-products, this mainly means non-starch polysaccharides such as arabinoxylans (present in bran), β-glucan (located in aleurone layer), and arabinogalacta. A good source of dietary fiber overall is bran and husk, and arabinoxylans and β-glucan are considered particularly important for nutrition among dietary fibers [154]. While arabinoxylans can occur in soluble and insoluble forms, β-glucans are water-soluble. Arabinoxylans (AX) and arabinoxylooligosaccharides (AXOS) present in the by-products of cereal production have beneficial effects on the human digestive system (improve the dietary fiber content) and may also improve the viscosity, texture, sensory properties, and shelf life of foods [155]. Due to their prebiotic nature, oligosaccharides derived from AX have scientific and commercial importance. The most important functional property of xylooligosaccharides (XOS) and AXOS is their ability to stimulate the growth of beneficial microflora in the gut [156]. Another relevant approach to valorise cereal by-products AX and AXOS is their complete hydrolysis to monosaccharides to produce other fine chemicals, e.g., production of xylitol from AX, which is suitable for diabetics [157]. Arabinoxylans are present in the bran of many cereals (between 10 and 25% of the total bran), but in wheat bran, arabinoxylans make up a significant proportion (52–70%) [154]. β-glucans are water-soluble fibers found in the aleurone layer of cereals and their proportion depends on the cereal variety from which the by-product was produced [158]. Since they have a positive effect on human health, β-glucans are used in the prevention of diabetes and control of blood cholesterol levels, as well as in the manufacture of products for people with lactose intolerance and cholesterol problems [159]. In food manufacturing, beta-glucans are used as stabilizers, thickeners, as fat replacers, as viscosity agents to improve texture and viscosity, etc. [160].

Most phytochemicals are found in the outer layers of the grain, such as the hull and bran, and in the germ of the grain. Cereal by-products are a rich source of compounds that contribute to antioxidant capacity, such as phenolic acids, flavonoids, and tannins [161]. Phenolic acids are present in various cereal by-products such as bran, aleurone, germ and hull (husk). Some authors reported the presence of several phenolic acids in cereal by-products such as *p*-hydroxybenzoic acid, chlorogenic acid, syringic acid, vanillic acids, vanillin, *p*-coumaric acid, ferulic acid, salicylic acid, and caffeic acids [65, 162, 163]. Another important group of polyphenols in cereal by-products are flavonoids, which are found in the pericarp and aulerone layer of the grain. Among the flavonoids, the anthocyanins should be mentioned, especially the peonidin and cyanidin glycosides, which are found in purple and blue wheat varieties [164]. In cereal by-products, tannins are present in the bran fraction of cereal grains (corn), germ (wheat, rice, corn), and husk (rice). Since human enzymes cannot digest tannins, they are fermented by the microbiota in the large intestine [165, 166].

Carotenoids are present in the endosperm and bran of cereal by-products (wheat, corn, rice) [164, 166].

Lipids present in cereal by-products (mainly in wheat brewers' spent grains) are triglyceride, while among fatty acids there are linoleic acid, palmitic acid, and oleic acid, as well as linoleic acid, stearic acid, myristic acid, vaccenic acid, arachidic acid, 11-eicosenoic acid, behenic acid, and lignoceric acid [167]. A good source of fatty acids is wheat brewers' spent grains and rice husk [168] and wheat bran [45]. Protein derived from cereal by-products is mainly concentrated in the bran [41].

Humankind today is facing several problems, such as the increasing number of various diseases and stress levels, as well as the appearance of an increasingly aging population. In addition, the number of well-educated people in the population is growing and the mentality of disease prevention as well as the awareness of a healthier and more individualized diet. The COVID-19 pandemic has driven up spending on multivitamins to boost the immune system. This trend is predicted to continue in the future. The above facts have led to a growth in demand for nutraceuticals (prediction is of about 8.9% per year in the period 2020-2028). Actually, this has led many scientists to intensively research the market and possible applications of nutraceuticals. Due to their potential nutritional, safety, and therapeutic effects, nutraceuticals are becoming increasingly important to consumers. According to the type of source, nutraceuticals are as follows:

- Probiotics
- Proteins & amino acids
- Phytochemicals & plant extracts
- Fibers & specialty carbohydrates
- Omega-3 fatty acids
- Vitamins
- Prebiotics
- Carotenoids
- Minerals
- Others (glucosamine, chondroitin, and polyols)

Obtaining nutraceuticals from cereal processing by-products is a challenge for scientists. The commercial implementation of the production of nutraceuticals from these sources is a very complex approach and depends on several parameters that should be considered, such as economic profitability and environmental impact, as well as the stability, applicability and quality of the product obtained. In order to commercialize nutraceuticals from food waste, the process of obtaining and characterizing nutraceuticals must be researched under laboratory conditions, taking into account the functional properties of the final product. Once this phase has been successfully carried out, the product must be patented in accordance with legal requirements. Once the product is patented, the process is carried out on a semi-industrial scale, followed by the calculation of its use and economic viability, as well as its application in the food sector and confirmation of the product on the market [169]. Typical nutraceuticals available to consumers on the market are seen in the form of snacks (protein bars), various energy drinks and juices, dietary supplements (vitamin and mineral supplements), various tablets, protein powders, probiotics, polyunsaturated fatty acids, etc. Primary (e.g., bran, germ, hulls) and secondary by-products (e.g., fiber, oil, protein, phytochemicals, nutrients) from cereal processing could be converted into various commodities. Some of these products can be found on the market:

- PROMITOR® Soluble Fiber (manufacturer Tate & Lyle) is corn-based soluble fiber suitable for restoring the amount and viscosity of sweeteners in reduced-sugar products, allowing manufacturers to improve viscosity and mouthfeel.
- RiBran® (manufacturer RiceBran Technologies)—A good source of vitamins and vital amino acids, dietary fiber, fatty acids and protein. It is easily incorporated into various granulations for special formulations and applications.
- Organic RiBalance® (manufacturer RiceBran Technologies)—powder, source of vitamins, amino acids and fiber. It has many uses, as a substitute meal in the form of a drink, as a substitute for cellulose and inulin in the preparation of food bars and cereal formulations.
- Organic RiSolubles® (manufacturer RiceBran Technologies)—Water-soluble powder, a source of vitamins, amino acids and antioxidants. It is used in the preparation of beverage powders, emulsifiers, carriers, icings, ice cream and yogurt. It is a natural substitute for gums, lecithin, silicon dioxide and magnesium stearate.
- Organic RiFiber® (manufacturer RiceBran Technologies)—the source of natural fiber, protein and essential amino acids. Use to increase water binding capacity, cooking efficiency and improved emulsification. It is used in the preparation of powders for athletes, energy bars, snacks and gluten-free pastries.

- Proryza® Gold (manufacturer RiceBran Technologies)—contains vegetable proteins and dietary fiber. Excellent as a functional and nutritious base ingredient to add texture, adhesive properties and a neutral flavour profile.
- AgriFiber MFO (manufacturer AgriFiber Solutions) are three cereal-based by-products made from corn bran and Oat Hull.

  - *Multi-functional Corn Bran Fiber* (MFC)—its water-binding properties make it a perfect emulsifier, stabilizer or yield improver in meat, sauces and baking.
  - *Multi-functional Oat Hull Fiber* (MFO)—natural, low-calorie oat fiber for a variety of applications in baked goods and fillings.
  - *Soluble Fiber Corn* (SFC)—Soluble corn fiber with emulsifying and binding properties for beverages, snacks and plant-based alternatives.

- Soluble Corn Arabinoxylan Fiber (manufacturer ClonBio Group Ltd)—corn bran arabinoxylans is a slow and completely fermented fiber with gut health benefits. They are used in the production of snacks and baked goods, where this fiber can replace up to 50% of the sugar in snacks such as granola bars.
- Arrabina™ (manufacturer CometBio) is an arabinoxylan prebiotic clinically proven to help maintain healthy blood sugar levels, support immune health, and promote the growth of beneficial bifidobacteria in the gut.
- Naxus® (manufacturer BioActor) is an arabinoxylan derived from wheat that systemically downregulates glucose and insulin level, improves intestinal barrier integrity, and strengthens adaptive and innate immune cells.
- PromOat oat beta-glucan (manufacturer Lantmännen Oats)—is a high concentration of soluble oat bran fiber that can be used alone or in foods and beverages to maintain healthy blood sugar and cholesterol levels, or used for its beneficial effects on digestion and gut health.
- Stabilized Rice Bran (manufacturer NutraBio)—a natural source of B vitamins, fiber, phytonutrients, antioxidants, omega acids, tocopherols and tocotrienols, which have a

positive effect on maintaining normal cholesterol and lipid levels and contribute to optimal immune health.

- Gama Oryzanol (manufacturer Swanson Health Products)—is a mixture of sterile and triterpenyl esters of ferulic acid extracted from rice bran and rice germ. It has high antioxidant capacity, helps maintain physical and emotional well-being during the transition to menopause, and provides favorable lipid balance for healthy cardiovascular function.
- Oat Fiber Plus (manufacturer Now Foods)—are a combination of fiber from oats, brown rice and apple pectin in tablet form, useful to balance fiber intake.
- NOW Real Food® Organic Oat Bran—oat bran is rich in nutrients and a good source of fiber and protein, which can reduce the risk of heart disease. NOW Real Food® Organic Oat Bran is an excellent addition to most baked goods and is mainly used to increase fiber content, regulate colour and improve flavour.
- Life Extension NK Cell Activator™ (manufacturer Swanson Health Products)—rice bran that influences the maintenance of the immune system.
- All-Bran Original (manufacturer Kellogg)—cereal rich in wheat bran fiber (86%).
- Betaglucare (manufacturer Prorsum Healthcare AB)—a natural source of beta-glucan from oats, acts on lowering cholesterol and blood sugar levels after a meal.
- Beta-Glucan 500mg (manufacturer Fair & Pure)—food supplement in capsule form, contains 70% oat polysaccharides, contributes to the maintenance of normal cholesterol levels.

6.7 Conclusion and Future Perspective

The by-products of cereal processing are usually disposed of into the environment or used as livestock feed. Since cereal processing produces several by-products with different chemical compositions, they represent a valuable potential of various desirable nutrients such as carbohydrates, proteins, fiber, lipids, vitamins, polyphenols, and minerals that can be recycled and reused in the production of new value-added products like nutraceuticals.

Because the process of recycling by-products is challenging, measures are needed to manage large variations in bio-resource content (biomaterials are subject to change), collect waste at the source (so additional transportation costs should be factored in), and avoid microbial decomposition of the original bio-resource. The current way of dealing with industry by-products, where agricultural waste is first disposed of and then recycled, leads to production losses, and such an approach to nutraceutical production is not economically viable. Therefore, it is necessary to find cost-effective, environmentally friendly and efficient production solutions that justify the production of nutraceuticals from cereal by-products, and to deepen scientific and technological efforts to effectively utilize cereal by-products and residues.

In the production of nutraceuticals from cereal by-products, one should consider the operating costs and efficiency of the process itself, which question the justification of the production itself. The above problems are currently the main economic and technical obstacles that need to be solved in the future. It is well known that a project that focuses on recovery technologies without researching and determining the ultimate use of the final product as food (which is also economically viable) is doomed to failure. In addition, the new product may not be as useful as expected at the research level, or it may be difficult to commercialize as a nonspecific functional ingredient. Research should focus on developing reliable biotechnology processes and systems that are commercially viable. For this to be the case, marketable products of high value and scale must be produced. Economic viability is only possible through increased competitiveness achieved by developing and optimizing more efficient, sustainable, and cost-effective downstream processes and lower initial capital costs.

The scope of nutraceuticals is very broad, so they are used to improve health and prevent chronic diseases, as well as to delay the aging process and prolong life. Due to their potential nutritional, safety and therapeutic effects, they are becoming more and more interesting on the market today. Nutraceuticals help maintain good health and they can be taken without consulting a doctor. There are also some disadvantages in the use of nutraceuticals, namely the fact that their mode of action is very slow (because they are more food than medicine) and that there is no strict quality control of nutraceuticals in terms of concentration, quantity and origin of ingredients (which sometimes vary). For the above reasons, stronger control of the production and declaration of the products is needed. Future research should focus on target groups of nutritional products, such as elderly people, people concerned with weight control, people following vegan or vegetarian diets, people interested in alternative and natural health, athletes and recreational athletes, people who want to improve and maintain their health (memory, immune system or other body functions).

Moreover, research should focus on the development and production of appropriate packaging solutions for several reasons. First, because nutraceuticals are available not only as pills but also in powder or liquid form, and second because the sensitive nature of nutraceuticals (susceptible to interaction with moisture, oxygen, and light from the environment) requires strict safety and protection standards for packaging. Furthermore, when developing appropriate solutions for packaging nutraceuticals, it is necessary to follow the trend of using sustainable materials in the packaging industry and use environmentally friendly packaging materials, materials with a natural composition and full recycling capabilities.

References

1. Laskowski, W., Górska-Warsewicz, H., Rejman, K., Czeczotko, M., Zwolińska, J., How important are cereals and cereal products in the average polish diet?, (in eng). *Nutrients*, 11, 3, Mar 21 2019.
2. Poole, N., Donovan, J., Erenstein, O., Continuing cereals research for sustainable health and well-being. *Int. J. Agric. Sustain.*, 20, 1–12, 2021.
3. Henry, A.G., Brooks, A.S., Piperno, D.R., Plant foods and the dietary ecology of Neanderthals and early modern humans. *J. Hum. Evol.*, 69, 44–54, 2014/04/01/.
4. Pankin, A. and von Korff, M., Co-evolution of methods and thoughts in cereal domestication studies: A tale of barley (Hordeum vulgare). *Curr. Opin. Plant Biol.*, 36, 15–21, 2017/04/01/, doi: https://doi.org/10.1016/j.pbi.2016.12.001.
5. Food and Agriculture Organization (FAO), FAOSTAT, FAO Statistics Division, 2020, http://www.fao.org/faostat/en/#data/QC.
6. Food and Agriculture Organization (FAO), Sustainable development goals, https://www.fao.org/sustainable-development-goals/indicators/1231/en/#:~:text=this%20sub%2Dindicator.-,Target%2012.3,%2C%20including%20post%2Dharvest%20losses.
7. Munekata, P.E.S. *et al.*, Valorization of by-products from Prunus genus fruit processing: Opportunities and applications, (in eng). *Crit. Rev. Food Sci. Nutr.*, 1–16, Mar 14 2022.
8. Food and Agriculture Organization (FAO), Food wastage footprint: Impacts on natural resources, https://www.fao.org/news/story/en/item/196402/icode/.
9. World Counts Statistics, https://www.theworldcounts.com/challenges/consumption/foods-and-beverages/food-waste-facts.
10. Skendi, A., Zinoviadou, K.G., Papageorgiou, M., Rocha, J.M., Advances on the valorisation and functionalization of by-products and wastes from

cereal-based processing industry. *Foods*, 9, 9, 1243, 2020. [Online]. Available: https://www.mdpi.com/2304-8158/9/9/1243.

11. Xue, L. *et al.*, Missing food, missing data? A critical review of global food losses and food waste data. *Environ. Sci. Technol.*, 51, 12, 6618–6633, 2017/06/20 2017.
12. Chauhan, C., Dhir, A., Akram, M.U., Salo, J., Food loss and waste in food supply chains. A systematic literature review and framework development approach. *J. Cleaner Prod.*, 295, 126438, 2021/05/01/, doi: https://doi.org/10.1016/j.jclepro.2021.126438.
13. Luithui, Y., Baghya Nisha, R., Meera, M.S., Cereal by-products as an important functional ingredient: Effect of processing, (in eng). *J. Food Sci. Technol.*, 56, 1, 1–11, Jan 2019.
14. Dapčević-Hadnađev, T., Hadnađev, M., Pojić, M., The healthy components of cereal by-products and their functional properties, in: *Sustainable Recovery and Reutilization of Cereal Processing by-Products*, pp. 27–61, Elsevier, Woodhead Publishing, Sawston, United Kingdom, 2018.
15. Moreau, R.A., Singh, V., Eckhoff, S.R., Powell, M.J., Hicks, K.B., Norton, R.A., Comparison of yield and composition of oil extracted from corn fiber and corn bran. *Cereal Chem.*, 76, 3, 449–451, 1999.
16. Prueckler, M. *et al.*, Wheat bran-based biorefinery 1: Composition of wheat bran and strategies of functionalization. *LWT-Food Sci. Technol.*, 56, 2, 211–221, 2014.
17. Salazar-López, N.J., Ovando-Martínez, M., Domínguez-Avila, J.A., Cereal/grain by-products, in: *Food Wastes and By-products*, pp. 1–34, 2020.
18. DeFelice, S.L., The nutraceutical revolution: Its impact on food industry R&D. *Trends Food Sci. Technol.*, 6, 2, 59–61, 1995/02/01/, doi: https://doi.org/10.1016/S0924-2244(00)88944-X.
19. Blaze, J., A comparison of current regulatory frameworks for nutraceuticals in Australia, Canada, Japan, and the United States, (in eng). *Innov. Pharm.*, 12, 2, 1–8, 2021.
20. Santini, A. *et al.*, Nutraceuticals: Opening the debate for a regulatory framework. (in eng). *Br. J. Clin. Pharmacol.*, 84, 4, 659–672, Apr 2018.
21. Gupta, P.K., Regulatory guidelines for nutraceuticals and food supplements in India, in: *Nutraceuticals in Veterinary Medicine*, R.C. Gupta, A. Srivastava, R. Lall (Eds.), pp. 807–814, Springer International Publishing, Cham, 2019.
22. Zeisel, S.H., Regulation of "nutraceuticalsv" (in eng). *Science*, 285, 5435, 1853–5, Sep 17 1999.
23. Brower, V., Nutraceuticals: Poised for a healthy slice of the healthcare market?, (in eng). *Nat. Biotechnol.*, 16, 8, 728–31, Aug 1998.
24. H. Canada, Nutraceuticals/functional foods and health claims on foods, https://www.canada.ca/en/health-canada/services/food-nutrition/food-labelling/health-claims/nutraceuticals-functional-foods-health-claims-foods-policy-paper.html.

25. M.-W. O. Dictionary, Merriam-webster online dictionary, Merriam-Webster Inc., P.O. Box 281, Springfield, MA 01102, United States, 2015.
26. European Nutraceutical Association (ENA), *Science behind nutraceuticals*, vol. 2016, E.N. Association (Ed.), p. 594, Switzerland, Basel, 2016.
27. Kalra, E.K., Nutraceutical–definition and introduction, (in eng). *AAPS PharmSci*, 5, 3, E25, 2003.
28. Gul, K., Singh, A.K., Jabeen, R., Nutraceuticals and functional foods: The foods for the future world, (in eng). *Crit. Rev. Food Sci. Nutr.*, 56, 16, 2617–27, Dec 9 2016.
29. McClements, D.J., Li, F., Xiao, H., The nutraceutical bioavailability classification scheme: Classifying nutraceuticals according to factors limiting their oral bioavailability, (in eng). *Annu. Rev. Food Sci. Technol.*, 6, 299–327, 2015.
30. Regulation (EC), Regulation (EC) No 178/2002 of the European Parliament and of the Council of 28 January 2002 laying down the general principles and requirements of food law, establishing the European Food Safety Authority and laying down procedures in matters of food safety, https://eur-lex.europa.eu/legal-content/EN/TXT/HTML/?uri=CELEX:32002R0178.
31. Sauer, S. and Plauth, A., Health-beneficial nutraceuticals-myth or reality?, (in eng). *Appl. Microbiol. Biotechnol.*, 101, 3, 951–961, Feb 2017.
32. OECD, Food, and A. O. o. t. U. Nations, *OECD-FAO agricultural outlook 2016-2025*, 2016.
33. Arzami, A.N., Ho, T.M., Mikkonen, K.S., Valorization of cereal by-product hemicelluloses: Fractionation and purity considerations. *Food Res. Int.*, 151, 110818, 2022/01/01/, doi: https://doi.org/10.1016/j.foodres.2021.110818.
34. Papageorgiou, M. and Skendi, A., 1 - Introduction to cereal processing and by-products, in: *Sustainable Recovery and Reutilization of Cereal Processing By-Products*, C.M. Galanakis (Ed.), pp. 1–25, Woodhead Publishing, Sawston, United Kingdom, 2018.
35. Luo, Y. *et al.*, The production of furfural directly from hemicellulose in lignocellulosic biomass: A review. *Catal. Today*, 319, 14–24, 2019.
36. Schiere, J. *et al.*, Grain and straw for whole plant value: Implications for crop management and genetic improvement strategies. *Exp. Agric.*, 40, 3, 277–294, 2004.
37. Gares, M., Hiligsmann, S., Kacem Chaouche, N., Lignocellulosic biomass and industrial bioprocesses for the production of second generation bio-ethanol, does it have a future in Algeria? *SN Appl. Sci.*, 2, 10, 1680, 2020/09/12.
38. Kim, S. and Dale, B.E., Global potential bioethanol production from wasted crops and crop residues. *Biomass Bioenergy*, 26, 4, 361–375, 2004/04/01/, doi: https://doi.org/10.1016/j.biombioe.2003.08.002.
39. Verni, M., Rizzello, C.G., Coda, R., Fermentation biotechnology applied to cereal industry by-products: Nutritional and functional insights, (in English). *Front. Nutr., Rev.*, 6, 1–13, 2019-April-12.

40. Salazar-López, N.J., Ovando-Martínez, M., Domínguez-Avila, J.A., Cereal/grain by-products, in: *Food Wastes and By-products: Nutraceutical and Health Potential*, pp. 1–34, 2020.
41. Balandrán-Quintana, R.R., Recovery of proteins from cereal processing by-products, in: *Sustainable Recovery and Reutilization of Cereal Processing by-Products*, pp. 125–157, Elsevier, Woodhead Publishing, Sawston, United Kingdom, 2018.
42. Slavin, J.L., Jacobs, D., Marquart, L., Grain processing and nutrition. *Crit. Rev. Food Sci. Nutr.*, 40, 4, 309–326, 2000.
43. Wronkowska, M., Wet-milling of cereals. *J. Food Process. Preserv.*, 40, 3, 572–580, 2016, doi: https://doi.org/10.1111/jfpp.12626.
44. Mazza, G., Gao, L., Oomah, B.D., O'connor, D., Crowe, B., Functional, water-soluble protein-fibre products from grains, Google Patents, 2001.
45. De Brier, N., Hemdane, S., Dornez, E., Gomand, S.V., Delcour, J.A., Courtin, C.M., Structure, chemical composition and enzymatic activities of pearlings and bran obtained from pearled wheat (Triticum aestivum L.) by roller milling. *J. Cereal Sci.*, 62, 66–72, 2015/03/01/, doi: https://doi.org/10.1016/j.jcs.2014.12.009.
46. Gupta, M., Abu-Ghannam, N., Gallaghar, E., Barley for brewing: Characteristic changes during malting, brewing and applications of its by-products. *Compr. Rev. Food Sci. Food Saf.*, 9, 3, 318–328, 2010.
47. Moongngarm, A., Daomukda, N., Khumpika, S., Chemical compositions, phytochemicals, and antioxidant capacity of rice bran, rice bran layer, and rice germ. *APCBEE Proc.*, 2, 73–79, 2012/01/01/, doi: https://doi.org/10.1016/j.apcbee.2012.06.014.
48. Mule, T.A., Sawant, S.S., Odaneth, A.A., Maize bran as a potential substrate for production of β-glucosidase. *Biomass Convers. Biorefin.*, 1–11, 2022.
49. Ghafoor, K. *et al.*, Nutritional composition, extraction, and utilization of wheat germ oil: A review. *Eur. J. Lipid Sci. Technol.*, 119, 7, 1600160, 2017.
50. Mussatto, S.I., Brewer's spent grain: A valuable feedstock for industrial applications. *J. Sci. Food Agric.*, 94, 7, 1264–1275, 2014.
51. Ed Nignpense, B., Francis, N., Blanchard, C., Santhakumar, A.B., Bioaccessibility and bioactivity of cereal polyphenols: A review. *Foods*, 10, 7, 1595, 2021. [Online]. Available: https://www.mdpi.com/2304-8158/10/7/1595.
52. Horvat, D. *et al.*, Phenolic acid profiles and antioxidant activity of major cereal crops. *Antioxidants*, 9, 6, 527, 2020. [Online]. Available: https://www.mdpi.com/2076-3921/9/6/527.
53. Strygina, K.V., Synthesis of flavonoid pigments in grain of representatives of poaceae: General patterns and exceptions in N.I. Vavilov's homologous series. *Russ. J. Genet.*, 56, 11, 1345–1358, 2020/11/01.
54. Dykes, L. and Rooney, L.W., Phenolic compounds in cereal grains and their health benefits. *Cereal Foods World*, 52, 105–111, 2007.
55. Zhao, X., Liang, K., Zhu, H., Carotenoids in cereals and related foodstuffs: A review of extraction and analysis methods. *Food Rev. Int.*, 1–16, 2022.

56. Panfili, G., Fratianni, A., Irano, M., Improved normal-phase high-performance liquid chromatography procedure for the determination of carotenoids in cereals. *J. Agric. Food Chem.*, 52, 21, 6373–6377, 2004.
57. Kean, E.G., Hamaker, B.R., Ferruzzi, M.G., Carotenoid bioaccessibility from whole grain and degermed maize meal products. *J. Agric. Food Chem.*, 56, 21, 9918–9926, 2008/11/12.
58. Zhu, F., Dietary fiber polysaccharides of amaranth, buckwheat and quinoa grains: A review of chemical structure, biological functions and food uses. *Carbohydr. Polym.*, 248, 116819, 2020/11/15/, doi: https://doi.org/10.1016/j.carbpol.2020.116819.
59. Biel, W. and Maciorowski, R., Evaluation of chemical composition and nutritional quality of buckwheat groat, bran and hull (fagopyrum esculentum Möench L.). *Ital. J. Food Sci.*, 25, 384–389, 01/01 2013.
60. Skrabanja, V. *et al.*, Nutrient content in buckwheat milling fractions. *Cereal Chem.*, 81, 2, 172–176, 2004, doi: https://doi.org/10.1094/CCHEM.2004.81.2.172.
61. Xiao, Y., Liu, Y., Wang, X., Li, M., Lei, H., Xu, H., Cellulose nanocrystals prepared from wheat bran: Characterization and cytotoxicity assessment. *Int. J. Biol. Macromol.*, 140, 225–233, 2019/11/01/, doi: https://doi.org/10.1016/j.ijbiomac.2019.08.160.
62. Zhang, Z., Smith, C., Li, W., Extraction and modification technology of arabinoxylans from cereal by-products: A critical review. *Food Res. Int.*, 65, 423–436, 2014/11/01/, doi: https://doi.org/10.1016/j.foodres.2014.05.068.
63. Niño-Medina, G., Carvajal-Millán, E., Rascon-Chu, A., Marquez-Escalante, J.A., Guerrero, V., Salas-Muñoz, E., Feruloylated arabinoxylans and arabinoxylan gels: Structure, sources and applications. *Phytochem. Rev.*, 9, 1, 111–120, 2010/03/01.
64. Rattan, O., Izydorczyk, M.S., Biliaderis, C.G., Structure and rheological behaviour of arabinoxylans from Canadian bread wheat flours. *LWT-Food Sci. Technol.*, 27, 6, 550–555, 1994.
65. Apprich, S. *et al.*, Wheat bran-based biorefinery 2: Valorization of products. *LWT - Food Sci. Technol.*, 56, 2, 222–231, 2014/05/01/, doi: https://doi.org/10.1016/j.lwt.2013.12.003.
66. Comino, P., Collins, H., Lahnstein, J., Beahan, C., Gidley, M.J., Characterisation of soluble and insoluble cell wall fractions from rye, wheat and hull-less barley endosperm flours. *Food Hydrocolloids*, 41, 219–226, 2014.
67. Izydorczyk, M.S. and Biliaderis, C.G., Structural and functional aspects of cereal arabinoxylans and β-glucans, in: *Developments in Food Science*, vol. 41, G. Doxastakis and V. Kiosseoglou (Eds.), pp. 361–384, Elsevier, 2000.
68. Lazaridou, A. and Biliaderis, C.G., Molecular aspects of cereal β-glucan functionality: Physical properties, technological applications and physiological effects. *J. Cereal Sci.*, 46, 2, 101–118, 2007/09/01/, doi: https://doi.org/10.1016/j.jcs.2007.05.003.

69. Cozzolino, D. and Degner, S., An overview on the role of lipids and fatty acids in barley grain and their products during beer brewing. *Food Res. Int.*, 81, 114–121, 2016/03/01/, doi: https://doi.org/10.1016/j.foodres.2016.01.003.
70. Wang, J. *et al.*, A comprehensive review of cereal germ and its lipids: Chemical composition, multi-objective process and functional application. *Food Chem.*, 362, 130066, 2021, doi: https://doi.org/10.1016/j.foodchem.2021.130066. 2021/11/15/.
71. Saldivar, X., Wang, Y.-J., Chen, P., Hou, A., Changes in chemical composition during soybean seed development. *Food Chem.*, 124, 4, 1369–1375, 2011.
72. Balandrán-Quintana, R.R., Mercado-Ruiz, J.N., Mendoza-Wilson, A.M., Wheat bran proteins: A review of their uses and potential. *Food Rev. Int.*, 31, 3, 279–293, 2015.
73. Balindong, J.L., Liu, L., Ward, R.M., Barkla, B.J., Waters, D.L., Optimisation and standardisation of extraction and HPLC analysis of rice grain protein. *J. Cereal Sci.*, 72, 124–130, 2016.
74. Collins, H.M. *et al.*, Genes that mediate starch metabolism in developing and germinated barley grain. *Front. Plant Sci.*, 12, 641325, 2021.
75. Geng, L., Li, M., Zhang, G., Ye, L., Barley: A potential cereal for producing healthy and functional foods. *Food Qual. Saf.*, 6, 1–13, 2022.
76. Zhuang, X., Zhao, C., Liu, K., Rubinelli, P.M., Ricke, S.C., Atungulu, G.G., Cereal grain fractions as potential sources of prebiotics: Current status, opportunities, and potential applications, in: *Food and Feed Safety Systems and Analysis*, pp. 173–191, Elsevier, Academic Press, Cambridge, Massachusetts, United States, 2018.
77. Mollakhalili-Meybodi, N., Arab, M., Nematollahi, A., Mousavi Khaneghah, A., Prebiotic wheat bread: Technological, sensorial and nutritional perspectives and challenges. *LWT*, 149, 111823, 2021/09/01/, doi: https://doi.org/10.1016/j.lwt.2021.111823.
78. Martinez-Soberanes, E.E., Mustafa, R., Reaney, M.J.T., Zhang, W.J., Seed hull utilization, in: *Food Wastes and By-products*, pp. 291–326, 2020.
79. Palma, M. *et al.*, Chapter 2 Extraction of natural products: Principles and fundamental aspects, in: *Natural Product Extraction: Principles and Applications*, pp. 58–88, The Royal Society of Chemistry, Royal Society of Chemistry (RSC) Publishing, Cambridge, United Kingdom, 2013.
80. Galanakis, C.M., Recovery of high added-value components from food wastes: Conventional, emerging technologies and commercialized applications. *Trends Food Sci. Technol.*, 26, 2, 68–87, 2012/08/01/, doi: https://doi.org/10.1016/j.tifs.2012.03.003.
81. Mussatto, S.I., Dragone, G., Roberto, I.C., Brewers' spent grain: Generation, characteristics and potential applications. *J. Cereal Sci.*, 43, 1, 1–14, 2006/01/01/, doi: https://doi.org/10.1016/j.jcs.2005.06.001.
82. Galitsky, C., Worrell, E., Ruth, M., Energy efficiency improvement and cost saving opportunities for the Corn Wet Milling Industry: An ENERGY STAR Guide for Energy and Plant Managers, Lawrence Berkeley National

Laboratory, 2003, [Online]. Available: https://escholarship.org/uc/item/34h4p4fr.
83. Galanakis, C., The universal recovery strategy, in: *Food Waste Recovery*, pp. 59–81, 12/31 2015.
84. Zhang, Q.-W., Lin, L.-G., Ye, W.-C., Techniques for extraction and isolation of natural products: A comprehensive review. *Chin. Med.*, 13, 1, 20, 2018/04/17.
85. Arshadi, M. *et al.*, Pre-treatment and extraction techniques for recovery of added value compounds from wastes throughout the agri-food chain. *Green Chem.*, 18, 23, 6160–6204, 2016.
86. Zannini, E., Bravo Núñez, Á., Sahin, A.W., Arendt, E.K., Arabinoxylans as functional food ingredients: A review. *Foods*, 11, 7, 1026, 2022. [Online]. Available: https://www.mdpi.com/2304-8158/11/7/1026.
87. Radoiu, M., Kaur, H., Bakowska-Barczak, A., Splinter, S., Microwave-assisted industrial scale cannabis extraction. *Technologies*, 8, 3, 45, 2020. [Online]. Available: https://www.mdpi.com/2227-7080/8/3/45.
88. Mustafa, A. and Turner, C., Pressurized liquid extraction as a green approach in food and herbal plants extraction: A review, (in eng). *Anal. Chim. Acta*, 703, 1, 8–18, Oct 3 2011.
89. Sanchai, K., Wasawat, K., Jaruwan, D., Kajornsak, F., Hydro-fractionation for biomass upgrading, in: *Renewable Resources and Biorefineries*, J.-L. Eduardo and Z. Leila Queiroz (Eds.), p. Ch. 5, IntechOpen, Rijeka, 2018.
90. Mendiola, J.A., Herrero, M., Castro-Puyana, M., Ibáñez, E., Chapter 6 Supercritical fluid extraction, in: *Natural Product Extraction: Principles and Applications*, pp. 196–230, The Royal Society of Chemistry, Cambridge, United Kingdom, 2013.
91. Moreno-González, M. and Ottens, M., A structured approach to recover valuable compounds from agri-food side streams. *Food Bioprocess Technol.*, 14, 8, 1387–1406, 2021/08/01.
92. Renger, A. and Steinhart, H., Ferulic acid dehydrodimers as structural elements in cereal dietary fibre. *Eur. Food Res. Technol.*, 211, 6, 422–428, 2000/11/01.
93. Cocero, M.J. *et al.*, Understanding biomass fractionation in subcritical & supercritical water. *J. Supercrit. Fluids*, 133, 550–565, 2018/03/01/, doi: https://doi.org/10.1016/j.supflu.2017.08.012.
94. Carvalheiro, F., Duarte, L.C., Gírio, F., Moniz, P., Chapter 14 - Hydrothermal/liquid hot water pretreatment (Autohydrolysis): A multipurpose process for biomass upgrading, in: *Biomass Fractionation Technologies for a Lignocellulosic Feedstock Based Biorefinery*, S.I. Mussatto (Ed.), pp. 315–347, Elsevier, Amsterdam, 2016.
95. Bastos, R., Coelho, E., Coimbra, M.A., 8 - Arabinoxylans from cereal by-products: Insights into structural features, recovery, and applications, in: *Sustainable Recovery and Reutilization of Cereal Processing By-Products*, C.M. Galanakis (Ed.), pp. 227–251, Woodhead Publishing, Sawston, United Kingdom, 2018.

96. Harmsen, P., Huijgen, W., Bermudez, L., Bakker, R., *Literature review of physical and chemical pretreatment processes for lignocellulosic biomass*, Wageningen UR - Food & Biobased Research, 2010.
97. Rissanen, J.V., Grénman, H., Willför, S., Murzin, D.Y., Salmi, T., Spruce hemicellulose for chemicals using aqueous extraction: Kinetics, mass transfer, and modeling. *Ind. Eng. Chem. Res.*, 53, 15, 6341–6350, 2014/04/16.
98. Cabeza, A., Piqueras, C.M., Sobrón, F., García-Serna, J., Modeling of biomass fractionation in a lab-scale biorefinery: Solubilization of hemicellulose and cellulose from holm oak wood using subcritical water. *Bioresour. Technol.*, 200, 90–102, 2016/01/01/, doi: https://doi.org/10.1016/j.biortech.2015.09.063.
99. Bender, D. *et al.*, Optimization of arabinoxylan isolation from rye bran by adapting extraction solvent and use of enzymes. *J. Food Sci.*, 82, 11, 2562–2568, 2017, doi: https://doi.org/10.1111/1750-3841.13920.
100. Li, W., Cui, S.W., Kakuda, Y., Extraction, fractionation, structural and physical characterization of wheat β-d-glucans. *Carbohydr. Polym.*, 63, 3, 408–416, 2006/03/03/, doi: https://doi.org/10.1016/j.carbpol.2005.09.025.
101. Maheshwari, G., Sowrirajan, S., Joseph, B., Extraction and isolation of β-glucan from grain sources—A review. *J. Food Sci.*, 82, 7, 1535–1545, 2017, doi: https://doi.org/10.1111/1750-3841.13765.
102. Sibakov, J., Abecassis, J., Barron, C., Poutanen, K., Electrostatic separation combined with ultra-fine grinding to produce β-glucan enriched ingredients from oat bran. *Innov. Food Sci. Emerg. Technol.*, 26, 445–455, 2014/12/01/, doi: https://doi.org/10.1016/j.ifset.2014.10.004.
103. Zhang, L. *et al.*, Arabinoxylans-enriched fractions: From dry fractionation of wheat bran to the investigation on bread baking performance. *J. Cereal Sci.*, 87, 1–8, 2019/05/01/, doi: https://doi.org/10.1016/j.jcs.2019.02.005.
104. Wang, J., Sun, B., Cao, Y., Tian, Y., Li, X., Optimisation of ultrasound-assisted extraction of phenolic compounds from wheat bran. *Food Chem.*, 106, 2, 804–810, 2008/01/15/, doi: https://doi.org/10.1016/j.foodchem.2007.06.062.
105. Das, A.B., Goud, V.V., Das, C., Extraction of phenolic compounds and anthocyanin from black and purple rice bran (Oryza sativa L.) using ultrasound: A comparative analysis and phytochemical profiling. *Ind. Crops Prod.*, 95, 332–341, 2017/01/01/, doi: https://doi.org/10.1016/j.indcrop.2016.10.041.
106. Alonso-Riaño, P., Sanz Diez, M.T., Blanco, B., Beltrán, S., Trigueros, E., Benito-Román, O., Water ultrasound-assisted extraction of polyphenol compounds from brewer's spent grain: Kinetic study, extract characterization, and concentration. *Antioxidants*, 9, 3, 265, 2020. [Online]. Available: https://www.mdpi.com/2076-3921/9/3/265.
107. Rebolleda, S., González-San José, M.L., Sanz, M.T., Beltrán, S., Solaesa, Á. G., Bioactive compounds of a wheat bran oily extract obtained with supercritical carbon dioxide. *Foods*, 9, 625, 2020. [Online]. Available: https://www.mdpi.com/2304-8158/9/5/625.
108. Benito-Román, O., Varona, S., Sanz, M.T., Beltrán, S., Valorization of rice bran: Modified supercritical CO2 extraction of bioactive compounds.

J. Ind. Eng. Chem., 80, 273–282, 2019/12/25/, doi: https://doi.org/10.1016/j.jiec.2019.08.005.

109. Macias-Garbett, R., Serna-Hernández, S.O., Sosa-Hernández, J.E., Parra-Saldívar, R., Phenolic compounds from brewer's spent grains: Toward green recovery methods and applications in the cosmetic industry, (in English). *Front. Sustain. Food Syst., Mini Rev.*, 5, 1–10, 2021-June-14.
110. Chan, K.W., Khong, N.M.H., Iqbal, S., Ismail, M., Isolation and antioxidative properties of phenolics-saponins rich fraction from defatted rice bran. *J. Cereal Sci.*, 57, 3, 480–485, 2013/05/01/, doi: https://doi.org/10.1016/j.jcs.2013.02.002.
111. Pazo-Cepeda, V., Benito-Román, Ó., Navarrete, A., Alonso, E., Valorization of wheat bran: Ferulic acid recovery using pressurized aqueous ethanol solutions. *Waste Biomass Valorization*, 11, 9, 4701–4710, 2020/09/01.
112. Buranov, A.U. and Mazza, G., Extraction and purification of ferulic acid from flax shives, wheat and corn bran by alkaline hydrolysis and pressurised solvents. *Food Chem.*, 115, 4, 1542–1548, 2009/08/15/, doi: https://doi.org/10.1016/j.foodchem.2009.01.059.
113. Liu, L. *et al.*, Effect of steam explosion-assisted extraction on phenolic acid profiles and antioxidant properties of wheat bran. *J. Sci. Food Agric.*, 96, 10, 3484–3491, 2016, doi: https://doi.org/10.1002/jsfa.7532.
114. Castro-Muñoz, R., Barragán-Huerta, B.E., Yáñez-Fernández, J., The use of nixtamalization wastewaters clarified by ultrafiltration for production of a fraction rich in phenolic compounds. *Waste Biomass Valorization*, 7, 5, 167–1176, 2016.
115. Zheng, Y., Gao, N., Wu, J., Yin, B., Chapter 11 - Rice bran protein: Extraction, nutraceutical properties, and potential applications, in: *Rice Bran and Rice Bran Oil*, L.-Z. Cheong and X. Xu (Eds.), pp. 271–293, AOCS Press, Urbana, Illinois, 2019.
116. Zhang, H.J., Zhang, H., Wang, L., Guo, X.N., Preparation and functional properties of rice bran proteins from heat-stabilized defatted rice bran. *Food Res. Int.*, 47, 2, 359–363, 2012/07/01/, doi: https://doi.org/10.1016/j.foodres.2011.08.014.
117. Gadalkar, S.M., Gogate, P.R., Rathod, V.K., Recovery of proteins from rice mill industry waste (rice bran) using alkaline or NaCl-assisted alkaline extraction processes. *J. Food Process Eng.*, 40, 3, e12430, 2017, doi: https://doi.org/10.1111/jfpe.12430.
118. Fabian, C. and Ju, Y.-H., A review on rice bran protein: Its properties and extraction methods. *Crit. Rev. Food Sci. Nutr.*, 51, 9, 816–827, 2011/10/01 2011.
119. Phongthai, S., Lim, S.-T., Rawdkuen, S., Ultrasonic-assisted extraction of rice bran protein using response surface methodology. *J. Food Biochem.*, 41, 2, e12314, 2017, doi: https://doi.org/10.1111/jfbc.12314.

120. Phongthai, S., Lim, S.-T., Rawdkuen, S., Optimization of microwave-assisted extraction of rice bran protein and its hydrolysates properties. *J. Cereal Sci.*, 70, 146–154, 2016/07/01/, doi: https://doi.org/10.1016/j.jcs.2016.06.001.
121. Idris, W.H., Babiker, E.E., El Tinay, A.H., Fractionation, solubility and functional properties of wheat bran proteins as influenced by pH and/or salt concentration. *Food / Nahrung*, 47, 6, 425–429, 2003, doi: https://doi.org/10.1002/food.200390094.
122. Tang, D.-S. *et al.*, Recovery of protein from brewer's spent grain by ultrafiltration. *Biochem. Eng. J.*, 48, 1, 1–5, 2009/12/15/, doi: https://doi.org/10.1016/j.bej.2009.05.019.
123. Lizardi-Jiménez, M.A. and Hernández-Martínez, R., Solid state fermentation (SSF): Diversity of applications to valorize waste and biomass. *3 Biotech.*, 7, 1, 44, 2017/04/25.
124. Özdemir, S., Matpan, F., Okumus, V., Dündar, A., Ulutas, M.S., Kumru, M., Isolation of a thermophilic Anoxybacillus flavithermus sp. nov. and production of thermostable α-amylase under solid-state fermentation (SSF). *Ann. Microbiol.*, 62, 4, 1367–1375, 2012/12/01.
125. Abbasi, H., Mortazavipour, S.R., Setudeh, M., Polygalacturonase (PG) production by fungal strains using agro-industrial bioproduct in solid state fermentation. *Chem. Eng. Res. Bull.*, 15, 1, 1–5, 08/08 2011.
126. Saba, I., Mukhtar, H., Haq, I., Production of alkaline protease by Bacillus subtilis using solid state fermentation. *Afr. J. Microbiol. Res.*, 7, 1558–1568, 04/16 2013.
127. McKinney, K., Combs, J., Becker, P., Humphries, A., Filer, K., Vriesekoop, F., Optimization of phytase production from escherichia coli by altering solid-state fermentation conditions. *Fermentation*, 1, 1, 13–23, 2015. [Online]. Available: https://www.mdpi.com/2311-5637/1/1/13.
128. Saratale, G.D. *et al.*, Cellulolytic enzymes production by utilizing agricultural wastes under solid state fermentation and its application for biohydrogen production, (in eng). *Appl. Biochem. Biotechnol.*, 174, 8, 2801–17, Dec 2014.
129. Yang, S.-S. and Ling, M.-Y., Tetracycline production with sweet potato residue by solid state fermentation. *Biotechnol. Bioeng.*, 33, 8, 1021–1028, 1989, doi: https://doi.org/10.1002/bit.260330811.
130. Adinarayana, K., Ellaiah, P., Srinivasulu, B., Bhavani Devi, R., Adinarayana, G., Response surface methodological approach to optimize the nutritional parameters for neomycin production by Streptomyces marinensis under solid-state fermentation. *Process Biochem.*, 38, 11, 1565–1572, 2003/06/30/, doi: https://doi.org/10.1016/S0032-9592(03)00057-8.
131. El-Naggar, M., El-Aassar, S.A., Abdul-Gawad, S., Solid-state fermentation for the production of meroparamycin by streptomyces sp. strain MAR01. *J. Microbiol. Biotechnol.*, 19, 468–73, 06/01 2009, doi: 10.4014/jmb.0807.457.
132. Reddy, D.S.R., Latha, D.P., Latha, K., Production of lovastatin by solid state fermentation by Penicillium funiculosum NCIM 1174. *Drug Invent. Today*, 3, 6, 75–77, 2011.

133. Naveena, B.J., Altaf, M., Bhadriah, K., Reddy, G., Selection of medium components by Plackett–Burman design for production of l(+) lactic acid by Lactobacillus amylophilus GV6 in SSF using wheat bran. *Bioresour. Technol.*, 96, 4, 485–490, 2005, doi: https://doi.org/10.1016/j.biortech.2004.05.020.
134. Wildman, R.E.C. and Bruno, R.S., *Handbook of nutraceuticals and functional foods*, CRC Press, Boca Raton, 2019.
135. Gulati, O.P. and Berry Ottaway, P., Legislation relating to nutraceuticals in the European Union with a particular focus on botanical-sourced products, (in eng). *Toxicology*, 221, 1, 75–87, Apr 3 2006.
136. FAO, World food situation, 2022, [Online]. Available: http://www.fao.org/worldfoodsituation/csdb/en.
137. FAO, Global food losses and food waste – extent, causes and prevention, Rome, 2011.
138. Ravindran, R. and Jaiswal, A.K., Exploitation of food industry waste for high-value products, (in eng). *Trends Biotechnol.*, 34, 1, 58–69, Jan 2016.
139. Balandrán-Quintana, R.R., 5 - Recovery of proteins from cereal processing by-products, in: *Sustainable Recovery and Reutilization of Cereal Processing By-Products*, C.M. Galanakis (Ed.), pp. 125–157, Woodhead Publishing, Sawston, United Kingdom, 2018.
140. Chandrasekara, A., Naczk, M., Shahidi, F., Effect of processing on the antioxidant activity of millet grains. *Food Chem.*, 133, 1, 1–9, 2012/07/01/, doi: https://doi.org/10.1016/j.foodchem.2011.09.043.
141. Mullen, A.M., Álvarez, C., Pojić, M., Hadnadev, T.D., Papageorgiou, M., Chapter 2 - Classification and target compounds, in: *Food Waste Recovery*, C.M. Galanakis (Ed.), pp. 25–57, Academic Press, San Diego, 2015.
142. Delcour, J.A. and Hoseney, R.C., *Principles of cereal science and technology*, AACC International, St Paul, MN, USA, 2010.
143. Anson, N.M., Hemery, Y.M., Bast, A., Haenen, G.R.M.M., Optimizing the bioactive potential of wheat bran by processing. *Food Funct.*, 3, 4, 362–375, 2012.
144. Zhu, K.-X., Zhou, H.-M., Qian, H.-F., Proteins extracted from defatted wheat germ: Nutritional and structural properties. *Cereal Chem.*, 83, 1, 69–75, 2006, doi: https://doi.org/10.1094/CC-83-0069.
145. Liukkonen, K.-H. *et al.*, Process-induced changes on bioactive compounds in whole grain rye. *Proc. Nutr. Soc.*, 62, 1, 117–122, 2003.
146. Patel, S., Cereal bran: The next super food with significant antioxidant and anticancer potential. *Mediterr. J. Nutr. Metab.*, 5, 2, 91–104, 2012/08/01.
147. Ross, A.B., Kamal-Eldin, A., Åman, P., Dietary alkylresorcinols: Absorption, bioactivities, and possible use as biomarkers of whole-grain wheat- and rye-rich foods. *Nutr. Rev.*, 62, 3, 81–95, 2004, doi: https://doi.org/10.1111/j.1753-4887.2004.tb00029.x.
148. de Vasconcelos, M.C.B.M., Bennett, R., Castro, C., Cardoso, P., Saavedra, M.J., Rosa, E.A., Study of composition, stabilization and processing of wheat

germ and maize industrial by-products. *Ind. Crops Prod.*, 42, 292–298, 2013/03/01/, doi: https://doi.org/10.1016/j.indcrop.2012.06.007.
149. ElMekawy, A., Diels, L., De Wever, H., Pant, D., Valorization of cereal based biorefinery by-products: Reality and expectations. *Environ. Sci. Technol.*, 47, 16, 9014–9027, 2013/08/20.
150. McCarthy, A.L. *et al.*, The hydroxycinnamic acid content of barley and brewers' spent grain (BSG) and the potential to incorporate phenolic extracts of BSG as antioxidants into fruit beverages. *Food Chem.*, 141, 3, 2567–2574, 2013/12/01/, doi: https://doi.org/10.1016/j.foodchem.2013.05.048.
151. Kitrytė, V., Šaduikis, A., Venskutonis, P.R., Assessment of antioxidant capacity of brewer's spent grain and its supercritical carbon dioxide extract as sources of valuable dietary ingredients. *J. Food Eng.*, 167, 18–24, 2015/12/01/, doi: https://doi.org/10.1016/j.jfoodeng.2014.12.005.
152. Izydorczyk, M.S. and Dexter, J.E., Barley β-glucans and arabinoxylans: Molecular structure, physicochemical properties, and uses in food products–a review. *Food Res. Int.*, 41, 9, 850–868, 2008/11/01/, doi: https://doi.org/10.1016/j.foodres.2008.04.001.
153. van Nierop, S.N., Axcell, B.C., Cantrell, I.C., Rautenbach, M., Quality assessment of lager brewery yeast samples and strains using barley malt extracts with anti-yeast activity," (in eng). *Food Microbiol.*, 26, 2, 192–6, Apr 2009.
154. Prückler, M. *et al.*, Wheat bran-based biorefinery 1: Composition of wheat bran and strategies of functionalization. *LWT - Food Sci. Technol.*, 56, 2, 211–221, 2014/05/01/, doi: https://doi.org/10.1016/j.lwt.2013.12.004.
155. Mudgil, D. and Barak, S., Composition, properties and health benefits of indigestible carbohydrate polymers as dietary fiber: A review, (in eng). *Int. J. Biol. Macromol.*, 61, 1–6, Oct 2013.
156. Rastall, R.A., Functional oligosaccharides: Application and manufacture, (in eng). *Annu. Rev. Food Sci. Technol.*, 1, 305–39, 2010.
157. Prakasham, R.S., Rao, R.S., Hobbs, P.J., Current trends in biotechnological production of xylitol and future prospects. *Curr. Trends Biotechnol. Pharm.*, 3, 8–36, 2009.
158. Burton, R.A. and Fincher, G.B., Evolution and development of cell walls in cereal grains, (in English). *Front. Plant Sci., Rev.*, 5, 1–15, 2014-September-11.
159. Patsioura, A., Galanakis, C.M., Gekas, V., Ultrafiltration optimization for the recovery of β-glucan from oat mill waste. *J. Membr. Sci.*, 373, 1, 53–63, 2011/05/01/, doi: https://doi.org/10.1016/j.memsci.2011.02.032.
160. Lee, Y.T., Puligundla, P., Schwarz, P.B., Molecular weight, solubility and viscosity of β-Glucan preparations from barley pearling by-products. *Sains Malays.*, 46, 713–718, 2017.
161. Pihlava, J.-M., Nordlund, E., Heiniö, R.-L., Hietaniemi, V., Lehtinen, P., Poutanen, K., Phenolic compounds in wholegrain rye and its fractions. *J. Food Compos. Anal.*, 38, 89–97, 2015/03/01/, doi: https://doi.org/10.1016/j.jfca.2014.10.004.

162. Fares, C., Platani, C., Baiano, A., Menga, V., Effect of processing and cooking on phenolic acid profile and antioxidant capacity of durum wheat pasta enriched with debranning fractions of wheat. *Food Chem.*, 119, 3, 1023–1029, 2010/04/01/, doi: https://doi.org/10.1016/j.foodchem.2009.08.006.
163. Gao, L., Wang, S.D., Oomah, B.D.D., Mazza, G., Wheat quality: Antioxidant activity of wheat millstreams, in: *Wheat Quality Elucidation*, P. Ng and C.W. Wrigley (Eds.), pp. 219–231, AACC International, St. Paul, 2002.
164. Giordano, D. *et al.*, Bioactive compound and antioxidant activity distribution in roller-milled and pearled fractions of conventional and pigmented wheat varieties, (in eng). *Food Chem.*, 233, 483–491, Oct 15 2017.
165. Tan, B.L. and Norhaizan, M.E., Scientific evidence of rice by-products for cancer prevention: Chemopreventive properties of waste products from rice milling on carcinogenesis in *vitro* and in *vivo*, (in eng). *Biomed. Res. Int.*, 2017, 9017902, 2017.
166. Smuda, S.S., Mohsen, S.M., Olsen, K., Aly, M.H., Bioactive compounds and antioxidant activities of some cereal milling by-products. *J. Food Sci. Technol.*, 55, 3, 1134–1142, 2018/03/01.
167. Fărcaş, A.C., Socaci, S.A., Dulf, F.V., Tofană, M., Mudura, E., Diaconeasa, Z., Volatile profile, fatty acids composition and total phenolics content of brewers' spent grain by-product with potential use in the development of new functional foods. *J. Cereal Sci.*, 64, 34–42, 2015/07/01/, doi: https://doi.org/10.1016/j.jcs.2015.04.003.
168. Kim, Y. *et al.*, Composition of corn dry-grind ethanol by-products: DDGS, wet cake, and thin stillage. *Bioresour. Technol.*, 99, 12, 5165–5176, 2008/08/01/, doi: https://doi.org/10.1016/j.biortech.2007.09.028.
169. Galanakis, C., *Concluding remarks and future perspectives*, pp. 319–327, Springer Cham, Cham, Switzerland, 2018.

7

Mustard By-Products: A Promising Sustainable Natural Source of Bioactive Compounds

Salima Acherar[1], Hayate Haddadi-Guemghar[1,2]*, Khodir Madani[1,3] and Lila Boulekbache-Makhlouf[1]

[1]Laboratory of Biomathematics, Biochemistry, Biophysics and Scientometrics, Abderrahmane Mira University of Bejaia, Bejaia, Algeria
[2]Department of Food Sciences, Faculty of Natural Sciences, University of Bejaia, Bejaia, Tharga Ouzemour Road, Algeria
[3]National Agri-Food Technology Research Center, Bejaia, Algeria

Abstract

During the few recent decades, consumers food behavior has been changed, they focus more on their health, so they pay more attention for natural products, which are rich in biomolecules that we can find in all plant parts. Following the recent food trends, sustainable agro-industrial food valorization has gained attention of scientists and activists around the world. Furthermore, they could be a raw material for good production thanks to its high potential content of bioactive compounds. Mustard is considered as a low-cost natural product; however, it is rich in nutrients especially the seed part. Mustard can be processed as a condiment preparation or for oil production. After processing of mustard an important quantities of bran seed reaching up to 60% (w/w) are generated. The concept of management of such by-products requires the employment of innovative and eco-friendly extraction methods each carrying their own advantages and disadvantages, these technics include microwaves, ultrasounds, pulsed electric extraction, and membrane-based processes in the purpose of recovering the most content of biocompounds existing in these residues and which of these methods serve more efficient. In this present chapter, the possibilities of optimizing high value biologically active compounds will be discussed also we will have an outlook of future implication of agro-industrial mustard waste in other sectors like pharmaceutical, cosmetic or as a source of alternative energy; this discussion will based on the analysis of the different studies and reviews have been realized in relation with this topic.

**Corresponding author*: hayate.haddadi@univ-bejaia.dz

Umile Gianfranco Spizzirri (ed.) Nutraceutics from Agri-Food By-Products, (223–244) © 2023 Scrivener Publishing LLC

***Keywords*:** Mustard seed, valorization, by-products, bioactive compounds, extraction methods, protein, glucosinolates, fibers

7.1 Introduction

Actually, researchers and industrials realized that chemical and synthetic ingredients, which were useful in the past in several applications, become no longer aligned with the new concepts of green chemistry. Global interest is heading towards natural biocompounds derived from plants and vegetables. Natural sources can confer the same properties that can provide synthetic ones [1]. Moreover, with less effect to the environment and human health.

In addition, the fast growing of population implicates a huge consumption of industrial food products which their process generates every year large amounts of by-products that are considered as unusable due to their low value importance or sometimes are used for animal feed. However, these residues conceal in their composition valuable bioactive compounds with, antioxidant, antibacterial and nutraceutical effects which have attracted the attention of scientists around the world to investigate them [2]. The big deal with the use of these agri-food wastes and their application residues is founding a sustainable strategy, which can minimize cost and maximize the value of these invaluable by-products.

To reach such a sustainable management of food waste, the use of green and novel extraction techniques is recommended which use of green and less harmful solvents [3], also implicating the use of less energy [4]. These natural substances may provide a source of preservatives that could be used to extend food shelf life [5].

This chapter will provide informations related to mustard seed, composition, by-products generated after seed processing, properties of the biocompounds inside these residues and we will talk about the emerging and novel techniques used through the literature for the recovery of the biologically active compounds. Finally, we will talk about the different ways that mustard waste products were valorized in various fields through the different previous studies realized within thie concept of valorization.

7.2 Mustard

Mustard belongs to *Cruciferae* family which includes canola, cabbage, rapeseed, horseradish, broccoli, kale, Brussels sprouts and cauliflower which are known by their economic importance [6]. There are three varieties of mustard that human usually use black mustard (*Brassica nigra*), yellow or white mustard (*Sinapis alba*), and brown mustard (*Brassica junceae*). The seed is characterized with round shape of a small size [7]. Mustard seed was described by the literature as a higher versatile food due to its composition on biologically active compounds proteins [8], phenolic compounds [9], glucosinolates [10], polyunsaturated fatty acids [11], fibers [12], and hydrocolloids [13].

Mustard whole seed is used commonly as a condiment preparation or extracted to produce oilseed [14]. Mustard bran produced is the by-product of mustard condiment processing, but actually it becomes valuable due to its emulsifying and stabilizing properties [15]. Mustard bran consists generally of seed coats and other seed fractions like cotyledons.

Mustard bran fraction is rich in dietary fibers 15% while the oil and protein fractions are lower 7% and 13% to 16%, respectively [15]. This composition varies according to the type of the treatments used [16]. The specific property of mustard bran is his ability to hold up to nine times his weight in water due to the mucilage contained in the coat [17].

The other fraction obtained from the oil extraction of mustard seed is the oilseed cake or meal, it represents the other by-product of mustard, it is generated during mustard oil extraction, the meal makes about 60% of the seed [16].

This residue contains principally proteins which their composition varies from 33-40%, and 80-83% represents true protein with an appreciable proportion of albumine, glutelin, and globulin [18, 19].

The composition of mustard cake on bioactive compounds includes phenolics with 77-81 mg/kg, glucosinolates and phytates. phenolics can be present both in free and esterified forms [16, 19]. Mustard cake fraction is also containing carbohydrates, lipids with 21%, crude fibers with 8.5% and 8% of ashes [20].

7.2.1 Mustard Bioactive Compounds Composition and their Properties

Mustard composition is divided on primary metabolites like proteins, lipids and hydrocolloids and secondary metabolites such as phenolic acids nd glucosinolates as principal components.

7.2.2 Primary Metabolites

7.2.2.1 Proteins

Mustard protein composition and functional properties varies within the spice and the plant family [21, 22]. The high nutraceutical quality of mustard proteins is due to their composition on lysine with a strong amino acid profile with an acceptable amounts of methionine and cysteine which make it a great substituent to most cereals and oilseed proteins [23]. However, the reason which explain the non-use of mustard proteins in human feed is the presence of anti-nutritional components, like glucosinolates, phenolics, and phytates [24]. The reduction of glucosinolates with some method before it could become a food ingredient was proposed by the same author [24].

7.2.2.2 Hydrocolloids

Hydrocolloids, gums or mucilages are the different appellations for the ingredients that used for the enhancement of food structure. Thus, these water-soluble polymers improve viscosity and gel formations in food products as reported by Dickinson [25].

Mucilage as a linear hydrocolloid located in the epidermal layer of mustard has the ability to catch both water and fat in the same time due the unique capacity of mucilage to reduce the interfacial tension in water-oil and water-air emulsions [26, 27].

Mucilage is also composed of 84% pectin polysaccharides which provide the formation of gels in the extracellular matrix, and it is composed of 16% of non-pectin polysaccharides, which form gels when exposed to heat, acid, and alkaline [28, 29].

Yellow mustard bran contains about 5.5% of mucilage compared with other varieties of mustard like brown or oriental mustard [30]. It was reported that mustard mucilage protects emerging seedling during germination and imbibition as a biological function [31].

7.2.2.3 Fatty Acids

Mustard is known as an oilseed crop, the seed contains high levels of different fatty acids, and their composition varies within the spice [32]. Saturated fatty acids are lower in mustard seed while the non-saturated ones are in high levels, which make the oil extracted healthy with a nutritional quality [33]. Among the most essential mustard seed fatty acids are the omega-3 ones which contain more than one carbon double bound in their structure [34].

Mustard lipids especially polyunsaturated and monounsaturated fatty acids can reduce the risk of cardiovascular diseases [35].

7.2.3 Secondary Metabolites

7.2.3.1 Phenolic Compounds

Phenolic compounds are natural compounds which contain at least one aromatic ring in their chemical structure and a number of different hydroxyl substituents [36]. Phenolic function in plant is the protection from soil pathogens or insects [37].

The composition of mustard seed of the three varieties known in phenolic compounds has been studied through several studies [38–40]. The results obtained revealed that mustard seed contains large amounts of phenolics. Thus, involve that mustard seed is rich in natural antioxidants that gained the attention of scientists around the world during the recent years.

Phenolic existing in mustard seed is represented by two principal components, p-hydroxybenzoic acid and sinapic acid, which exists both in free and soluble bond forms, in addition to hydrocinnamic, hydroxycinnamolyquinic, p-coumaric, and ferulic acid [40, 41]. Sinapic acid constitutes over 85.4% and 72.6% in free and esterified forms respectively of the phenolic acid fraction [42].

Antioxidant effect of phenolic compounds of mustard was carried out by a team of researchers (Amarozwics *et al.* [38]) and the results obtained showed that mustard phenolics have a high antioxidant activity. Antioxidant activity of mustard seed could be affected by several parameters like processing method and variety thus, the radical scavenging activity in black mustard seed which ranged from 5.9 to 16.1 mg GAE/g DW was higher than mustard phenolic radical scavenging in brown mustard seed which ranged from 8.1 to 13.8 mg GAE/g DW as reported by the authors [43].

Mustard phenolics are also characterized by a high antimicrobial power as reported by several studies [5, 44] a study performed by Khatib and Al-Makky showed that a methanolic extract of white mustard has an antibacterial effect on *Staphylococcus aureus* bacteria type [44].

7.2.3.2 Glucosinolates

Cruciferous family are characterized by the presence of glucosinolates, these compounds give to mustard his pungent special odour, principally during the growth stage for the reason of protection from herbivores [45].

Glucosinolates are known as mustard oil glucosides [46] and three compartments build up glucosinolates composition: thioglucose, thiohydroxymate-o-sulfonate, and an α-amino acid [34].

Most of the hydrophilic glucosinolates are chemically and thermally stable; however, they can be converted into diverse breakdown products in presence of water; β-thioglucosidases known as myrosinases enzymes into isothiocyanates and thiocyanates or nitriles [47, 48].

Glucosinolates profile varies within the variety, thus, the main glucosinolate in brown mustard is sinigrin, while the main glucosinolate in white mustard is sinalbin as reported by Cools and Terry [49].

The anti-carcinogen effect of glucosinolates breakdown products of mustard seed have been proven through several studies, thus, mustard glucosinolates derivates, such as allylisothiocyanates, could inhibit cancer cell proliferation by increasing detoxification activity of carcinogens as reported by Razis *et al.* [50] and Ahmed *et al.* toward human lung cancer cells [51].

7.3 Bioactive Compounds Recovery from Mustard Waste Products

Biobased compounds have become more attractive and that thanks to their biological activity especially scavenging properties that characterize these compounds also their anti-inflammatory, anti-age, and anti-*uv* properties [52].

Several studies have reported and analyzed the content of mustard on phenolic compounds in mustard by-products, thus the capacity of sinapate and antioxidant power of canola and mustard seed hulls and cotyledons using ultrasound-assisted extraction and HPLC-DAD for the identification of the targeted compounds have been studied [53]. It has been demonstrated that the content of sinapates was higher in mustard seed fractions while there was a small amount of sinapic acid and sinigrin with a significant variation between the two seed fractions. The authors mentioned also that the use of HPLC-DAD system provides an efficient identification of phenolic compounds.

Reungoat *et al.* [54] optimized the extraction of sinapine from mustard bran fraction using an ethanol/water mixed solvent. The authors used HPLC-DAD for the quantification instead of using Folin-Ciocalteu assay. The results shown revealed that the highest yield was 8.8±0.1 mg/g (of dry matter) of sinapine during an extraction for 30 minutes, which carried out at 75°C with the use of 70% ethanol.

Mahoonak and Swamylingappa [55] studied the effects of different treatments on mustard cake protein recovery by the use of activated carbon, heat coagulation and alkalase enzyme modification. The results showed a higher yield for the three treatments. However, alkalase enzyme modification treatment seems that it enhances protein functional properties like solubility.

Engels *et al.* [56] investigated antioxidant and antibacterial activity of sinapic acid and its derivates derived from defatted mustard meal using ultrasound for the extraction of the targeted compounds and UHPLC-DAD-ESI-MS for their identification. The antibacterial results showed that sinapic acid inhibits *Escherichia coli* and salmonella bacteria with higher activity also this bioactive component it was selective toward good bacteria like Lactobacillus, which showed a resistance toward sinapic acid inhibition power.

Hydrocolloid properties of mustard were evaluated through a comparative study performed by Wu *et al.* [57] based on the water solubility property of yellow mustard mucilage compared to two other commercial polysaccharides gum Arabic and citrus pectin, yellow mustard mucilage showed a higher surface activity compared to the two other gums studied.

Coagulant power of mustard residues was studied by Ram and Raw [58], in this study, mustard seed powder and cake used to coagulate municipal wastewater derived from sanitary system of commercial housing, industrials and institutions, pH, coagulant dose were studied with the use of digital flocculator apparatus. Results obtained showed that mustard cake and powder were able to coagulate wastewater with a turbidity reduction of 23% and 25% with the following optimum conditions a dose of 25 g for 300 ml and pH of 8 and 6.5.

Lett *et al.* [17] induced 5 g of yellow mustard bran into potato and leek soup in order to study the effect of mustard bran of glycaemic response and satiety in a category of 10 males in a good health, moderated and they do not smoke. Results obtained revealed that after the addiction of YMB, the post-prandial glycemic response was reduced, and the satiety was increased.

7.4 Emerging Methods Used for the Extraction of Biobased Elements from Mustard By-Products

Sample preparation cannot be realized without being applying a good and efficient extraction method, thus choosing extraction method is considered as the most crucial step that we must care of it [59, 60], and it should be less harmful for the environment and the economy [61].

Recovery of bioactive compounds from biowastes depends generally on the source, functionality, chemical properties, and destiny of these compounds [62]. Chadni *et al.* [63] have also studied the effect of pH on the of extraction efficiency of phenolic compounds, which the results obtained showed that the higher yield of sinapic acid was obtained at basic pH of 12 with an ethanol proportion under 50%.

The extraction of bioactive compounds from plants materials is difficult because these substances are surrounded by rigid and insoluble structures like vacuoles, bilayers, lignins, or cell walls [64] thus, their extraction implicates intensive and powerful methods and material to release and extract the biologically active compounds inside these natural materials.

7.4.1 Ultrasound-Assisted Extraction

Sonication process involves the use of energy to move particles around a sample in an aqueous solution, above a sonication of 20 kHz, it will be considered as an ultra-sonication [65].

Ultrasound-assisted extraction (UAE) is classified as a cell disruption technique, this principle function is based on the phenomena of acoustic cavitation, thus the penetration of the reagent into the sample which increases the surface contact between the solvent and the sample matrix is assisted due to the ultrasonic waves propagation [66, 67]. However, Chemat *et al.* [68] have mentioned that solvent physical parameters affect cavitation of ultrasound, such as vapour pressure, surface tension, and viscosity. UAE is versatile; it can be used both on small and large scales [67].

UAE can increase cell wall destruction, cause leakage of cellular material, enhance the penetration of solvent, which facilitate hydration and swelling, also it can improve mass transfer [69, 70].

Within the framework of the use of eco-friendly extractions methods, some studies have investigated the combination of both hydroethanolic extractions with physical treatments like ultrasound to ensure an efficient recovery of phenolic compounds in mustard waste [53, 71–74].

The use of an innovative technology as high intensity ultrasound-assisted extraction can lead to a similar sinapic acid yield 7.83 ± 0.07 and 8.81 ± 0.17 mg, respectively, compared to a conventional hot ethanolic extraction method using water and methanol V/V at 80°C. More of that, this modern technique can be realized at low temperatures (40°C) and during less time also this kind of emerging techniques conserve the properties of the antioxidants and avoid their degradation reported as by Mayengbam *et al.* [53].

The efficiency of ultrasound-assisted extraction is related to the physical effects of cavitation as reported by Wang *et al.* [75].

Ultrasound-assisted extraction was used also in the study performed by Nadeem *et al.* [74], which aimed to study the nematicidal effect of defatted mustard meal on root-knot nematode, *Meloidogyne javania*; in this study, UAE showed a great effectiveness for the extraction of phenolic compounds from mustard meal fraction with a higher proportion for sinapine and gluconapin, respectively.

An extraction of glucosinolates mainly sinigrin compound from mustard seed was optimized using ultrasound-assisted extraction and response surface methodology, ethanol concentration, extraction time, and temperature were studied; Results showed that with an ethanol concentration of 70% and at a temperature of 80°C during 60 minutes, the extraction yield obtained is 70.67% [76].

Another study was performed in order to optimize the extraction of proteins from mustard meal fraction, defatted mustard used and the optimum conditions were studied are pH, particle size, extraction amplitude, and extraction time. Results showed that the highest yield obtained for proteins isolates was 44.87% at a pH of 11, particle size was 375μm, an extraction amplitude of 90% during 10 minutes. The authors have mentioned also that UAE conserve mustard protein properties [77].

7.4.2 Mechanical Cell Disruption Technique by High-Pressure Homogenization

High-pressure homogenization (HPH) technique is the commonly way used to disrupt plant cells, based on turbulence formation due to the high pressure applied [78, 79]. Pressure is used in this method for the reason of protection of food properties [80].

A noticeable characteristic of mechanical high pressure homogenization cell disruption is scalability, this propriety allows to HPH to be used for both huge and small volumes, also without any interruption [81].

To achieve a maximal recovery of bioactive compounds enclosed in mustard bran cells Donsi and Velikov [82] used an emerging technique based on the application of a shear mixing at first followed by a high-pressure homogenization of mustard bran material at different pressure levels (50-150) and varying the number of passes (1-10) and homogenizing systems. The results obtained showed that mechanical cell disruption technique improve the release of protein from mustard bran cells, which allow an enhancement of protein digestibility, also this eco-friendly method conserves rheological properties of mustard as reported by the authors [82].

7.4.3 Microwave-Assisted Extraction Technique

Microwaves extraction method consists on the use of microwaves energy ranges from 300 kHz to 300 MHz for heating polar solvents which are in direct contact with the sample [83, 84]. MAE's main advantage is rapidity; it could be realized at less than 30 minutes [85], also the specificity of heating from inside to outside compared to the other heating extraction methods [86].

The another factor that influences microwave extraction yield is viscosity of the matrix, higher viscosity refrains sample molecules to be aligned with the microwave field [87].

Ronak [88] studied the effect of microwave-assisted extraction on the phenolic profile of yellow mustard bran, the results showed that MAE conserve sinapine content of mustard bran extracted at a power of 300 W during 20 minutes using 70% of methanol.

Omirou *et al.* [86] used MAE for the extraction of glucosinolates from brassica seed spices (*Eruca sativa*), the results showed that MAE higher efficiency applied at a temperature of 80°C during 10 minutes with a power of 250 W, which allowed the extraction of 90% of glucosinolates.

7.4.4 Enzyme-Assisted Extraction

Extraction using enzymes consists the use of specific enzymes to disrupt and hydrolyze cell walls compounds and allowing to the targeted compounds to become more accessible for their extraction [89].

Reungoat *et al.* [90] used enzyme-assisted extraction for the recovery of sinapic acid from mustard bran through two preliminary extraction phases; thus, an aqueous extraction was applied at first followed by an enzymatic hydrolysis carried out using commercial enzymatic cocktails of cellulases. The results showed that EAE was efficient in the extraction of sinapic acid from mustard bran.

Another study performed by Tabtabaei [91] in order to recover proteins and oil from deshelled mustard flour using nonenzymatic and enzymatic-assisted extractions methods, results showed that this technique provides a high oil yield especially when it has been associated with alkaline treatment.

7.4.5 Pulsed Electric Field-Assisted Extraction

Pulsed electric field is an eco-extraction technique used for the biologically active compounds. PEF is based on the application of number of electric field pulses ranging from 100–300 V/cm to 20–80 V/cm in a short duration (nanoseconds to milliseconds, which create a such permeability of cell membranes). Each increase of the number of pulses leads to an increase of inactivation activity of microorganisms [92–94]. Using low temperatures is not the only advantage of this method. However, it could reduce solvent proportion which will serve the economy [90]. Also, PEF can preserve food rheological and physical properties as reported by Picart and Cheftel [95].

Pulsed electric field extraction used by Zhang *et al.* [96] in order to extract oil from a plant belongs to Brassica family, canola seeds were used in this study; the extraction yield was higher using PEF also he has improved protein properties of canola like solubility, water binding capacity, emulsifiability, and foamiabilty; however, this retreatment has caused a change in the structure of secondary proteins of canola resulting an aggregation of proteins [96].

7.5 Valorization of Mustard By-Products Properties in Various Fields

To protect the environment and serve the economy too, recycling and valorizing industrial food waste is necessary, especially mustard production wastes; the common way of their application is their use for animal feed like cows, poultry, aquaculture as a natural protein plant source [97–100].

7.5.1 Edible Applications

Numerous studies have investigated the different ways to valorize mustard production wastes, thus, some studies have adopted the concept of the use of mustard biomass as food preservative in food preparations.

Torrijos *et al.* [5] investigated the possibility of the application of white mustard bran and mustard flour as preservative ingredient in bread preparation, the author reported that mustard bran can preserve bread preparation against mytotoxic fungi based on the results obtained, mustard bran stored at low temperatures 4°C showed a noticeable stability and high antifungal activity, besides of that, the organoleptic characteristics of the bread prepared were preserved too, what make mustard bran a good natural and low cost preservative, which can be used as a substituting of chemical preservative ingredient in bakery products.

The use of yellow mustard mucilage or gum as a stabilizer and thickener in food preparations was investigated by Wu *et al.* [31], this low-cost material showed special and individual rheological properties.

Hendrix *et al.* [101] developed a biopolymer with good rheological and physical properties using a sonicated and homogenized defatted mustard meal with 0.6% glycerol.

Goswami *et al.* [102] performed a study in order to investigate the effect of substituting wheat flour with mustard meal in cookies. MAE was used for the extraction of phenolics and glucosinolates from mustard meal fraction.

7.5.2 Inedible Applications

The non-food application of mustard waste has been inspected by different studies, thus, the application of mustard husk ash to produce silica have studied by Faris [103] in this study mustard husk was transformed into ash by heating at 600°C during 60 minutes then the produced mustard husk ash was leached using citric acid as leaching agent at a concentration of 2%. The silica produced from the leached mustard ash showed a high purity of 92%. Silica obtained from mustard MHA could be applied in various fields like rubber production, paper, and composites and also to produce silicon as reported by the author.

The use of low-cost biosorbents has been grown tremendously nowadays, and they applied for depolluting contaminated wastewater. The use of mustard waste especially husk fraction as natural sorbent for the elimination of heavy metals for depolluted water has been carried out by many studies.

Nemeş and Bulgariu [104] have performed a study in batch system and the results obtained have revealed a higher biosorbent power for both Pb (II), Zn (II), and Cd (II), respectively, which give the mustard waste biomass the possibility to be used as a cheap biosorbent and also it will offer a natural source within the concept of green chemistry and also from an economic point of view.

The same results have been shown by Meena *et al.* [105], Pb (II) and Cd (II) metal ions have been removed by the use of mustard husk and it seems a great future biosorbent for heavy metal recovery than other current process technologies.

In the same concept with the first study, Humelnicu *et al.* [106] have analyzed the adsorptive performance of soy bran and mustard husk towards arsenic (V) ions from synthetic aqueous solutions, the different parameters studied were pH, initial arsenic ion concentration, contact time, sorbent mass and temperature. The results have shown an excellent removal power for both soy bran and mustard husk, respectively.

Another study performed by Negrila *et al.* [107] has also examined the possibility of using mustard waste biomass obtained after oil extraction with n-hexane for Zn(II) ions removal from aqueous solution a bioadsorbent agent, pH, adsorbent dose, initial concentration of Zn (II) and contact time parameters were evaluated, results showed that optimum condition required for a best adsorption acquired with an amount of 0.125g of mustard residues for the removal of 40.67 mg L^{-1} of Zn (II) ions in a solution of 25 ml during 60 minutes at an initial pH of 5.5 at room temperature.

The absorption power of mustard husk for the elimination of anthraquinones dyes has been studied by Gautam *et al.* [108], Alizarin red S has been absorbent by mustard husk from an aqueous solution at pH ranged from 2 to 3 using a batch system.

The production of a biofuel from mustard oilseed cake was investigated by Kumar *et al.* [109], in this experience, a pyrolysis at high temperature 450°C to 600°C in a semibatch system was applied to mustard cake. Results showed a yield of 51 % of oil, which make the use of mustard cake useful for the production of biofuel.

Mustard de-oiled cake has been also used as a natural adsorbent for metal cutting elimination from industrial wastewater by Katiyar *et al.* [110], and it showed that activated carbon of mustard cake forms oil film or coating appeared on the surface of adsorbents.

Mustard de-oiled cake material contains a number of pores on a rough and snare type structure of surface where the oil droplets may be trapped and adsorbed easily as explained by the authors [110].

Pal *et al.* [111] studied the possibility of using mustard stalk and straw as libnocellulolytic enzyme as substrate for sugar saccharification for the production of bioethanol, the results obtained showed that mustard residues could be used as a cheap material serving to generate bioethanol especially when it has been pretreated with salt.

Pramanik *et al.* [112] investigated a simple synthesis of porous spherical carbon nanostructure from mustard husk to produce energy, and the results showed that mustard husk waste can serve as a good source of bio and cost-effective energy for electrochemical energy storage applications.

Within the concept of the application of mustard residues as biosorbents Jadhav[a] and Jadhav[b] [113] used mustard husk ash for the removing of fluoride from water, response surface methodology was applied with the optimization of pH, adsorbent dose, contact time and stirring rate; the results showed a high efficiency for fluoride removal 85% from water using mustard husk ash material.

Tinoi *et al.* [70] investigated the use of mustard waste as a substrate for growth and astaxantin production and the results obtained showed that mustard waste could be a good medium for astaxantin production compared to the other materials wastes.

These all results indicate the possibility to replace chemicals preservatives and adsorbents by natural ones especially from mustard residues, which allow a protection of the environment, and they will serve as economic issues as low cost materials.

7.6 Conclusion

From data cited in this chapter, we conclude that mustard seed is considered as a functional food due to its composition on biologically active compounds, which also present in its by-products.

However, the investigation of these residues and their use as substituents of synthetic ingredients requires a sustainable management of agrifood waste, means acquiring the most knowledge about this residue their composition, properties and functionalities.

In, addition this knowledge cannot be useful without applying green, novel, and efficient techniques for the recovery of the targeted compounds.

Finally, the last step is their valorization in numerous fields, like pharmacology, cosmetics, food preparations, and feedstock, will be complete. The information acquired about these by-products and their adequation with each application will be used in the future.

References

1. Tringali, C., *Bioactive compounds from natural sources: Isolation, characterization and biological properties*, CRC Press, London/New York, 2000, https://doi.org/10.1201/9781482289268.
2. Shirahigue, L.D. and Ceccato-Antonini, S.R., Agro-industrial wastes as sources of bioactive compounds for food and fermentation industries. *Ciênc. Rural*, 50, 1–17, 2020.
3. Płotka-Wasylka, J., Rutkowska, M., Owczarek, K., Tobiszewski, M., Namieśnik, J., Extraction with environmentally friendly solvents. *TrAC - Trends Anal. Chem.*, 91, 12–25, 2017.

4. Chaves, J.O., De Souza, M.C., Da Silva, L.C., Lachos-Perez, D., Torres-Mayanga, P.C., Machado, A.P.D.F., Rostagno, M.A., Extraction of flavonoids from natural sources using modern techniques. *Front. Chem.*, 8, 507887, 2020.
5. Torrijos, R., Nazareth, T., Quiles, J.M., Mañes, J., Meca, G., Application of white mustard bran and flour on bread as natural preservative agents. *Foods*, 10, 431, 2021.
6. Anjum, N.A., Gill, S.S., Ahmad, I., Pacheco, M., Duarte, A.C., Umar, S., Pereira, M.E., *The plant family Brassicaceae: An introduction. The plant family Brassicaceae: Contribution towards phytoremediation*, pp. 1–33, Springer, Dordrecht: The Netherlands, 2012.
7. Chan, K., Zhang, H.W., Lin, Z.X., Treatments used in complementary and alternative medicine, in: *Side Effects of Drugs Annual*, vol. 35, pp. 889–898, Elsevier, Amsterdam, The Netherlands, 2014.
8. Abul-Fadl, M.M., El-Badry, N., Ammar, M.S., Nutritional and chemical evaluation for two different varieties of mustard seeds. *World Appl. Sci. J.*, 15, 9, 1225–1233, 2011.
9. Sosulski, F.W., Decyne formation in cottonseed oil fried foods. *J. Am. Oil Chem.*, 60, 6, 1119–1123, 1983.
10. Cools, K. and Terry, L.A., The effect of processing on the glucosinolate profile in mustard seed. *Food Chem.*, 252, June, 343–348, 2018.
11. Mukherjee, K.D. and Kiewitt, I., Changes in fatty acid composition of lipid classes in developing mustard seed. *Phytochemistry*, 23, 2, 349–352, 1984.
12. Bala, M. and Singh, M., Non destructive estimation of total phenol and crude fiber content in intact seeds of rapeseed-mustard using FTNIR. *Ind. Crops Prod.*, 42, 1, 357–362, 2013.
13. Weber, F.E., Taillie, S.A., Stauffer, K.R., Functional characteristics of mustard mucilage. *J. Food Sci.*, 39, 3, 461–466, 1974.
14. NIIR Board of Consultants & Engineers, *The complete book on spices and condiments (with cultivation, processing and uses) 2nd revised edition:* p. 880, Asia Pacific Business Press Inc., New Delhi, India, 2006, [Online]. Available: https://books.google.com/books?id=039ZCwAAQBAJ&pgis=1.
15. Cui, S.W., *Polysaccharides gums from agricultural products: Processing, structures and functionality*, Tech Pub Co, Lancaster, Pennsylvania USA, 2001.
16. Sehwag, S., Das, M. *et al.*, A brief overview: Present status on utilization of mustard oil and cake. *J. Tradit. Knowl.*, 14, 2, 244–250, 2015.
17. Lett, A.M., Thondre, P.S., Rosenthal, A.J., Yellow mustard bran attenuates glycaemic response of a semi-solid food in young healthy men. *Int. J. Food Sci. Nutr.*, 64, 2, 140–146, 2013.
18. Cilly, V.K., Lodhi, G.N., Ichhponani, J.S., Nutritive value of mustard cake derived from seeds of different genetic make up for growing chicks. *J. Agric. Sci.*, 91, 3, 543–550, 1978.

19. Thiyam, U., Claudia, P., Jan, U., Alfred, B., De-oiled rapeseed and a protein isolate: Characterization of sinapic acid derivatives by HPLC-DAD and LC-MS. *Eur. Food Res. Technol.*, 229, 5, 825–831, 2009.
20. Islam, S., Shelly, N.J., Ahmed, K.U., Chowdhury, S.R., Islam, S., Shelly, N.J., Chowdhury, S.R., Study on the nutritional composition of oil cakes of different released and line cultivars of mustard and rapeseed (Brassica spp.). *Int. J. Bio-Resour. Stress Manage.*, 11, 5, 437–444, 2020.
21. VanEtten, C.H., Kwolek, W.F., Peters, J.E., Barclay, A.S., Plant seeds as protein sources of food or feed. Evaluation based on amino acid composition of 379 species. *J. Agric. Food Chem.*, 15, 6, 1077–1089, 1967.
22. Aluko, R.E. and Mcintosh, T., Electrophoretic and functional properties of mustard seed meals and protein concentrates. *J. Am. Oil Chem.*, 81, 7, 679–683, 2004.
23. Thompson, L.U., Potential health benefits and problems associated with antinutrients in foods. *Food Res. Int.*, 26, 2, 131–149, 1993.
24. Diosday, L.L., Tzeng, Y.-M., Rubin, L.J., Preparation of rapeseed protein concentrates and isolates using ultrafiltration. *J. Food Sci.*, 49, 3, 768–770, 1984.
25. Dickinson, E., *Introduction to food colloids*, vol. 36, p. 514, Oxford University Press, Cambridge: United Kingdom, 1992.
26. Wolcott, M.P., Davin, L.B., Lewis, N.G., Patten, A.M., Vassa, D.G., *Trees: A remarkable biochemical bounty*, pp. 1173–1175, Elsevier, Oxford, UK, 2010.
27. Weber, F.E., Taillie, S.A., Stauffer, K.R., Functional characteristics of mustard mucilage. *J. Food Sci.*, 39, 3, 461–466, 1974.
28. Wu, Y., Eskin, N.A.M., Cui, W., Pokharel, B., Emulsifying properties of water soluble yellow mustard mucilage: A comparative study with gum Arabic and citrus pectin. *Food Hydrocoll.*, 47, 191–196, 2015.
29. Western, T.L., Skinner, D.J., Haughn, G.W., Differentiation of mucilage secretory cells of the Arabidopsis seed coat. *Plant Physiol.*, 122, 2, 345–355, 2000.
30. Of, S., Interactions, S., Galactomannans, B., From, P., Soluble, W., Mustard, Y., *Study of synergistic interactions between galactomannans and non-pectic polysaccharides from water soluble yellow mustard mucilage*, Doctoral dessertation: University of Guelph, 2009.
31. Wu, Y., Cui, W., Eskin, N.A.M., Goff, H.D., Food hydrocolloids fractionation and partial characterization of non-pectic polysaccharides from yellow mustard mucilage. *Food Hydrocoll.*, 23, 6, 1535–1541, 2009.
32. Grygier, A. and Grygier, A., Mustard seeds as a bioactive component of food mustard seeds as a bioactive component of food. *Food Rev. Int.*, 17, 1–14, 2022.
33. Mawlong, I., Kumar, M.S., Kandpal, B.K., Premi, O.P., Gurung, B., Singh, D., Meal and oil quality among genotypes of Indian mustard (Brassica juncea) varies under recommended dose of nitrogen fertilizer. *Appl. Ecol. Environ. Res.*, 15, 4, 1427–1445, 2017.

34. Das, G., Tantengco, O.A.G., Tundis, R., Robles, J.A.H., Loizzo, M.R., Glucosinolates and omega-3 fatty acids from mustard seeds: Phytochemistry and pharmacology. *Plants*, 11, 17, 1–22, 2022.
35. Kumar, V., Thakur, A.K., Barothia, N.D., Chatterjee, S.S., Therapeutic potentials of Brassica juncea: An overview. *Cell Medecine*, 1, 1, 2–1, 2011.
36. Ogidi, O.I., Omu, O., Ezeagba, P.A., Ethno pharmacologically active Components of Brassica Juncea (Brown Mustard) Seeds. *Int. J. Pharm. Res. Dev.*, 1, 1, 9–13, 2019.
37. Boscariol, G., Hermenegildo, M., Matias, S., Alencar, D., Biologically active compounds from white and black mustard grains: An optimization study for recovery and identification of phenolic antioxidants. *Ind. Crops Prod.*, 135, March, 294–300, 2019.
38. Karamak, M. and Shahidi, F., Antioxidant activity of ethanolic extract of mustard seed. *Food/Nuhrung*, 40, 5, 261–263, 1996.
39. Das, R., Bhattacherjee, C., Ghosh, S., Preparation of Mustard (Brassica juncea L.) protein isolate and recovery of phenolic compounds by ultrafiltration, *Ind. Eng. Chem. Res.*, 48, 10, 2807, 4939–4947, 2009.
40. Martinovi, N., Polak, T., Abramovi, H., Mustard seed: Phenolic composition and effects on lipid oxidation in oil, oil-in-water emulsion and oleogel. *Ind. Crops Prod.*, 156, 112851, November 2019, 2020.
41. Gök, V., Çağlar, M.Y., Tomar, O., Comparison of chemical properties, antioxidant capacity, and phenolic acids of autoclaved and unautoclaved ground mustard seeds. *Food Sci. Technol.*, 2061, 6–11, 2020.
42. Shahidi, F. and Naczk, M., *Phenolics in food and nutraceuticals*, CRC Press, Boca Raton, FL, USA, 2003.
43. Semeniuc, C.A., Socaci, S., Vlaic, R., Tofana, M., The influence of variety and processing on the total phenolic content and antioxidant activity of mustard. *Rom. Biotechnol. Lett.*, 22, 3, 12514–12519, July, 2017.
44. Aziz, S.S., El-zayat, M.M., El-khateeb, A.Y., Phytochemical characterization, antioxidant and antimicrobial activities of Brassica juncea (L.) mustard seeds aqueous and ethanolic extracts. *J. Plant Prod.*, 11, 2, 85–88, 2020.
45. Ladak, Z., Khairy, M., Armstrong, E.A., Yager, J.Y., *Glucosinolates: Paradoxically beneficial in fighting both brain cell death and cancer*, Elsevier, Cambridge, MA, USA, 2021.
46. Halkier, B.A. and Gershenzon, J., Biology and biochemistry of glucosinolates. *Annu. Rev. Plant Biol.*, 57, 303–333, 2006.
47. Barba, F.J., Bioavailability of glucosinolates and their breakdown products: Impact of Processing. *Front. Nutr.*, 3, August, 1–12, 2016.
48. Borek, V., Morra, M.J., Brown, P.D., Allelochemicals produced during sinigrin decomposition in soil. *J. Agric. Food Chem.*, i, 208, 1030–1034, 1994.
49. Cools, K. and Terry, L.A., The effect of processing on the glucosinolate profile in mustard seed. *Food Chem.*, 252, January, 343–348, 2018.

50. Faizal, A., Razis, A., Arumugam, A., Konsue, N., Glucosinolates and isothiocyanates: Cancer preventive effects 11, in: *2 Mechanisms of Cancer Prevention: In Vivo Studies*, vol. I, pp. 199–210, 2018.
51. Ahmed, A.G., Hussein, U.K., Ahmed, A.E., Kim, K.M., Mahmoud, H.M., Hammouda, O., Bishayee, A., Mustard seed (Brassica nigra) extract exhibits antiproliferative effect against human lung cancer cells through differential regulation of apoptosis, cell cycle, migration, and invasion. *Molecules*, 25, 9, 2069, 2020.
52. Ni, N., Sinapic acid and its derivatives: Natural sources and bioactivity. *Compr. Rev. Food Sci. Food Saf.*, 13, Figure 1, 34–51, 2014.
53. Mayengbam, S., Aachary, A., Thiyam-Holländer, U., Endogenous phenolics in hulls and cotyledons of mustard and canola: A comparative study on its sinapates and antioxidant capacity. *Antioxidants*, 3, 3, 544–558, 2014.
54. Reungoat, V., Gaudin, M., Flourat, A.L., Isidore, E., Mouterde, L.M., Allais, F., Ioannou, I., Optimization of an ethanol/water-based sinapine extraction from mustard bran using Response Surface Methodology. *Food Bioprod. Process.*, 122, 322–331, 2020.
55. Mahoonak, A.S. and Swamylingappa, B., *A new method for preparation of non-toxic, functional protein hydrolysate from commercial mustard cake*, in: *Proceeding of the 12th International Congress of Rapeseed*, 15, pp. 142–145, Wuhan, 2006.
56. Engels, C., Schieber, A., Gänzle, M.G., Sinapic acid derivatives in defatted Oriental mustard (Brassica juncea L.) seed meal extracts using UHPLC-DAD-ESI-MS n and identification of compounds with antibacterial activity. *Eur. Food Res. Technol.*, 234, 535–542, 2012.
57. Wu, Y., Eskin, N.A.M., Cui, W., Pokharel, B., Emulsifying properties of water soluble yellow mustard mucilage: A comparative study with gum Arabic and citrus pectin. *Food Hydrocolloids*, 47, 191–196, 2015.
58. Ram, B.A.A. and Rao, J.H., Effectiveness of the mustard seed powder and cake for the coagulation of synthetic municipal wastewater. *Res. J. Pharm. Biol. Chem. Sci.*, 9, 4, 1422–1428, 2018.
59. Aly, A.A. and Górecki, T., Green approaches to sample preparation based on extraction techniques. *Molecules*, 25, 7, 1719, 2020.
60. Alexovič, M., Andruch, V., Balogh, I.S., Šandrejová, J., A single-valve sequential injection manifold (SV-SIA) for automation of air-assisted liquid-phase microextraction: Stopped flow spectrophotometric determination of chromium (VI). *Anal. Methods*, 5, 10, 2497–2502, 2013.
61. Carpentieri, S., Soltanipour, F., Ferrari, G., Pataro, G., Dons, F., Emerging green techniques for the extraction of antioxidants from agri-food by-products as promising ingredients for the food industry, *Antioxidants*, 10, 9, 1417, 2021.
62. Dini, I., Bio discarded from waste to resource. *Foods*, 10, 11, 2652, 2021.
63. Chadni, M., Flourat, A.L., Reungoat, V., Mouterde, L.M., Allais, F., Ioannou, I., Selective extraction of sinapic acid derivatives from mustard seed meal by

acting on pH: Toward a high antioxidant activity rich extract. *Molecules*, 26, 1, 212, 2021.
64. Corrales, M., Toepfl, S., Butz, P., Knorr, D., Tauscher, B., Extraction of anthocyanins from grape by-products assisted by ultrasonics, high hydrostatic pressure or pulsed electric fields: A comparison. *Innov. Food. Sci. Emerg. Technol.*, 9, 85–91, 2008.
65. Chung, D.L., 6 - Cement-matrix composites, Second Edi, Elsevier Inc., Butterworth-Heineman, UK, 2017.
66. Duarte, K., Justino, C.I.L., Gomes, A.M., Rocha-Santos, T., *Green analytical methodologies for preparation of extracts and analysis of bioactive compounds*, vol. 65, Elsevier B.V., Kidlington, UK, 2014.
67. Chemat, F., Rombaut, N., Sicaire, A., Meullemiestre, A., Abert-vian, M., Ultrasound assisted extraction of food and natural products. Mechanisms, techniques, combinations, protocols and applications. *Ultrason. – Sonochem.*, 34, 540–560, 2016.
68. Vinatoru, M., An overview of the ultrasonically assisted extraction of bioactive principles from herbs. *Ultrason Sonochem.*, 8, 303–313, 2001.
69. Vilkhu, K., Mawson, R., Simons, L., Bates, D., Applications and opportunities for ultrasound assisted extraction in the food industry: A review. *Innov. Food Sci. Emerg. Technol.*, 9, 2, 161–169, 2007.
70. Tinoi, J., Rakariyatham, N., Deming, R.L., Utilization of mustard waste isolates for improved production of astaxanthin by Xanthophyllomyces dendrorhous. *J. Ind. Microbiol. Biotechnol.*, 33, 4, 309–314, 2006.
71. Dubie, J., Stancik, A., Morra, M., Nindo, C., Antioxidant extraction from mustard (Brassica juncea) seed meal using high-intensity ultrasound. *J. Food Sci.*, 78, 4, 542–548, 2013.
72. Flourat, A.L., Willig, G., Teixeira, A.R.S., Allais, F., Eco-friendly extraction of sinapine from residues of mustard production. *Front. Sustain. Food Syst.*, 3, March, 1–10, 2019.
73. Nadeem, H., Malan, P., Khan, A., Asif, M., Siddiqui, M.A., New insights on the utilization of ultrasonicated mustard seed cake: Chemical composition and antagonistic potential for root-knot nematode, Meloidogyne javanica. *Journal of the Jiang University*, 22, 7, 563–574, 2021.
74. Kentish, S. and Ashokkumar, M., *The physical and chemical effects of ultrasound*, pp. 1–12, Springer, New York, 2011.
75. Wang, T., Liang, H., Yuan, Q., Optimization of ultrasonic-stimulated solvent extraction of sinigrin from indian mustard seed (Brassica Juncea L.) using response surface methodology. *Phytochem. Anal.*, 22, 3, May 2010 205–213, 2011.
76. Jahan, K., Ashfaq, A., Islam, R.U., Younis, K., Yousuf, O., Optimization of ultrasound-assisted protein extraction from defatted mustard meal and determination of its physical, structural, and functional properties. *J. Food Process. Preserv.*, 46, e16764, 2022. https://doi.org/10.1111/jfpp.16764.

77. Diels, A.M. and Michiels, C.W., High-pressure homogenization as a non-thermal technique for the inactivation of microorganisms. *Crit. Rev. Microbiol.*, 32, 4, 201–16, 2006 Oct-Dec.
78. Shah, A.A., Seehar, T.H., Sharma, K., Toor, S.S., Biomass pretreatment technologies, in: *Hydrocarbon Biorefinery*, pp. 203–228, Elsevier, Amsterdam, The Netherlands, 2022.
79. Yong, S.X.M., Song, C.P., Choo, W.S., Impact of high-pressure homogenization on the extractability and stability of phytochemicals. *Front. Sustain. Food Syst.*, 4, 593259, 2021.
80. Donsì, F. and Velikov, K.P., Mechanical cell disruption of mustard bran suspensions for improved dispersion properties and protein release. *Food Funct.*, 11, 7, 6273–6284, 2020.
81. Yap, B.H., Dumsday, G.J., Scales, P.J., Martin, G.J., Energy evaluation of algal cell disruption by high pressure homogenisation. *Bioresour. Technol.*, 184, 280–285, 2015.
82. Twarużek, M., Błajet-Kosicka, A., Kosicki, R., Grajewski, J., Mycotoxin analytical methods, in: *Environmental Mycology in Public Health*, pp. 363–386, Academic Press, Amsterdam The Netherlands, 2016.
83. Mandal, V., Mohan, Y., Hemalatha, S., Microwave assisted extraction: An innovative and promising extraction tool for medicinal plant research. *Pharmacogn. Rev.*, 1, 1, 7–18, 2007.
84. Veggi, P.C., Martinez, J., Meireles, M.A.A., *Microwave-assisted extraction for bioactive compounds*, p. 4, Theory and Practice Springer Science & Business Media, New York, USA, 2013.
85. Chemat, F. and Cravotto, G. (Eds.), *Microwave-assisted extraction for bioactive compounds: Theory and practice*, vol. 4, Springer Science & Business Media, New York, USA, 2012.
86. Omirou, M., Papastylianou, I., Iori, R., Papastephanou, C., Papadopoulou, K.K., Ehaliotis, C., Karpouzas, D.G., Microwave-assisted extraction of glucosinolates from Eruca sativa seeds and soil: Comparison with existing methods. *Phytochem. Anal.*, 20, 3, 214–220, 2009.
87. Cheng, X., Bi, L., Zhao, Z., Chen, Y., Advances in enzyme assisted extraction of natural products, in: *3rd International Conference on Material, Mechanical and Manufacturing Engineering (IC3ME 2015)*, Atlantis Press, pp. 371–375, 2015.
88. Fahmi, R., Antioxidant and antibacterial properties of endogenous phenolic compounds from commercial mustard products (Master's thesis), University of Manitoba: Manitoba, 2016.
89. Cheng, X., Bi, L., Zhao, Z., Chen, Y., Advances in enzyme assisted extraction of natural products. *IC3ME*, pp. 371–375, 2015.
90. Reungoat, V., Mouterde, L.M., Chadni, M., Couvreur, J., Isidore, E., Allais, F., Ioannou, I., Simultaneous extraction and enzymatic hydrolysis of mustard bran for the recovery of sinapic acid. *Food Bioprod. Process.*, 130, 68–78, 2021.

91. Tabtabaei, S., *Production of biodiesel and protein isolates from dehulled yellow mustard flour*, University of Toronto, Toronto, 2015, [Online]. Available: https://hdl.handle.net/1807/77702.
92. Vega-Mercado, H., Gongora-Nieto, M.M., Barbosa-Canovas, G.V., Swanson, B.G., Pulsed electric fields in food preservation, in: *Handb. Food Preserv.*, pp. 823–842, 2020.
93. Barba, F.J., Parniakov, O., Pereira, S.A., Wiktor, A., Grimi, N., Boussetta, N., Saraiva, J.A., Raso, J., Martin-Belloso, O., Witrowa-Rajchert, D., Lebovka, N., Vorobiev, E., Current applications and new opportunities for the use of pulsed electric fields in food science and industry. *Food Res. Int.*, 77, 773–798, 2015.
94. Carullo, D., Pataro, G., Donsì, F., Ferrari, G., Redondo, L.M., Pulsed electric fields-assisted extraction of valuable compounds from arthrospira platensis: Effect of pulse polarity and mild heating. *Front. Bioeng. Biotechnol.*, 8, September, 1–15, 2020.
95. Picart, L. and Cheftel, J.C., Pulsed electric fields, in: *Food Preservation Techniques*, pp. 360–427, Woodhead Publishing, Boca Raton, FL, USA, 2003.
96. Zhang, L., Wang, L., Jiang, W., Qian, J., Effect of pulsed electric field on functional and structural properties of canola protein by pretreating seeds to elevate oil yield. *LWT - Food Sci. Technol.*, 84, 73–81, 2017.
97. Newkirk, R.W., Classen, H.L., Tyler, R.T., Nutritional evaluation of low glucosinolate mustard meals (Brassica juncea) in broiler diets. *Poult. Sci.*, 76, 9, 1272–1277, 1997.
98. Rao, S.V., Raju, M.V.L.N., Panda, A.K., Shashibindu, M.S., Utilization of low glucosinalate and conventional mustard oilseed cakes in commercial broiler chicken diets. *Asian-Australas. J. Animal Sci.*, 18, 8, 1157–1163, 2005.
99. Maiga, H.A., Bauer, M.L., Dahlen, C.R., Badaruddin, M., Scholljegerdes, E.J., Mustard bran in lactating dairy cow diets. *J. Dairy Sci.*, 94, 6, 3054–3062, 2011.
100. Kumar, S., Kumar, S., Kumar, B., Sinha, M., Comparative evaluation of 'Azolla' with wheat bran & mustard cake in cross breed cow. *J. AgriSearch*, 6, 112–114, 2002.
101. Hendrix, K.M., Morra, M.J., Lee, H., Min, S.C., Food hydrocolloids defatted mustard seed meal-based biopolymer film development. *Food Hydrocoll.*, 26, 1, 118–125, 2012.
102. Goswami, D., Sharma, S., Singh, B., Tyagi, S.K., Effect of microwave-assisted processing of mustard meal on physicochemical and sensory properties of cookies. *Int. J. Curr. Microbiol. Appl. Sci.*, 9, 9, 2439–2446, 2020.
103. Faris, M., Production of silica from mustard husk ash. *Modern Approaches on Materials Science (MAMS)*, 4, 3, 525–529, 2021.
104. Paper, I., Bulgariu, L., Access, O., Optimization of process parameters for heavy metals biosorption onto mustard waste biomass. *Open Chemistry*, 14, 1, 175–187, 2016.

105. Kumar, A., Kadirvelu, K., Mishraa, G.K., Rajagopal, C., Nagar, P.N., Adsorption of Pb (II) and Cd (II) metal ions from aqueous solutions by mustard adsorption of Pb (II) and Cd (II) metal ions from aqueous solutions by mustard husk. *J. Hazard. Mater.*, 150, 3, 619–625, March, 2008.
106. Humelnicu, D., Soroaga, L.V., Arsene, C., Humelnicu, I., Olariu, R.I., Adsorptive performance of soy bran and mustard husk towards arsenic (V) ions from synthetic aqueous solutions. *Acta Chim. Slov.*, 66, 2, 326–336, July, 2019.
107. Bulgariu, L., Biosorption of zn (ii) ions from aqueous solution onto mustard biosorption of zn (ii) ions from aqueous solution onto mustard waste biomass. *Bulletin of the Polytechnic Institute of Jassy, Constructions Architecture Section*, 62, 9–19, june, 2017.
108. Gautam, R.K., Chattopadhyaya, M.C., Sharma, S.K., *Biosorption of heavy metals: Recent trends and challenges*, Springer, Dordrecht, October, 2013.
109. Kumar, A., Gandhi, S.K., Singh, S., Pyrolysis of mustard de-oiled cake. *Int. J. Sci. Technol. Res.*, 2, 5, 230–232, 2013.
110. Katiyar, A., Singh, A.K., Joshi, A., Utilization of waste material (Part II): DMAC (De-oiled mustard cake), as an efficient adsorbent for the removal of metal cutting fluids from aqueous medium/industrial wastewater. *Int. J. Eng. Res. Technol.*, 0869, 12, 131–137, 2015.
111. Pal, S., Prosad, S., Khowala, S., Mustard stalk and straw: A new source for production of lignocellulolytic enzymes by the fungus Termitomyces clypeatus and as a substrate for saccharification. *Ind. Crops Prod.*, 41, 283–288, 2013.
112. Pramanik, A. and Chattopadhyay, S., Materials advances efficient energy storage in mustard husk derived porous spherical carbon nanostructures †. *Mater. Adv.*, 2, 22, 7463–7472, 2021.
113. Jadhav, A.S. and Jadhav, M.V., Materials today: Proceedings use of response surface methodology for optimization of fluoride removal efficiency by adsorption on black mustard husk ash. *Mater. Today Proc.*, 61, 150–157, 2022.

8

Trends in Utilization of Whey and Buttermilk—Valuable By-Products of the Dairy Industry

Katarina Lisak Jakopović*, Irena Barukčić Jurina and Rajka Božanić

University of Zagreb, Faculty of Food Technology and Biotechnology, Department of Food Engineering, Laboratory for Milk and Milk Products, Zagreb, Croatia

Abstract

Whey, as well as buttermilk, represents the main by-products of the dairy industry. They abound with high nutritional value, and their usage is increasingly present in modern men's daily consumption. Since their consumption is more represented in human nutrition, dairy industries put more effort into producing highly nutritional products with specific health benefits. Thanks to the nutritional value and composition of whey and buttermilk, there are many ways for their usage. The most common method of whey utilization is beverage production, whey cheese production, whey protein isolation, lactose production, and many others. Buttermilk is usually processed by fermentation with lactic acid bacteria or it is used for milk fat globule membrane (MFGM) and phospholipids isolation. This chapter highlights the need to increase the usage of whey and buttermilk-based products and presents current achievements focused on whey and buttermilk-based products and the isolation of high nutritional components for technological and functional usage.

***Keywords*:** Buttermilk, whey, utilization, composition, biologically active compounds, dairy by-products

**Corresponding author*: klisak@pbf.hr

Umile Gianfranco Spizzirri (ed.) Nutraceutics from Agri-Food By-Products, (245–284) © 2023 Scrivener Publishing LLC

8.1 Whey

8.1.1 Introduction

Whey is green-yellowish translucent liquid fraction remaining after milk clotting and casein removal during cheese or casein manufacturing. Total solids of whey are about 6% to 6.5%. Whey comprises 80% to 90% of the total volume of milk and contains about 50% of the nutrients in the original milk: such as lactose (~70%; i.e. depending on the acidity of the whey), whey proteins (~10%), minerals (~12%), vitamins and some fat (Figure 8.1). Calcium, potassium, sodium and magnesium salts make up the bulk of these minerals (of >50% NaCl and KCl, calcium salts) with trace amounts of metals such as zinc and copper.

Whey has been viewed until recently as one of the major disposal problems of the dairy industry. The biological oxygen demand (BOD) of whey is very high (40 000 mg kg^{-1} or more), constituting a major ecological burden if disposal practices of the past, including drainage into waste treatment facilities or spraying onto fields, are currently seldom practiced. The use of whey as cattle or pig feed is still one of the significant alternatives to utilization in the human food chain, now being predominantly favored due to the economic opportunities provided by some of the milk nutrients contained in the whey [1].

8.1.2 Whey Types and Composition

The type of whey depends upon the processing technique used to remove casein from the milk, with the two main types being sweet whey and acid whey. Whey, as a by-product from the manufacture of cheese and rennet casein, is known as sweet whey with a pH of 5.9 to 6.6. Manufacture of mineral-acid precipitated casein or cheese produced by bacterial culture

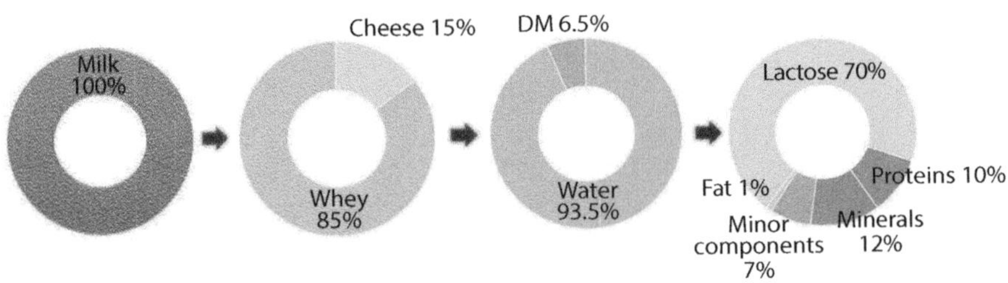

Figure 8.1 The volume ratio of milk and whey during cheese production and the composition of whey.

yields acid whey with a pH of 4.3 to 4.6. Whey's composition and sensory characteristics vary depending on the kind of the whey (acid or sweet), the source of the milk (cow, sheep, bovine milk etc.) and the feed of the animal, which produced the milk, the cheese processing used, the time of the year and the stage of lactation. But the chemical composition of whey varies the most in relation to method used for its production (acid whey or sweet whey). The main differences are in the calcium, phosphate, lactic acid and lactate contents, which are higher in acid whey than in sweet whey (Table 8.1). Whey also contains small quantities of other components like lactic and citric acids, non-protein nitrogen compounds (urea and uric acid) and B group vitamins.

Colloidal calcium becomes more soluble in acid environment and, consequently, by the acid coagulation of casein, part of the calcium dissolves and passes into the whey. In contrast, sweet whey, except whey proteins, contains glycomacropeptides formed by the enzymatic hydrolysis of κ-casein. Thus, the glycomacropeptides constitute approximately 20% of the whey protein fraction of sweet, rennet-bases whey but are not found

Table 8.1 Approximate composition (%) of whey [2].

Ingredient	Sweet whey	Acid whey
Total solids	6.0	6.4
Water	94	93.6
Fat	0.05	0.05
True protein	0.60	0.60
NPN (non-protein nitrogen)	0.20	0.20
Lactose	4.5	4.6
Ash (minerals)	0.5	0.8
Calcium	0.035	0.12
Phosphorus	0.040	0.065
Sodium	0.045	0.050
Potassium	0.14	0.16
Chloride	0.09	0.11
Lactic acid	0.05	0.4

in the acid whey unless renneting was included in the fresh cheese manufacturing process. In addition, the proportion of whey protein is slightly lower in whey obtained from ultrafiltered (UF) milk, and in whey obtained from milk heated at high temperature [1]. The main reason is that part of the whey proteins is retained in curd of the cheese and not seeped into the whey, i.e. an approach in increase the yield of cheese. In the conventional processes of cheese making, proteins, which are insensitive to the action of enzymes and/or acids, pass into the whey. Consequently, this group of proteins were termed as whey proteins [3]. Whey proteins are globular molecules with substantial contents of α-helix motifs, in which acidic/basic and hydrophobic/hydrophilic amino acids are balanced throughout their sequences. The whey proteins include β-lactoglobuline, α-lactalbumine, immunoglobulins, blood serum albumin, bovine lactoferrin, and lactoperoxidase, in addition to other minor proteinaceous components, such as glycomacropeptide, which is released from κ-casein in the first step of enzymatic cheese making [4]. Although whey protein content is similar in sweet and acid whey (Table 8.2), free amino acids content varies a lot and is mostly dependent on the level of casein hydrolysis in cheese making process. Free amino acids content is about four times higher in sweet whey and about ten times higher in acid whey than in milk [5].

However, the largest constituent of whey is lactose (~70% based on dray matter basis). Incidentally, most milk carbohydrates pass into the whey after cheese making of which 90% is lactose including some glucose, galactose, oligosaccharides and amino sugars [3]. Lactose is a very important source of energy, and has multiple roles. Some of the beneficial effects of lactose are stimulation of intestinal peristalsis, facilitate calcium and phosphorus absorption [6]; establishment of a mildly acid reaction in the intestines and thereby preventing the growth and multiplication of harmful bacteria; ensures optimal magnesium levels and improves digestion of fat and other nutrients in the human body; participate in the development of dental plaque as well as heat treatment of whey converts lactose into lactulose, which is one of the growth promoters of the bifidobacteria.

Table 8.2 Content of amino acids (mg/L) in whey [5].

	Free amino acids		**Aminoacids in proteins**	
Whey	**Total**	**Essential**	**Total**	**Essential**
Sweet whey	132.7	51.0	6.490	3.326
Acid whey	450.0	356.0	5.590	2.849

Also, during the manufacture of cheese or casein, some water-soluble vitamins permeate from milk into the whey, but their amounts are very variable and greatly dependent on whey handling. The most important of these vitamins are riboflavin, folic acid and cobalamine. However, the latter two vitamins are bonded to whey proteins, what is the reason why they pass into the whey during cheese making. It is of interest to note that whey contains higher amounts of vitamin B2 than milk due to the activity of some starter cultures (i.e. lactic acid bacteria [LAB]) during the manufacture of cheese [3, 7].

Whey is an excellent source of functional proteins and peptides, lipids, vitamins, minerals, and lactose that could be exploited by the agri-food, biotechnology, medical, and related industries and in the last several decade major research efforts have seen whey and whey proteins transformed from 'gutter-to-gold'.

8.1.3 Whey Processing and Utilization for Food Applications

The main problem associated to large quantities of whey remaining in cheese manufacture comes from its high potential to pollute the environment. Whey has a very high biochemical oxygen demand (BOD) and chemical oxygen demand (COD), both arising from the high lactose content [8]. As already previously highlighted, apart from lactose, whey also contains whey proteins and non-protein nitrogen components, vitamins, minerals and very small amounts residual milk fat. Such composition makes it an excellent medium for microbial growth, demanding in turn obligate processing either by applying thermal treatments or some of novel nonthermal processing methods. Thereby, a special focus has been put on high-quality whey proteins, their isolation and processing into different powdered products such as whey protein concentrates (WPC) or whey protein isolates (WPI). Besides, a lot of research has been performed with the objective of developing and suggesting possible processing methods to enhance whey valorization in the production of riboflavin, lactic acid, starter and probiotic cultures as well as their metabolites, whey cheese, whey beverages or different food supplements. Nowadays, scientific research in the area of food processing targets to meet the requirements of zero-waste principles and establishment sustainable technological processes. Accordingly, enhancing whey utilization has become one of the key points of the dairy industry sector. Some of the most investigated approaches will be described in more detail in this chapter.

8.1.3.1 Condensing and Drying for Whey Powder Production

Due to a high susceptibility to microbial spoilage and the tendency of whey proteins to thermal denaturation, as well as the undesirable sensory properties, whey is most often processed into whey powder. However, prior to drying whey needs to be concentrated, which is most often achieved by a multiple-effect evaporator, where two or more units are connected and operate at pressures, which become progressively lower. Accordingly, the boiling temperatures become lower too, as the process moves on, and the condensate from the previous section is being removed. In the dairy industry, a well-known installation is the falling film evaporator that consists of several tubes inside of which whey is flowing in a thin film, while the outer layer of the tubes is covered by a steam heating jacket under vacuum. During the process, water and condensed vapor are removed as condensate in the condenser placed at the bottom of the installation. At the outlet of the evaporator a vapor cyclone is placed which separates concentrated whey from the vapor [9]. The main disadvantages of this method are the high energy consumption and the high content of minerals and lactose that remain in the concentrate. Besides evaporation, which is commonly used, some more recent concentration technologies rely on membrane separation techniques as appropriate tools for whey concentration without applying high temperatures that decrease the nutritional value of whey proteins. These are primarily based on ultrafiltration as the most applied membrane process in the production of whey powder and whey protein concentrates and isolates, which will be described in more detail in the next section 1.3.2. Drying of whey can be regarded as a continuation of whey concentration with the main objective of producing a stable low moisture highly functional product, which will be commercially available.

Spray drying has been recognized as the most appropriate method of whey drying since it results in a powder of good quality and is applicable for liquids with low dry matter content. Besides, an important advantage of spray drying is that water can be removed from the concentrated whey at the lowest temperature and in the shortest time, which reduces undesired heat-induced changes, especially on whey proteins, as well as energy consuming process costs. Spray drying of whey consists of entering the concentrated whey into the spray-dryer whereby very fine droplets are generated by using a nozzle or a rotary atomizer into a hot dry air stream typically set at 150-250 °C [9, 11]. At the bottom of the chamber the produced powder exits the process. A certain amount of the produced powder is blown away by the outlet warm air, but in order to reduce the production loss arising from this phenomenon, several cyclones are usually placed outside the dryer to remove dry whey residues from humid air through vortex separation. The final drying and cooling occurs in a fluidized bed where the powders coming from the end of the chamber and the cyclones mix together [11]. Spray drying can be carried out in installations of different capacities of one—or multistage dryers. This method has many advantages such as rapid drying, large processing capacities and continuous operating conditions but there are also few disadvantages like the heat induced denaturation of proteins and high costs of operation [1, 11]. Also, there is a lack of data on influence of spray drying on functional and antioxidative characteristics of whey. Some of more recent studies indicate changes in activity and functional properties of whey proteins during the drying process [12, 13]. It has even been demonstrated that the presence of some whey protein (WP) fractions is dependent on thermal treatment of milk prior to cheese making [14]. Thus, there is an ongoing research focused on improving drying techniques in order to overcome quality issues related to the produced whey powders. Some of the implemented key dryer developments include fluid bed improvements and enhanced approaches to agglomeration resulting in highly dispersible, soluble and free-flowing powders, developing energy saving strategies leading to more cost-effective operations and implementing novel approaches for drying of remaining effluents (e.g., permeate) resulting from whey processing. Whey powder is successfully used in manufacturing ice cream, fermented milks, cereal, confectionary and meat products, ice cream, yoghurt, flavored milk-based drinks, as well as in animal feed, mostly for being a cheap replacement for skim milk powder [9, 10]. Whey powder has also a high potential for application in probiotic, pharmaceutical and similar functional products that are used as food supplements.

8.1.3.2 Membrane Processing

Processes based on membrane separation technology are well implemented in the dairy industry for numerous applications. This approach refers to the use of semi-permeable membrane filters with the aim to concentrate or fractionate a liquid, whereby a hydrostatic pressure gradient is the driving force responsible for the sieving effect. Depending on the applied working pressures and membrane pore sizes or molecular weight cut offs, membrane processes are utilized for different purposes in the dairy industry, such as skimming and microbial stabilization, demineralization, concentration of the total dry matter, concentration and/or fractionation of casein and whey proteins. The two major types of membrane separation techniques employed by the dairy industry for the separation of whey proteins are ultrafiltration and microfiltration [15]. Ultrafiltration (UF) is the most commonly used industrial method for the separation and fractionation of whey proteins since it implies the recovery of whey proteins in their native form. The typical range of processing parameters refers to molecular weight cut-offs between 1 and 100 kDa, working pressures 0.1-1 MPa and temperatures below 55°C [15, 16]. The application of UF can increase the protein content of whey up to 85%, thus, it most commonly applied to obtain whey protein concentrates (WPCs) and whey protein isolates (WPIs), or concentrates of specific whey protein fractions, such as ß-lactoglobulin (β-LG), α-lactalbumin (α -LA) or glycomacropeptide (GMP), but also to reduce the lactose and ash content of whey concentrate prior to drying [17]. An inevitable phenomenon occurring during UF is membrane fouling, which is a result of deposition and accumulation of microorganisms, residual fat, proteins and minerals at the membrane surface as well as their mutual interactions. Consequently, UF efficiency in whey processing is reduced [16]. In order to minimize the fouling intensity, a lot of modifications have been investigated and made to conventional ultrafiltration with the main objective to increase the yield of whey protein recovery. Some authors have suggested different pre-treatments of whey prior to UF, such as microfiltration (MF) to reduce microbial load and remove larger particles like casein or fat residues, which in results in minimizing fouling and its adverse effects on UF performance [17, 18]. For instance, Barukčić *et al.* [16] found that pre-treatment of sweet whey by applying MF with 0.5 μm ceramic membrane prior to UF resulted in a significant flux increase (4.5 times higher), while the fouling intensity was up to 10 times lower in comparison to conventionally applied UF conditions. It has also been found that proteins of larger size have less influence on the membrane and are easier to clean in comparison to the small sized proteins [19–21], so it has been proposed to employ enzymatic crosslinking, most often by transglutaminase, in order to obtain larger molecules and thus enhance whey protein retention in the UF concentrate [22]. Also, some other approaches have been investigated like the membrane back-flushing, introducing electric field or application of ultrasound to avoid the adherence of particles reducing fouling and/or might be responsible for the removal of particles from membrane surface giving a cleaning effect for the already fouled membrane [22–24]. Khaire *et al.* [24] was demonstrated that whey pre-treatment by thermos-sonication at 60°C at 140 W/25min and the subsequent ultrasound assisted UF at varying power levels from 50W to 120 W, reduced the fouling by about 32% compared to the conventional approach. Conclusively, ultrasound assisted ultrafiltration was demonstrated as an effective approach to improve membrane operation in the context of the reduced fouling. Besides UF, also other membrane processes like nanofiltration (NF) are applied for whey processing. NF is particularly useful for separation of peptides, which are gaining increased interest due to their bioactive nature [25]. Butylina *et al.* [26] found that nanofiltration through a sulphonated polyether sulphone membrane (cut-off 1000 g/mol, nominal) could be successfully applied for separation of polypeptides (molar mass range of 500–1800 g/mol) from lactose. However, NF is most often used for demineralization and removal of lactic acid from whey prior to spray drying. As a result, stickiness during drying and hygroscopicity of the obtained powder are reduced [27, 28]. Another benefit of NF application is the capability to concentrate whey up to a total dry matter content of 20–22%, which contributes to economization of the evaporation costs. Also, NF membranes are also to a large extent permeable for organic acids like lactic and acetic acid, which can be used for deacidification of acid whey and thus contribute to its wider utilization. Some authors proposed the combination of several membrane processes like reverse osmosis (RO), MF, UF and NF as efficient protocols for recovery of whey proteins with maintained nutritional and functional properties [29–32].

8.1.3.3 Demineralization in Whey Processing

The relatively high content (8–10% on a dry weight basis) of minerals and salts remaining in whey, especially in acid and salty types, make it difficult for further utilization. Besides contributing to an excessively salty taste, minerals and salts also tend to precipitate at higher temperatures applied during obligate thermal treatments, which results in sediment formation Both, the salty taste as well as the formed sediment, are undesirable attributes in all kinds of food products, especially in dietetic and baby foods, so a reduction of the mineral content is needed [33]. Also, the high mineral content can be problematic in producing whey powder, when processing steps like concentrating, lactose crystallization and spray drying. In order to obtain non-hygroscopic whey powder with mineral composition similar to human milk, which is then a suitable ingredient for certain foods, especially for infant formulas, a 90% to 95% reduction in minerals is needed. A smaller overall reduction in minerals (50–70%) is often enough reduction for other applications, like for ice cream or bakery products [34]. The three main technologies for whey demineralization are electrodialysis, ion exchange and nanofiltration (NF), as well as their combinations.

Electrodialysis is a selective demineralization method, which relies on the transport of mono-valent ions through semipermeable membranes. The driving forced for this separation process is the electric field caused by an applied direct current (DC) source. It consists of a number of compartments separated by alternate cation and anion exchange membranes, while the electrodes are located at the end compartments. Alternate cells in the electrodialysis stack act as concentration and dilution cells. When DC is applied across the cells, cations migrate to the cathode and anions to the anode. Whey circulates through the dilution cells and whey salts are discharged through a 5% brine solution, which circulates through the concentration cells. The electrodialysis plant can be run continuously or in batches. A batch system can consist of one membrane stack over which the process liquid, for example, whey, is circulated until a certain ash level is reached. It is often used for demineralization levels above 70% whereby the holding time can be very long (5–6 h) for 90% demineralization at 30–40°C. So, in order to assure maximum demineralization and lower consumption of electric power, it is desirable to concentrate the whey to 20–30% dry matter (DM) prior to batch operated electrodialysis. If such concentrate is generated, it has to be clarified and microbiologically stabilized by adding a bacteriostatic compound like hydrogen peroxide before it enters the electrodialysis unit. In a continuous plant, the holding time can be reduced to 10 to 40 minutes, depending on the number of membrane stacks, while the maximum demineralization level that can be reached is 60% to 70% [34–36]. Diblíková *et al.* [37] found that electrodialysis can be used for effective mineral salts removal even from highly salted and concentrated whey solutions, whereby the total salt content decreased in all model solutions by minimally 90% within 2 h, without encountering any operating difficulties. These results are very useful for expanding further utilization of acid and salty whey remaining after Cheddar or blueveined cheese production, which are often disposed as waste due to high ash content and unfavourable sensory properties. Similar to other membrane processes, electrodialysis is also susceptible to occurrence of membrane fouling caused by calcium phosphate precipitation of on the cation exchange membrane surfaces and deposition of proteins on the anion exchange membrane surfaces. Thus, it is necessary to regularly change membranes and other parts of the unit. Since those make about 35–40% of the total running costs of the plant, they are considered as a major limiting factor for using electrodialysis in dairy processing [34].

Ion exchange is regarded as the most efficient method for whey demineralization. It relies on similar principle like electrodialysis, i.e. this on the reversible interaction between target molecules to be separated and membrane functional groups [36, 38]. Industrial application of ion exchange often involves pumping the liquid through a fixed column filled with polymeric beads loaded with ions that can be exchanged with the ions in the process liquid. The capacity of the process is limited by the amount of ions on the resin in the fixed bed. When the ion exchange capacity has been reached, the column needs to be regenerated with appropriate regeneration solution to remove the adsorbed ions. After the regeneration was completed, the ion-exchange plant is again capable of treating the process liquid (whey) which needs to be demineralized. Similar to electrodialysis, ion exchange membranes can also be divided into two main groups—anion and cation exchange membranes. Nowadays, they are usually constituted of macromolecular porous plastic materials formed into beads, which are capable to exchange the mobile ions (counterions) that they contain for ions of the same charge sign contained in whey, or the solution to be treated, respectively. The functional groups of ion exchange resins for demineralization purposes vary. Strong cation exchange materials often have sulfonic groups bound to the matrix, while the most common weak acid exchange material has a carboxylic acid. A simple whey demineralization plant based on ion exchange processing involves whey entering a strong cation exchanger loaded in the hydrogen ion form, and continuing the flow to the weak base anion exchanger in its free base form. The ion exchange columns are rinsed and regenerated separately with diluted hydrochloric acid and sodium hydroxide (or ammonia). The columns need to be disinfected on a daily basis by rinsing with water containing active chlorine [35, 36, 38].

Nanofiltration is an interesting alternative to electrodialysis and ion exchange. As mentioned in the previous section, NF is a membrane pressure-driven process for separation of molecules in the molecular range of 100–1000 Da. The usual operating conditions involve membranes with the mean pore diameter of approximately 1 nm and the applied pressure is typically 2–3 MPa. Thus, NF membranes are highly permeable for monovalent ions like Na+, K+, and Cl– and low molecular weight organic compounds, but very lowly permeable for high molecular weight organic compounds like lactose or proteins. Such selective separation makes NF favourable for target removal of electrolytes is required, so NF has become a new technology for the demineralization and concentration of whey, and especially a good alternative to the combined process of evaporation (EV) and electrodialysis (ED) [34, 39, 40]. Thus, the demineralization efficiency is mainly restricted to removal of monovalent ions. For a 4 to 5-fold volume reduction achieved by whey NF, approximately 40–60% of the ash content can be removed, which corresponds to approximately 70–80% of monovalent co-ions (Na+, K+, and H+ for acid whey; and Cl– and OH– for sweet whey) and 40–70% of monovalent counterions (Na+, K+, and H+ for sweet whey; and Cl– and OH– for acid whey), respectively. The reduction of divalent ions ranges between 3 and 20%. Pan *et al.* [41] found that NF process was highly effective for mineral salt removal from whey in order to reuse lactose and concentrate whey protein. The best effect of demineralization was achieved when the feeding pH was adjusted to the whey protein isoelectric point (4.60). They also investigated the diafiltration process and achieved approximately 72% removal of ash with almost no lactose loss. Partial demineralization is often required in the manufacturing of dairy ingredients like whey protein concentrates in order to adjust the mineral composition [34, 36]. The advantages of NF are low investment costs and easy operating, simple installations. Further, the amount of effluents is significantly lower in comparison to the other demineralization processes, and the generated effluents (permeates) are characterized by a lower biochemical oxygen demand (BOD) values. In contrast to that, demineralization by electrodialysis and ion exchange usually generate high amounts of effluents, and their running costs are 25–55% higher than NF [32, 34].

For large demineralization installations, which are designed for processing more than 400m^3 per day, it would be reasonable to invest in combined technologies relying on electrodialysis or NF in combination with the classical ion exchange process. The benefits are reduced costs of chemicals but the capital costs are higher. Many modern demineralization plants are combinations of classical ion exchange with NF, which results in up to 30% reduced ionic load and the columns size of the ion exchangers as well as lower volumes to treat [36].

8.1.3.4 Whey Protein Concentrates and Isolates

Whey protein concentrates (WPC) and whey protein isolates (WPI) belong to functional whey ingredients that have expanded the applications spectrum of whey protein ingredients. As described in previous sections, membrane processing, including industrial applications of ultrafiltration, diafiltration and microfiltration have significantly contributed to the development of these high-protein and low-fat products. Nowadays, in a large-scale whey processing plants, membrane processes or ion-exchange chromatography in combination with spray drying are commonly used to produce WPCs and WPIs, both of which contain all the whey proteins in the same proportions as in the original whey [42, 43]. Thereby, WPC can be distinguished as low protein (24-45%), medium protein (45-60%) and high protein WPC (60-80%), but most WPCs on the market contain either 34–35 or 80% protein [32, 35, 42]. WPIs contain 90% or more whey proteins and are usually produced from whey defatted by MF. Ion exchange chromatography belongs to highly selective methods used in the manufacture of WPIs, since it provides an additional level of selectivity other than molecular size typical for membrane separation techniques.

Low protein WPCs are mainly used in the manufacture of yogurt, processed cheese, infant formulae, various bakery applications, but also in sauces, stews and meat patties due to their thickening properties and nutritional benefits. The impact of 35% WPCs in food applications relies on the combined effects of protein, lactose, and minerals in these products. WPCs containing 80% protein are intended for applications such as gelation, emulsification or foam formation, where proteins play a dominant functional role. Due to the low carbohydrate content, high protein WPC is an ideal ingredient for specialized products designed for athletes or weight management. This group of WPC is also very often applied in manufacture of meat products where high gel strength and good water-binding properties are required [35, 42].

WPIs contain more than 90% protein, and due to the high protein purity and solution clarity, they are extensively used in nutritional supplements, sports and health drinks, and protein-fortified beverages.

The constantly increasing load of scientific evidence on various health promoting effects of whey proteins, result in an increasing need to produce individual whey proteins on a large scale without damaging their structure or functional properties. Thus, membrane processing has undergone numerous improvements in membrane materials and construction, and nowadays allows performing whey ultrafiltration and/or diafiltration at low temperatures (<10 °C) that enable preservation of functional activity and minimize protein denaturation. Further, membrane separation techniques became more sophisticated with the development and application of membrane adsorbers for the isolation and fractionation of whey proteins. Ion exchange is a typical example of such separation techniques, whereby the applied anion exchange membranes have positively charged functional groups like diethylamine and will theoretically bind negatively charged fractions such as β-lactoglobulin, α-lactalbumin and BSA. On contrary, the cation exchange membranes have negatively charged functional groups like carboxylic and sulfonic acid and will, in theory, bind positively charged fractions i.e. lactoferrin and lactoperoxidase [38, 44, 45]. For example, a combination of electrodialysis and ultrafiltration (EDUF) was patented by the Bazinet group [46] for the separation of peptides, which due to smaller size allow the application of weaker electrical field (< 15 V/cm), so water splitting is excluded. Wang *et al.* [47] succeeded in purification and isolation of lactoferrin and immunoglobulins by applying electrodialysis with filtration membrane (EDFM) approach, whereby a polyvinyl alcohol

(PVA) membrane was prepared in house. Lactoferrin was aggregated in a salty buffer similar to that naturally available in whey The solution containing the aggregated lactoferrin (LF) and other proteins was the be passed through an electrodialysis with filtration membrane process, where the smaller proteins such as BSA, can pass through, while the larger LF aggregates and Ig. These findings are of enormous importance since isolation of high value minor proteins such as lactoferrin and immunoglobulins is one of major challenges for the dairy industry. Gel filtration techniques are also used to separate whey proteins from other whey constituents based on their size difference, as well as affinity chromatography, which relies on specific reversible interactions between two biologically active molecules similar to the antigen-antibody principle. Modern applications of continuous chromatography are being successfully used for commercial manufacture of whey protein ingredients like WPI containing >90% proteins and specific fractions of purified bioactive proteins such as lactoferrin, growth factors or immunoglobulins [44, 45].

Current approaches focus on employing WPIs as carriers for various bioactive substances like polyphenols, probiotic bacteria, vitamins or anthocyanins in order to improve their stability and bioavailability in different pharmaceutical and food applications [48, 49]. Such complexes have also demonstrated to act as excellent stabilizers of biphasic dispersions like emulsions and foams, as Nayak *et al.* [50] proved for curcumin-WPI complexes and Chen *et al.* [51] for WPIs used for production of various foam like materials.

Most recently, whey protein hydrolysates have gained a lot of attention, since peptides obtained by enzymatic hydrolysis of WPC or WPI are absorbed slightly better and more quickly than whole proteins. Additionally, hydrolysis results in peptides characterized by better nutritional quality and lower allergenicity. Thus, whey protein hydrolysates obtained by moderate-to-high degree of hydrolysis have a high content of short peptides and are currently used in hypoallergenic infant formulae, sports nutrition drinks, nutritional bars, enteral formulae, and infant formulae. Besides the impact on nutritional characteristics, hydrolysis can also enhance foaming ability by increasing the amount of air incorporated into a foam (overrun). Accordingly, whey protein hydrolysates obtained by low degree of hydrolysis can be used to increase emulsion stability or to change the conditions required for gelation, as well as physical properties of gels. Increased thermal stability is another application associated with whey protein hydrolysates. In this application, the hydrolysis process is designed to produce peptides that are less prone to aggregation when heat processed [42, 53, 54].

8.1.3.5 *Lactose Recovery*

Lactose is the main component in the why dry matter. It is conventionally recovered from sweet or acid whey by crystallization, whereby two main approaches can be distinguished depending on the source—concentrated skimmed and casein-free whey or UF whey permeate. The whole process involves concentration of the feed containing lactose to a high total solids content so that lactose becomes supersaturated and crystallizes. The next step is separation of the lactose crystals by centrifugation [35, 53, 54]. Technological operations as well as the used equipment are very similar to those employed for the manufacture of sucrose from sugar cane. Lactose crystallization is a very complex process, which involves the diffusion of lactose molecules to the surface of a crystal. It occurs spontaneously either due to the super saturation of whey or whey permeate, or as a result of nuclei addition, which is more common. Because of the super saturation, mainly α-monohydrate lactose crystals arise, containing one molecule of water. After the crystallization process was completed, the slurry proceeds into the decanter centrifuge for crystal separation. Thus, it is very important to obtain a large quantity of crystals of similar proportions (approximate diameter of 0.2 mm), which will enable a good separation in this phase of the process [35, 51]. The lactose crystals are then processed by washing and centrifugation in the second decanter, which increases the lactose content to approximately 99% of total solids. A subsequent spray drying reduces the moisture content below 0.5% and results in producing edible lactose. Edible lactose is mostly utilized in the food industry for applications such as the production of infant formulae or baby foods, production of modified and reconstituted dairy products, confectionary products, pastry, pudding, sauces, salad dressings, etc. Lactose can also be used as a carrier of volatile aroma compounds, for adsorption of the unwanted aromas or as a sweetening agent to enhance product aroma [51]. Edible lactose for pharmaceutical applications requires higher degrees of purity, so it has to be additionally refined. Over the past decades, techniques for enhanced crystallization, isolation and purification, hydrolysis, and transformation of lactose chemistry have been developed and implemented around the world. Lactose can also be efficiently separated from whey by strong cation exchange resins based on the different ion- and size-exclusion effects of minerals and proteins, and on the ability of lactose to form complexes with cations bound to strong cation exchange resins. Other efficient methods involve UF and a combination of UF and NF for the lactose separation of from whey [50].

8.1.3.6 *Whey Beverages and Whey Cheese*

Production of whey-based beverages and whey cheese represent the most economical and simplest approaches in whey utilization. Up to now, numerous research studies focused on investigating whey beverage production from sweet or acid whey, powdered, deproteinated, or diluted whey, as well as from permeates remaining after ultrafiltration of whey. More recently alternative food processing methods such high-intensity ultrasound, ohmic heating, ozone processing, pulsed electric field processing or supercritical carbon dioxide technology are intensively being investigated for implementation in whey processing in order to overcome problems like microbiological instability, sediment formation or whey protein denaturation, as well as to improve the properties of existing products [44–52]. All of these research has resulted in placing on the market numerous whey-based beverages as refreshing, value added, and functional foods. Thereby, two main groups can be distinguished—non-alcoholic and alcoholic whey beverages. Non-alcoholic whey beverages can be found as mixtures of whey with fruit or vegetable concentrates and/or extracts, grains, spices, seeds, and similar ingredients, fermented beverages and refreshing beverages with supplemented with carbon dioxide. Regardless of the type, the main idea standing behind developing a certain recipe is to produce a beverage similar to fruit juice, which could be consumed as a breakfast-type, or snack-type beverage or a value added drink rich in proteins, vitamins and minerals. It has been found that the addition of citrus and tropical fruits to whey was the most appropriate for efficient covering of undesirable sour salty taste and the odour of cooked milk [53, 54]. Thus, the addition of orange juice to liquid or powder whey or whey permeate has been extensively investigated by numerous research groups, and the resulting formulations very often involved the addition of CO_2. Thereby the obtained beverages were characterized by satisfactory sensory properties and shelf life [55–58]. Very popular appeared to be also formulations with the addition of tropic fruits such as mango, papaya and passion fruit as well as the addition of herbal extracts like Mentha arvensis, barely, guava, green tea, or cereals [59–62]. However, the formation of undesirable sediment, poor mouthfeel, milky and salty sour taste and odour are still the main issues, which need to be overcome in whey beverage manufacture. It is of crucial importance to define the addition of sufficient amounts of fruit base in order to obtain desirable sensory properties, but to avoid the precipitation of certain components of the fruit dry matter. Also, the application of hydrocolloids, metal gluconate, citric acid, etc. belong to some of the proposed solutions for overwhelming the mentioned hurdles during whey processing into beverages. More recently, it has also been suggested to employ some of non-thermal processing methods like membrane processes, high intensity ultrasound, pulsed electric field or ohmic heating to achieve satisfactory microbiological stabilization and homogenizing effect without applying high temperatures, which in turn result in reduced or no sediment formation [16, 63–66]. Fermented beverages have been recognized by consumers all over the world for their therapeutic value. Since lactose is the main constituent of whey dry matter, there is an ongoing trend in developing novel fermented whey-based to yogurt-like and kefir-like drinks. The main requirement for the selection of starter cultures used for whey fermentations is their ability to metabolize lactose, which most often includes strains of lactic acid bacteria such as *Lactobacillus acidophilus, Lactobacillus delbrueckii* ssp. *bulgaricus, Streptococcus thermophilus, Lactobacillus reuteri, Bifidobacterium bifidum, Lactobacillus rhamnosus, Lactobacillus casei, Lactobacillus plantarum, Lactobacillus helveticus, Bifidobacterium animalis* ssp. *lactis*, and *Lactobacillus paracasei* [54]. Some of the used strains are also probiotics, which additionally enhances the nutritional and therapeutic value of the produced beverage [67–69]. Most recently, the use of acid whey has been proposed for larger utilization in producing fermented beverages by mixing acid whey with milk which is technically and economically viable. However, this approach requires a lot of research in the area of optimizing sensory properties in order to mask the salty sour taste originating from acid whey, as well as investigating optimal ratios for obtaining acceptable rheological properties of the end product [70]. Relying on the same principle, alcoholic beverages are also a possible solution of whey utilization by applying fermentation. Thereby, some authors have investigated the use of yeasts like *Kluyveromyces fragilis, Kluyveromyces marxianus*, and *Saccharomyces lactis* were proposed as suitable strains for this process. The production of whey beverages with a low alcohol content includes removal of proteins and concentration, lactose fermentation, or addition of sucrose until reaching the desired alcohol content (0.5%–1%), flavouring and bottling. A certain amount of lactose is fermented into lactic acid which gives a refreshing and sour taste to the end product, while the rest ferments to alcohol [33, 71]. Whey beer can be produced with or without the addition of malt and can be fortified with minerals. Some of problems that can occur during whey beer production are reduced or complete absence of beer foam due to residual milk fat, undesirable odour and taste resulting from poor solubility of whey proteins, and the choice of yeast strain since many of usual brewing yeasts do not ferment lactose [53].

Whey cheeses have in general not achieved a significant marketing success, but in a few localities like in the Mediterranean, Scandinavian, east and south European countries, whey cheeses belong among the most traditional and most important foods. Two types of whey cheeses are recognized in the textbooks and by the International Dairy Federation—the Italian type or Ricotta –type and the Norwegian or mysost-type. Both of those are principally obtained either through the concentration of whey and the subsequent moulding of the concentrated product or by heat induced coagulation of whey with or without acid addition. Regardless of the type, whey is mainly pre-concentrated prior to the further concentration or coagulation of the whey proteins. The process may also include the addition of milk, cream, or other raw materials of milk origin before or after concentration or coagulation [72, 73]. Italian-type whey cheese Ricotta is essentially a heat–acid-coagulated whey protein paste, sometimes referred to as 'whey quark', and similar products are also popular in Portugal, Turkey, and other localities. The processing technology involves whey heating to at least 90 °C for 10-20 minutes, which results in heat coagulation of whey proteins. In order to promote the coagulation, a certain amount of salt and organic acid (citric, acetic, acid whey) can be added. When the precipitated curd begins to "break," it is removed from the cheese tub or cheese vat and transferred either into perforated plastic moulds and left for draining for the following 4 to 6 hours or into the cloth bags for 6 to 8 hours. Salting can be done during the heating process or after draining. Very often, small amount of milk or cream (up to 10% v/v) is added to the whey prior to heating, in order to achieve better quality and higher yield of whey cheese [51, 73]. This method is widely used for traditional home-made whey cheese, which is commonly produced from sheep whey, due to the significantly higher total solids content. However, the chemical composition of these types of whey cheeses differ from the industrial product because of the whey type (sheep, goat and cow) and the type of used additives (salt, vinegar, acid whey) during the processing [51, 73, 74].

In contrast to the Ricotta-type whey cheese, which contains mostly the protein fraction of whey, the Norwegian-type whey cheese, like Mysost, utilizes all whey components and leaves no residue besides the water vapour. The principle of the Mysost cheese process relies on whey concentration to a maximum. Thereby, whey is supplemented by addition of some other components like cream or goats' milk in the traditional production. By using the concentration method, lactose is also concentrated which results in distinctive characteristics of this cheese type, such as yellowish to brown color and sweet/cooked or caramelized flavour. However, the main problem encountered during this method is also related to lactose concentration—that is the occurrence of lactose crystallization as a result of low solubility of lactose at high concentrations [51, 73, 75]. For that reason, rapid cooling of the hot paste from approximately 95°C to below 65°C is applied, and has to involve constant stirring. There are two basic types of mysost—sliceable or spreadable, which differ according to the moisture and fat content, brown color intensity, addition of goat milk, sweetening agents, or non-dairy ingredients like hazelnuts or chocolate, all of which are applied for obtaining desirable flavor. The largest market of this cheese type is still the country of its origin—Norway. Since the technology of production is relatively simple and results in no waste, its principle is much more useful than that of the Ricotta process, especially for small cheese manufacturers [73, 75]. However, lot of efforts need to be invested into product development in order to optimize sensory profile of the basic Norwegian products to suit the tastes of markets elsewhere.

8.1.4 Bioactive Compounds

Whey is a rich source of nutrients. In addition to naturally present components (proteins, lipids, lactose and micronutrients and macronutrients) in whey, there is variety of other components that occur during processing or intestinal digestion and are called bioactive components since they provide various health benefits. The term bioactive compounds refer to compounds either naturally existing in food or the ones formed or formulated during food processing that may have physiological and biochemical functions when consumed by humans [76]. The most abundant and the most researched bioactive components are related to proteins and arise during their digestion. It is known that proteins have effects that promote health benefits. Peptides derived from whey proteins are currently widely studied. These bioactive peptides are amino acid sequences that are encrypted within the first structure of proteins, which required hydrolysis for their release. The hydrolysis could be through *in vitro* or *in vivo* enzymatic digestion and using microorganisms in fermented systems. The biological activities associated with bio-peptides include immunomodulatory properties, antibacterial, antihypertensive, antioxidant and opioid, etc. These functions are related to general conditions of health or reduced risk of certain chronic illnesses. Also, it is important to mention immunoglobulins and other glycoproteins (lactoferrine, transferrine) end enzymes (lysozyme, lactoperoxidase) which are very notable factors of immunoactive system of whey. They have antimicrobial properties and have ability to reduce or even inhibit allergic reactions [5].

8.1.4.1 *Whey Proteins*

Whey proteins are nutritionally the most valuable components in whey. They are composed of thermo sensitive fractions, such as β-lactoglobulin (β-LG), α-lactalbumin (α-LA), blood serum albumin, and immunoglobulin, as well as thermo stable proteose-peptone (Table 8.3). Another soluble protein found in small amounts is lactoferrin, lactoperoxidase and in case of cheese whey caseino-macropeptide. Whey proteins have a compact globular structure that accounts for their solubility (unlike caseins that exist as a micellar suspension, with a relatively uniform distribution of non-polar, polar and charged groups). These proteins have amino acid profiles quite different from caseins: they have a smaller fraction of Glu and Pro, but a greater fraction of sulfur-containing amino acid residues (i.e., Cys and Met). These proteins are dephosphorylated, easily denatured by heat, insensitive to Ca^{2+}, and susceptible to intramolecular bond formation via disulfide bridges between Cys sulfhydryl groups. Nutritionally

Table 8.3 The constituents of the total whey proteins [78].

Protein	Total whey protein (% w/w)	Molecular weight (Daltons)
β-lactoglobulin (β-LG)	55	18 600
α-lactalboumin (α-LA)	15	14 200
Immunoglobulin A (IgA)	Max 15%	420 000
Immunoglobulin G1 (IgG1)	-	163 000
Immunoglobulin G2 (IgG2)	-	150 000
Immunoglobulin M (IgM)	-	1 000 000
Bovine serum albumin (BSA)	5	66 000
Lactoferrin (Lf)	<1%	76 000
Lysozyme (Ly)	<1%	18 000
Lactoperoxidase (Lp)	<1%	78 000
Casein macropeptides	Less than 10%	7000–33 000
Sulphydryl oxidase	<1%	1 000 000
Superoxide dismutase	<1%	32 000

whey proteins fractions are most valuable proteins because they contain high concentration of essential amino acids (especially lysine, cysteine, and methionine) and high concentration of cystine. They are rich in essential, branched-chain amino acids including leucine, isoleucine, and valine that play crucial roles in metabolism, blood glucose homeostasis, and neural function. Leucine is known to regulate skeletal muscle protein synthesis and cysteine is building block of glutathione, the dietary antioxidant [76]. Because of the desirable amino acid composition, whey proteins have higher biological value when compared to casein, and other proteins of animal origin, including egg, which were considered for a long time as referent proteins. Utilization of proteins in the body is closely related to the ratio of cysteine/methionine, which is about 10 times higher in whey proteins than in casein. Thermally denatured α-lactalbumine is almost completely absorbed in the digestive system compared to casein which one is absorbed only 75% [77]. Daily requirements for the most essential amino acids may be obtained by consuming ~1.5 L of whey or 0.5 L of milk [7]. The protein content in acid and sweet whey is very similar; however, the amount of free amino acids can vary and depends on a degree of casein hydrolysis during the manufacture of cheese (acid or sweet). Thus, the amount of free amino acids in sweet whey is about four times higher, and in acid whey even 10 times higher than in milk [3].

8.1.4.2 Bioactive Peptides

Whey proteins are well documented as an essential source of amino acids, but recently they have been recognized that dietary proteins exert many other functionalities *in vivo* by means of biologically active peptide. Whey is pool of bioactive peptides. Bioactive peptides have been defined as specific protein fragments that have a positive impact on body functions or conditions and may ultimately influence health. Such peptides are inactive within the sequence of the parent protein and can be released by digestive enzymes during gastrointestinal transit or by fermentation or ripening during food processing (Figure 8.2). Upon oral administration, bioactive peptides, may affect the major body systems—namely, the cardiovascular, endocrine, digestive, immune and nervous systems. The beneficial health effects may be classified as antimicrobial, antioxidative, antithrombotic, antidiabetic, antihypertensive, antimicrobial, immunomodulatory, etc. The following peptides were identified in the β-lactoglobulin sequences: β-lactorfin which influences the smooth muscles, β-lactotensin which exhibits hypocholesterolemic and anti-stress activities, and in the α-lactalbumine sequences: α-lactorfin which displays effects similar to that

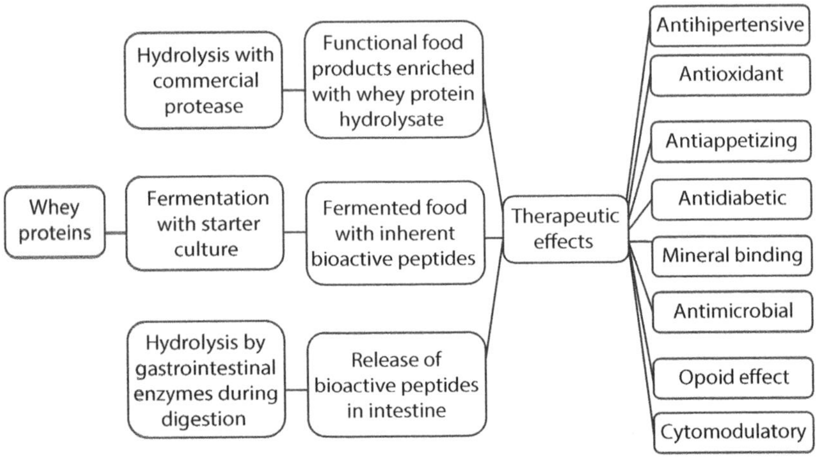

Figure 8.2 Possible action of bioactive peptides from whey proteins [77].

of morphine, namely blood pressure reduction [79]. For this reason, the potential of distinct dietary peptide sequences to promote human health by reducing the risk of chronic diseases or boosting natural immune protection has aroused increasing scientific and commercial interest over the past decade. The activity of these peptides is based on their inherent amino acid composition and sequence. The size of active sequences may vary from 2 to 20 amino acid residues, and many peptides are known to possess multifunctional properties [77].

8.1.4.2.1 Antihypertensive Peptides

Cardiovascular diseases are the major cause of death in humans globally [80]. High blood pressure is a modifiable risk factor for cardiovascular diseases. The angiotensin-converting enzyme (ACE, peptidyldipeptide hydrolase) has been associated with the renin-angiotensin system, which regulates peripheral blood pressure. Inhibition of this enzyme can exert an antihypertensive effect. ACE-inhibitory peptides are the most studied group of bioactive peptides and a large number of such peptides have been isolated from milk proteins hydrolysates. The ACE-inhibiting peptides derived from casein are termed casokinins, whereas those derived from whey (α-LA and β-LG) are called lactokinins [80].

8.1.4.2.2 Antioxidant Peptides

Oxidative stress, the increased production of reactive oxygen species and reactive nitrogen species, combined with overtaking endogenous antioxidant defence mechanisms, is a significant causative factor for the initiation or progression of several lifestyle-mediated diseases [80]. Dietary consumption of antioxidants appears to benefit endogenous antioxidant defence strategies in the fight against oxidative stress [81].

8.1.4.2.3 Antidiabetic Peptides

Diabetes is a serious chronic disease that occurs either when the pancreas does not produce enough insulin (a hormone that regulates blood sugar or glucose), or when the body cannot effectively use the insulin it produces. Diabetes is one of four noncommunicable diseases targeted for action by global researchers. Consumption of milk protein-derived bioactive peptides has been linked with serum glucose regulatory properties in humans. Different mechanisms may include an insulinotropic activity, incretin secretagogue action, as well as activity on different metabolic enzymes involved in the regulation of serum glucose. The insulinotropic properties of milk-derived components have been particularly attributed to whey-derived peptides. It has been shown that whey protein hydrolysates are able to induce an insulinotropic response in humans [77].

8.1.4.2.4 Mineral-Binding Peptides

Due to its prevalence worldwide, osteoporosis is considered a serious public health concern. According to International Osteoporosis Foundation, over 200 million women worldwide suffer from this disease. Worldwide, one in every three women and one in every five men older than 50 years have suffered from osteoporotic fractures. On the other side is anemia, which is also a widespread nutritional deficiency particularly common in children and women. Food supplementation with mineral salts is a challenge for the food industry, since various metal salts may result in changes in the physical and sensory properties of foods [82]. In addition, when ingested in the form of salts, they have low bioavailability, may promote the formation of ROS, and may be responsible for gastric mucosa damage [83]. Radicals can initiate peroxidation of lipids in biological membranes, and cause enzyme inactivation and damage to the DNA structure. Metal-chelating peptides have been identified as potential functional ingredients to improve bivalent mineral bioavailability [82]. Iron-peptide complexes have been considered as an alternative to mitigate the problems related to iron fortification, and have been considered as one of the best choices to replace iron supplements [84]. Whey protein-derived peptides, obtained from proteolytic digestion have shown considerable capacity for binding cations such as calcium, iron, and zinc.

8.1.4.2.5 Opioid Peptides

Opioid peptides are those peptides having pharmacological similarities to opium. Opioid peptide sequences typically contain Tyr–Gly–Gly–Phe at their N-terminal side, and in the case of atypical opioid peptides, a Tyr residue is found at the N-terminus [85]. These peptides have been reported to behave like morphine in the brain. These have been found within milk proteins, both in casein (αs1-, αs2-, β-, and κ-) (casomorphins) and whey proteins (lactorphins). α-Lactorphin, β-lactorphin, and β-lactotensin are opioid peptides that also have antihypertensive functions [4].

8.1.4.2.6 Antimicrobial Peptides

Milk is a rich source of antimicrobial proteins and peptides, capable of exerting antimicrobial activities comparable to antibiotics. This effect is due to the synergistic activity of naturally occurring peptides and defense proteins, such as lactoferrin, lactoperoxidase, and lysozyme and is greater than any individual contribution [86].

8.1.4.3 Lactoferrin

Lactoferrin is an iron-binding glycoprotein, which is less than 1% in total whey proteins. Several biological functions have been attributed to lactoferrin: improved bioavailability of iron, bacteriostatic effects, antioxidant, antiviral, anti-inflammatory, immunomodulatory, and anti-carcinogenic activity. Early research demonstrated lactoferrin to have bacteriostatic activity against intestinal pathogenes, including *Escherichia coli, Helicobacter pylori,* and *Vibrio cholera.* Lactoferrin adheres to the bacterial cell wall, and its strong iron-binding capacity allows it to compete with the bacteria for iron, ultimately depriving the bacteria of an essential nutrient. Iron-binding sites on lactoferrin are typically only approximately 30% saturated in milk allowing lactoferrin to retain significant capacity to bind additional iron. Also, lactoferrin binds directly to the cell surface of bacteria, damaging the external cell membrane through interaction with lipopolysaccharides, causing increased permeability and cell damage from lysozyme activity [87]. Lactoferrin is known to interfere with some of the receptors used by coronaviruses, it may contribute usefully to the prevention and treatment of coronavirus infections. In particular, there is evidence that it can bind to at least some of the receptors used by coronaviruses and thereby block their entry. Of importance are heparan sulfate proteoglycans (HSPGs) and the host receptor angiotensin-converting enzyme 2 (ACE2), as based on other activities lactoferrin might prevent severe acute respiratory syndrome coronavirus 2 (SARS-CoV-2) from attaching to the host cells. Lactoferrin (and more specifically enteric-coated LF because of increased bioavailability) may consequently be of preventive and therapeutic value during the present COVID-19 pandemic [88]. Hydrolysis of lactoferrin by pepsin yields peptides called lactoferricins with higher bacteriostatic activity than lactoferrin [89].

Many new products with bioactive peptides have been launched in the market. There is a growing body of evidence that whey peptides exhibit physiological activities on specific components of the immune response system. In addition, ACE-inhibitory whey peptides can play roles in blood pressure regulation and hypertension; unfortunately, a high ACE-inhibiting activity *in vitro* does not necessarily correlate with a high antihypertensive activity *in vivo*. Release of peptides with high ACE-inhibiting activity is not limited by the protein and enzyme sources, but the highest activities are typically found in peptides released by trypsin. For a candidate peptide to be labeled as bioactive, its resistance to gastrointestinal conditions must be determined in advance. The exact mechanisms by which whey peptides exert their bioactivities upon reaching the intestine need further elucidation; for example, whether their effect is mediated directly in the gut lumen or through receptors on the intestinal cell wall. Therefore, *in vivo* studies are essential not only to validate the physiological effects of tentative bioactive peptides, but also to confirm whether they will require protection from gastrointestinal enzymes when orally administered [4].

8.2 Buttermilk

8.2.1 Introduction

Buttermilk is a by-product that lags after the butter/ghee process making. Usually, buttermilk isn't a problem in the dairy industry since it is often considered as perfectly skimmed milk and after butter production most often it is returned to the consume milk processing. But still, buttermilk is a nutritionally worth product and thus some other perspective of its usage should be considered. Buttermilk contains all ingredients as milk except milk fat which lags in the butter, but milk fat fragments like phospholipids and milk fat globule membrane (MFGM) are present. Generally, buttermilk is rich in milk fat globule membrane, lecithin, valuable proteins, lactose and mineral content. Traditionally, buttermilk is obtained from fermented cream, and it is called sour buttermilk. Buttermilk can also be obtained from sweet cream and it is called sweet cream buttermilk. Sweet cream buttermilk undergoes a fermentation step after separation from the butter. Besides mentioned types, there are cultured buttermilk and whey buttermilk. Cultured buttermilk is produced by fermentation of sweet buttermilk, and some sources describe it as low-fat milk fermented with the type of culture used for fermented cream butter. Whey buttermilk can be produced by churning whey cream, also in the manufacturing process of whey butter.

Worldwide, about 6.5% to 7.0% of produced milk is used for the butter production and thus high (about 3.2 million tons per year) amounts of buttermilk is present as a by-product. The most common usage of buttermilk is beverage production, but it can be also concentrated and dried and used in food industry. Buttermilk has a lot of bioactive components like milk fat membrane globule and phospholipids that promote human health.

Since intensive consumer awareness about food intake and healthy lifestyle in the last decade is present, both researchers and the food industry develop a wide range of products with highlighted bioactive components and new health claims.

8.2.2 Buttermilk Types and Composition

Depends of the butter making process, apropos cream maturation there are several types of buttermilk. Sweet buttermilk, sour buttermilk, cultured

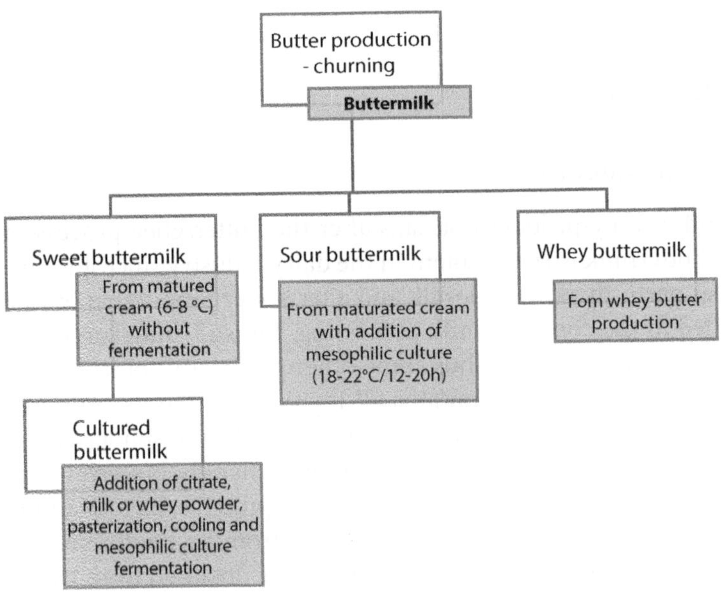

Figure 8.3 Types of buttermilk and obtaining methods.

buttermilk and whey buttermilk. Figure 8.3 shows the types of buttermilk as well as cream maturation obtaining parameters.

8.2.2.1 Sweet Buttermilk

Sweet buttermilk is obtained from the sweet cream which, before the churning process was subjected to physical maturation. Physical maturation is the obligatory step before the churning process [90]. This maturation process is obtained in the duplicator at low temperatures (6–8°C). The main goal of physical maturation is the crystallization and solidification of the milk fat globule which becomes suitable for destabilization during churning process. By the crystallization process, milk fat globule volume decreases due to the binding forces between membrane and water phase, and consequently milk fat membrane bursts. Thus, milk fat globule membrane becomes unstable and liquid fat from the inner side can migrate through membrane during churning [90]. Sweet buttermilk is abundant in MFGM components due to the transfer during churning to the buttermilk.

Sweet buttermilk is usually transferred to the consume milk production line and often is concentrated by evaporation or by spray drying and thus used in the food industry as an ingredient in confectionery and bakery industry as well as in ice cream production and processed cheese foods [57]. Composition on a dry matter basis of sweet buttermilk and other types is given in Table 8.4.

Furthermore, sweet buttermilk can be fermented by mesophilic culture and this type of buttermilk is called cultured buttermilk. In some countries, cultured buttermilk is sour buttermilk described in this chapter and in some it is a low-fat milk fermented with the same type of culture that

Table 8.4 Average composition (%) of different types of buttermilk (sweet, sour, and whey buttermilk) on a dry matter basis [57].

Buttermilk type	Total nitrogen	Fat	Phospholipids	Ash	Lactose	pH
Sweet buttermilk	31.5	13.1	1.27	6.7	48.7	6.61
Sour buttermilk	27.8	22.3	1.15	6.2	43.7	5.39
Whey buttermilk	14.1	15.5	1.87	7.0	63.4	5.98

is used for physical maturation of cream for butter production [91]. In South Asia, cultured buttermilk is called "Chhash" and it is known that has great therapeutic value. Then, in India, cultured buttermilk is most often consumed after the meal and sweetened version of cultured buttermilk is called "lassi" [92]. When sweet buttermilk obtains after the sweet cream churning, it is further processed by addition of milk or whey powder (to adjust dry matter), pasteurization, cooling and fermentation. For the fermentation, usually mesophilic starter cultures containing lactococci and leuconostocs is used [91].

8.2.2.2 *Sour Buttermilk*

Sour buttermilk (Table 8.4) occurs when before churning process cream goes to the biochemical and physical maturation. After the cream pasteurization (usually 85–95°C/10–30 seconds or 100–105°C) and cooling, maturation phase is a next step. First, biochemical maturation occurs and it is carried out by the mesophilic lactic acid bacteria (LAB) cream fermentation and thus sour cream obtains. Mesophilic LAB used for the biochemical maturation usually are homo-fermentative *Lactococcus lactis* subsp. *lactis* or *cremoris* as well as *Lactococcus lactis* subsp. *lactis* biovar. *diacetylactis*, which besides lactose ferment citrates and produce more aroma agents. Furthermore, hetero-fermentative *Leuconostoc mesenteroides* subsp. *cremoris* is also used in a cream fermentation [90]. Biochemical fermentation is carried out at the 18°C to 22°C until pH reaches 4.6 units, i.e., isoelectric point of casein. During the fermentation, lactic acid is produced by the lactose fermentation as well as aroma components like diacetyl and volatile acids. Depending on the fermentation temperature, biochemical maturation can last from 12 to 20 hours. After biochemical maturation physical maturation is a next step. Physical maturation is described in a sweet buttermilk section. Sour buttermilk is most often used as a beverage with or without addition of fruit base, or some functional ingredient addition.

8.2.2.3 *Whey Buttermilk*

Whey buttermilk lags after the whey cream churning in whey butter process obtaining. Composition (Table 8.4) and functional properties of whey buttermilk is different from sweet or sour buttermilk. Whey buttermilk has the lowest total nitrogen content on a dry matter basis compared to the sour and sweet buttermilk and the highest lactose content. Since ratio between proteins and phospholipids in whey buttermilk is the lowest, it has higher emulsification properties and lower foaming ability than sweet or sour buttermilk. Whey cream for the butter production is obtained from cheese whey fat separation, and afterward, whey cream is subjected to the butter production and a by-product whey buttermilk retains [93]. Whey buttermilk compared to the mentioned types of buttermilk is very different due to the lack of casein. Because of that, whey buttermilk is an excellent substrate for the MFGM isolation by membrane filtration [93].

8.2.3 Bioactive Components Isolated From Buttermilk

8.2.3.1 Milk Fat Globule Membrane (MFGM)

Supplementation with MFGM improves cognitive performance, and modulate metabolic functions. Also, it affects the gut microbiota [94], which is important for infant growth and development [95]. The milk fat globule membrane (MFGM) stabilizes and surrounds the milk fat globules, and releases in buttermilk during butter production (churning). MFGM has many health benefits due to its composition rich in membrane proteins and phospholipids [96]. Native milk fat globule membrane (MFGM) has unique composition as well as very complex structure. It contains proteins, glycoproteins, polar and nonpolar lipids, glycosphingolipids, cholesterol, cerebrosides, enzymes, and water. Besides, it contains carotenoids, vitamin A and minerals iron and cupper. MFGM accounts for about 2% to 6% of the total mass of milk fat, and proteins and polar lipids in the MFGM composition represents about 90% of MFGM dry matter. Furthermore, proteins in MFGM account around 1% of the milk fat globule mass, around 1–2% of the total milk protein mass and 25% to 70% of total MFGM mass [97]. Bioactive components from MFGM like phospholipids and glycosphingolipids (they account approximately half of membrane weight) have role in many biological processes in human body like cell growth regulation, adhesion, development and cross membrane trafficking. Due to its biological activity, which was proven *in vivo* and *in vitro* studies, MFGM possess numerous health benefits like inhibition of *Helicobacter pylori*, *Escherichia coli* and rotaviruses [98]. Furthermore, supplementation of food with MFGM has been reported to modulate immunity and metabolic function and improve cognitive performance and also it affects the gut microbiota, important for the infant development and growth [98].

There are many technics for MFGM separation from buttermilk, but they are very challenging, especially if they wanted to be expanded to the industrial scale. Very recently, group of authors [96] develop a new approach for MFGM separation from buttermilk. They have showed the efficacy of hydroxyapatite crystal interacting with MFGM isolates obtained from raw and pasteurized cream. Their results showed a very high affinity of the MFGM for hydroxyapatite crystal and they concluded that this method has a potential for MFGM separation from buttermilk in order to improve its valorization [96]. Furthermore, Ali and Wang [99] investigated the possible effect of pasteurization on the microbial transglutaminase as an enzyme-protein cross linking in order to isolate MFGM. Their investigation brought interesting results that increase of κ-casein in pasteurized sample was an indicator of a possible κ-casein association with MFGM and/or interaction with whey proteins (β-lactoglobulin) based upon heating. However, they have concluded that pasteurization might induce even more cross-linking of milk proteins and thermal treatment reduced microbial transglutaminase ability and show impact on MFGM and non-MFGM proteins interaction.

8.2.3.2 Phospholipids in Buttermilk

Generally, phospholipid content of buttermilk is greater than that in milk due to the migration of fragmented milk fat globule membranes disrupted during churning process to the aqueous phase [100]. The most common usage of phospholipids in food industry is its incorporation into the food processing as an emulsifier. Phospholipids have amphipathic nature due to the both a hydrophilic head and a hydrophobic tail. There are multitude researches that report about milk fat globule membrane (MFGM) phospholipids and they have received much attention, owing to their stability, composition, and health. Phospholipids (PL) in MFGM accounts about 25% of MFGM components. PL are mainly composed of phosphatidylcholine (PC), phosphatidylethanolamine (PE) and sphingomyelin (SM), and smaller amounts of phosphatidylinositol (PI), phosphatidylserine (PS) and lysophosphatidylcholine (LPC) [101]. PL, besides emulsifying properties, some components like SM have been associated with health benefits like improving cell regulation and growth, protecting against hypercholesterolemia and suppressing Alzheimer's disease. Considering the health benefits and technological use of MFGM phospholipids, efforts have been made for its concentration from buttermilk powder [102]. There are two main approaches that have been implemented for PL isolation. The first approach is based on the extraction of polar and nonpolar lipid from buttermilk powder using hexane, chloroform, ethanol, petroleum ether, methanol and N,N-dimethylcyclohexylamine as solvents. The other approach is based on increase of MFGM phospholipids concentration by removing other components like proteins, sugars and so on. Furthermore, membrane filtration in a combination with supercritical carbon dioxide

(SC-CO_2) extraction has been investigated for the concentration of phospholipids in buttermilk powder. Furthermore, Ubeyitogullari and Rizvi [101] have isolated phospholipid-rich lipid extract from buttermilk powder using a green technology by super critical (SC) CO_2. Their goal was to obtain the highest phospholipid recovery from buttermilk powder. The optimized process conditions for SC-CO_2 extraction conditions were 60°C, 30 MPa, and 15% ethanol where process resulted in a 6.3% total lipid yield. Environmental friendly method resulted in the extract which contained even 49% phospholipids [101]. Phospholipid extract can be used as a very good source of emulsifier in functional food production.

8.2.4 Buttermilk-Based Beverages

Recent and the widest usage of buttermilk is in a form of beverage. Buttermilk-based beverages are usually with a fruit addition or with some functional ingredient addition or they can be also carbonated. Some data shows that the global functional dairy beverage market reached US$ 13.9 billion in a period between 2016 and 2021 year [103].

The most common fruit addition or fruit aroma addition to the buttermilk are mango, orange, and banana juice bases. The amounts of added fruit bases into the buttermilk ranges between 10% and 35%. Mersham [104] develop mango buttermilk beverage composed with the 81% buttermilk, 12% mango juice, 7% sugar, and 0.05% pectin. Furthermore, some authors have developed buttermilk beverages with the addition of fruit juices or pulp like soursop and strawberry [57]. Sensory analysts have evaluated beverages with a high sensory score and consumers accept all produced beverages.

Since modern consumer tends to functional ingredient intake during food consumption, buttermilk presents great substrate for its fortification. Mudgil *et al.* [105] have developed product based on cultured buttermilk with *Aloe vera* juice addition. They have fortified buttermilk with *Aloe vera* juice in amounts from 5% to 20%. Results show that addition of *Aloe vera* juice decreased the phase separation and increased the viscosity of the beverage whereas viscosity increased proportionally with increased levels of juice. Sensory evaluation showed that addition of 10% *Aloe vera* obtained the highest sensory score as well as it improved nutritive and physicochemical characteristics. Furthermore, Mudgil and Barak [106] fortified buttermilk with soluble fiber, i.e., with partially hydrolyzed guar gum. Since guar gum is a very good source of soluble fibre, its addition in buttermilk can promote health benefits when consumed. They have optimized beverage with 4% of soluble fiber and thus improved both physicochemical and sensory properties. Fibers, in this case guar gum, are often used in food industry as a stabilizer and thickener and thus reduce the phase separation. Turkmen *et al.* [107] described probiotic buttermilk whey-based beverages. This type of beverages is not so common as cheese whey or milk probiotic drinks. It is successfully formulated a buttermilk-based probiotic beverage enriched with various flavours and carbohydrate sources and acceptable sensory quality whereas probiotic count was high after 28 days of storage.

Generally, by carbonation, beverage shelf life can be prolonged. Since buttermilk has all ingredients for microbial growth, carbonation can prolong its shelf life even up to 30 days [57]. The process for the preparation of carbonated fruit-based buttermilk has been developed and has highly acceptable sensory and physicochemical properties. The best beverage, scored by sensory panel, was fruit carbonated buttermilk with the addition of 24% pineapple juice and 12% sugar [108].

References

1. Jelen, P., Whey processing, in: *Encyclopedia of Dairy Sciences*, 2nd edition, vol. 4, Fuquay (Ed.), p. 2740, Academic Press - An Imprint of Elsevier, USA, 2003.
2. Bylund, G., *Dairy processing handbook*, Tetra Pak, Lund, Sweden, 2015.
3. Božanić, R., Barukčić, I., Lisak Jakopović, K., Tratnik, L., Possibilities of whey utilisation. *Austin J. Nutr. Food Sci.*, 2, 7, 1, 2014.
4. Madureira, A.R., Tavares, T., Gomes, A.M.P., Pintado, M.E., Malcata, F.X., Invited review: Physiological properties of bioactive peptides obtained from whey proteins. *J. Dairy Sci.*, 93, 2, 437, 2010.
5. Jeličić, I., Božanić, R., Tratnik, L., Whey-based beverages – A new generation of dairy products. *Mljekarstvo*, 58, 3, 275, 2008.
6. Kwak, H.S., Lee, W.J., Lee, M.R., Revisiting lactose as an enhancer of calcium absorption. *Int. Dairy J.*, 22, 2, 147, 2012.
7. Popović-Vranješ, A. and Vujičić, I., *Tehnologija surutke*, Poljoprivredni fakultet Novi Sad, Novi Sad, 1997.
8. Chatzipaschali, A.A. and Stamatis, A.G., Biotechnological utilization with a focus on anaerobic treatment of cheese whey: Current status and prospects. *Energies*, 5, 3492–3525, 2012.
9. Schuck, P., Milk powder: Types and manufacture, in: *Encyclopedia of Dairy Sciences*, 2nd Edition, H. Roginski, J.W. Fuquay, P.F. Fox, (Eds.), pp. 108–117, Academic Press, London, UK, 2011.

10. Henriques, M., Gomes, D., Rodrigues, D., Pereira, C., Gil, M., Performance of bovine and ovine liquid whey protein concentrate on functional properties of set yoghurts. *Proc. Food Sci.*, 1, 2007–2014, 2011.
11. Ganju, S. and Gogate, P.R., A review on approaches for efficient recovery of whey proteins from dairy industry effluents. *J. Food Eng.*, 215, 84–96, 2017.
12. Ma, J.-J., Xue-Ying Mao, X.-Y., Wang, Q., Yang, S., Dan Zhang, D., Shang-Wu Chen, S.-W., Li, Y.-H., Effect of spray drying and freeze drying on the immunomodulatory activity, bitter taste and hygroscopicity of hydrolysate derived from whey protein concentrate. *LWT*, 56, 296–30, 2014.
13. Nishanthi, M., Chandrapala, J., Vasiljevic, T., Compositional and structural properties of whey proteins of sweet, acid and salty whey concentrates and their respective spray dried powders. *Int. Dairy J.*, 74, 49–56, 2017.
14. Miloradovic, Z.N., Macej, O.D., Kljajevic, N.V., Jovanovic, S.T., Vucic, T.R., Zdravkovic, I.R., The effect of heat treatment of caprine milk on the composition of cheese whey. *Int. Dairy J.*, 58, 39–42, 2016.
15. Saxena, A., Tripathi, B.P., Kumar, M., Shahi, V.K., Membrane-based techniques for the separation and purification of proteins: An overview. *Adv. Colloid Interface Sci.*, 145, 1–22, 2009.
16. Barukčić, I., Božanić, R., Kulozik, U., Influence of process temperature and microfiltration pre-treatment on flux and fouling intensity during cross-flow ultrafiltration of sweet whey using ceramic membranes. *Int. Dairy J.*, 51, 1–7, 2015.
17. Kumar, P., Sharma, N., Ranjan, R., Kumar, S., Bhat, Z.F., Kee Jeong, D., Perspective of membrane technology in dairy industry: A review. *Asian-Australas. J. Anim. Sci.*, 26, 1347–1358, 2013.
18. Steinhauer, T., Schwing, J., Krauß, S., Kulozik, U., Enhancement of ultrafiltration-performance and improvement of hygienic quality during the production of whey concentrates. *Int. Dairy J.*, 45, 8–14, 2015.
19. Baldasso, C., Barros, T.C., Tessaro, I.C., Concentration and purification of whey proteins by ultrafiltration. *Desalination*, 278, 381–386, 2011.
20. Tolkach, A. and Kulozik, U., Transport of whey proteins through 0.1 mm ceramic membrane: Phenomena, modelling and consequences for concentration or diafiltration processes. *Desalination*, 199, 1–3, 340–341, 2006.
21. Norazman, N., Wu, W., Li, H., Wasinger, V., Zhang, H., Chen, V., Evaluation of chemical cleaning of UF membranes fouled with whey protein isolates via analysis of residual protein components on membranes surfaces. *Sep. Purif. Technol.*, 103, 241–250, 2013.
22. Tolkach, A. and Kulozik, U., Fractionation of whey proteins and caseinomacropeptide by means of enzymatic crosslinking and membrane separation techniques. *J. Food Eng.*, 67, 1–2, 13–20, 2005.
23. Qasim, M., Darwish, N.N., Mhiyo, S., Darwish, N.A., Hilal, N., The use of ultrasound to mitigate membrane fouling in desalination and water treatment. *Desalination*, 443, 143–164, 2018.

24. Khaire, R.A., Sunny, A.A., Gogate, P.R., Ultrasound assisted ultrafiltration of whey using dual frequency ultrasound for intensified recovery of lactose. *Chem. Eng. Process.*, 142, 107581, 2019.
25. Demers-Mathieu, V., Gauthier, S.F., Britten, M., Fliss, I., Robitaille, G., Jean, J., Antibacterial activity of peptides extracted from tryptic hydrolyzate of whey protein by nanofiltration. *Int. Dairy J.*, 28, 2, 94–101, 2013.
26. Butylina, S., Luque, S., Nyström, M., Fractionation of whey-derived peptides using a combination of ultrafiltration and nanofiltration. *J. Membr. Sci.*, 280, 418–426, 2006.
27. Bédas, M., Tanguy, G., Dolivet, A., Méjean, S., Gaucheron, F., Garric, G., Senard, G., Jeantet, R., Schuck, P., Nanofiltration of lactic acid whey prior to spray drying: Scaling up to a semi-industrial scale. *LWT*, 79, 355–360, 2017.
28. Chandrapala, J., Chen, G.Q., Kezia, K., Bowman, E.G., Vasiljevic, T., Kentish, S.E., Removal of lactate from acid whey using nanofiltration. *J. Food Eng.*, 177, 59–64, 2016.
29. Das, B., Sarkar, S., Sarkar, A., Bhattacharjee, A., Bhattacharjee, C., Recovery of whey proteins and lactose from dairy waste: A step towards green waste management. *Process Saf. Environ. Prot.*, 101, 27–33, 2016.
30. Yorgun, M.S., Akmehmet Balcioglu, I., Saygin, O., Performance comparison of ultrafiltration, nanofiltration and reverse osmosis on whey treatment. *Desalination*, 229, 1–3, 204–216, 2008.
31. Nguyen, M., Reynolds, N., Vigneswaran, S., By-product recovery from cottage cheese production by nanofiltration. *J. Cleaner Prod.*, 11, 7, 803–807, 2003.
32. Barukčić, I., *Optimizing efficiency of whey microfiltration and ultrafiltration by applying ceramic membranes—A dissertation*, University of Zagreb, Zagreb, Croatia, 2013.
33. Jeličić, I., Božanić, R., Tratnik, L., Whey-based beverages- A new generation of dairy products. *Mljekarstvo*, 58, 3, 257–274, 2008.
34. Gernigon, G., Schuck, P., Jeantet, R., Burling, H., Demineralisation, in: *Encyclopedia of Dairy Science*, 2nd edition, H. Roginski, J.W. Fuquay, P.F. Fox, (Eds.), pp. 738–744, Academic Press, London, UK, 2011.
35. de Witt, J.N., *Lecturer's handbook on whey and whey products*, European Whey products Association, Brussels, Belgium, 2011, Available at: http://ewpa.euromilk.org/fileadmin/user_upload/Public_Documents/EWPA_Publications/Lecturer_s_Handbook_on_Whey.pdf.
36. Strathmann, H., Electrodialysis, in: *Encyclopedia of Separation Science*, I.D. Wilson, E.R. Adlard, M. Cooke, C.F. Poole (Eds.), pp. 1707–1717, Academic Press, London, UK, 2000.
37. Diblíková, L., Curda, L., Kincl, J., The effect of dry matter and salt addition on cheese whey demineralisation. *Int. Dairy J.*, 31, 29–33, 2013.
38. Goodall, S., Grandison, A.S., Jauregi, P.J., Price, J., Selective separation of the major whey proteins using ion exchange membranes. *J. Dairy Sci.*, 91, 1–10, 2008.

39. Alkhatim, H.S., Alcaina, M.I., Soriano, E., Iborra, M.I., Treatment of whey effluents from dairy industries by nanofiltration membranes. *Desalination*, 199, 177–184, 1998.
40. Mazzoni, C., Bruni, L., Baudini, S., Nanofiltration: Role of the electrolyte and pH on Desal DK performances. *Ind. Eng. Chem. Res.*, 46, 2254–2262, 2007.
41. Pan, K., Song, Q., Wang, L., Cao, B., Study of demineralization of whey by nanofiltration membrane. *Desalination*, 267, 217–221, 2011.
42. Foegeding, E.A., Luck, P., Vardhanabhuti, B., Whey protein products, in: *Encyclopedia of Dairy Science*, 2nd edition, H. Roginski, J.W. Fuquay, P.F. Fox, (Eds.), pp. 873–878, Academic Press, London, UK, 2011.
43. Yee, K.W.K., Wiley, D.E., Bao, J., Whey protein concentrate production by continuous ultrafiltration: Operability under constant operating conditions. *J. Membr. Sci.*, 290, 1–2, 125–137, 2007.
44. Smithers, G.W., Whey and whey proteins—From 'gutter-to-gold'. *Int. Dairy J.*, 18, 695–704, 2008.
45. Smithers, G.W., Whey-ing up the options. Yesterday, today and tomorrow. *Int. Dairy J.*, 48, 2–14, 2015.
46. L. Bazinet, J. Amiot, J.-F. Poulin, D. Labbé, D. Tremblay, Process and system for separation of organic charged compounds, WO 2005/082495 A1, 2005.
47. Wang, Q., Chen, G.Q., Kentish, S.E., Isolation of lactoferrin and immunoglobulins from dairy whey by an electrodialysis with filtration membrane process. *Sep. Purif. Technol.*, 233, 115987, 2020.
48. Abbasi, A., Emam-Djomeh, Z., Mousavi, M.A.E., Davoodi, D., Stability of vitamin D3 encapsulated in nanoparticles of whey protein isolate. *Food Chem.*, 143, 379–383, 2014.
49. Chung, C., Rojanasasithara, T., Mutilangi, W., McClements, D.J., Enhanced stability of anthocyanin-based colour in model beverage systems through whey protein isolate complexation. *Food Res. Int.*, 76, 3, 761–768, 2015.
50. Nayak, A., Genot, C., Meynier, A., Dorlando, A., Capron, I., Impact of process and physico-chemical conditions on the formation of curcumin-whey protein composite particles capable to stabilize food-compatible oil in water emulsions. *LWT*, 153, 112421, 2022.
51. Chen, H., Wanga, Y., Schiraldi, D.A., Foam-like materials based on whey protein isolate. *Eur. Polym. J.*, 49, 3387–3391, 2013.
52. Rahmani-Manglano, N.E., Jones, N.C., Hoffmann, S.V., Guadix, E.M., Pérez-Gálvez, R., Guadix, A., García-Moreno, P.J., Structure of whey protein hydrolysate used as emulsifier in wet and dried oil delivery systems: Effect of pH and drying processing. *Food Chem.*, 390, 133169, 2022.
53. Mehra, R., Kumar, H., Kumar, N., Ranvir, Jana, A., Buttar, H.S., Telessy, I.G., Godswill Awuchi, C., Okpala, C.O.R., Korzeniowska, M., Guiné, R.P.F., Whey proteins processing and emergent derivatives: An insight perspective from constituents, bioactivities, functionalities to therapeutic applications. *J. Funct. Foods*, 87, 104760, 2021.

54. Harju, M., Kallioinen, H., Tossavainen, O., Lactose hydrolysis and other conversions in dairy products: Technological aspects. *Int. Dairy J.*, 22, 104–109, 2012.
55. Božanić, R., Barukčić, I., Jakopović, K.L., Tratnik, L., Possibilities of whey utilisation. *Austin J. Nutr. Food Sci.*, 2, 7, 7, 2014.
56. Barukčić, I., Lisak Jakopović, K., Božanić, R., Valorization of whey and buttermilk for production of functional beverages – An overview of current possibilities. *Food Technol. Biotechnol.*, 57, 4, 448–460, 2019.
57. Barukčić, I., Lisak Jakopović, K., Božanić, R., Whey and buttermilk—Neglected sources of valuable beverages, in: *Natural Beverages*, A.M. Grumezescu, and A.M. Holban, (Eds.), pp. 209–242, Academic Press Eslevier Inc., Cambridge, USA, 2019.
58. Chavan, R.S., Shraddha, R.C., Kumar, A., Nalawade, T., Whey based beverage: Its functionality, formulations, health benefits and applications. *J. Food Process. Technol.*, 6, 10, 2–8, 2015.
59. Sady, M., Jaworska, G., Grega, T., Bernas, E., Domagal, J., Application of acid whey in orange drink production. *Food Technol. Biotechnol.*, 51, 2, 266–277, 2013.
60. Pareek, N., Gupta, A., Sengar, R., Preparation of healthy fruit based carbonated whey beverages using whey and orange juice. *Asian J. Dairy Food Res.*, 33, 1, 5–8, 2014.
61. Chatterjee, G., De Neve, J., Dutta, A., Das, S., Formulation and statistical evaluation of a ready-to-drink whey based orange beverage and its storage stability. *Rev. Mex. Ing. Quım.*, 14, 2, 253–264, 2015.
62. Kumar, P.A. and Bangaraiah, P., Formulation of whey-sweet orange based ready-toserve fruit beverage. *Int. J. Pharma Bio. Sci.*, 5, 4, 101–1111, 2014.
63. Chavan, R.S., Nalawade, T., Kumar, A., Studies on the development of whey based mango beverage. *Res. Rev.: J. Food Dairy Technol.*, 3, 2, 1–6, 2015.
64. Valadao, N.K., Andrade, I.M.G., Jory, J.C., Fabio Augusto Gallo, F.A., Petrus, R.R., Development of a Ricotta cheese whey-based sports drink. *J. Adv. Dairy Res.*, 4, 3, 156–162, 2016.
65. Yadav, R.B., Yadav, B.S., Kalia, N., Development and storage studies on whey-based banana herbal (Mentha arvensis) beverage. *Am. J. Food Technol.*, 5, 2, 121–129, 2010.
66. Singh, D., Singh, R., Bhatt, F., Development, quality evaluation and shelf life studies of whey guava beverage. *Int. J. Curr. Eng. Technol.*, 4, 3, 2171–2175, 2014.
67. Barukčić, I., Božanić, R., Kulozik, U., Effect of pore size and process temperature on flux, microbial reduction and fouling mechanisms during sweet whey cross-flow microfiltration by ceramic membranes. *Int. Dairy J.*, 39, 8–15, 2014.
68. Jeličić, I., Božanić, R., Brnčić, M., Tripalo, B., Influence and comparison of thermal, ultrasonic and thermo-sonic treatments on microbiological quality

and sensory properties of rennet cheese whey. *Mljekarstvo*, 62, 3, 165–178, 2012.
69. Singh, R., Rathod, G., Meletharayil, G.H., Kapoor, R., Sankarlal, V.M., Amamcharla, J.K., Invited review: Shelf-stable dairy protein beverages—Scientific and technological aspects. *J. Dairy Sci.*, 105, 12, 9327–9346, 2022.
70. Ferreira, M.V.C., Cappato, L.P., Silva, R., Rocha, R.S., Guimarães, J.T., Balthazar, C.F., Esmerino, E.A., Freitas, M.Q., Rodrigues, F.N., Granato, D., Neto, R.P.C., Tavares, M.I.B., Silva, P.H.F., Raices, R.S.L., Silva, M.C., Cruz, A.G., Ohmic heating for processing of whey-raspberry flavored beverage. *Food Chem.*, 297, 125018, 2019.
71. Marsh, A.J., Hill, C., Rossa, R.P., Cotter, P.D., Fermented beverages with health-promoting potential: Past and future perspectives. *Trends Food Sci. Technol.*, 38, 2, 113–124, 2014.
72. Rocha-Mendoza, D., Kosmerl, E., Krentz, A., Zhang, L., Badiger, S., Miyagusuku-Cruzado, G., Mayta-Apaza, A., Giusti, M., Jiménez-Flores, R., García-Cano, I., Invited review: Acid whey trends and health benefits. *J. Dairy Sci.*, 104, 2, 1262–1275, 2021.
73. Dragone, G., Mussatto, S.I., Oliveira, J.M., Teixeira, J.A., Characterisation of volatile compounds in an alcoholic beverage produced by whey fermentation. *Food Chem.*, 112, 929–935, 2009.
74. Codex Alimentarius, *Milk and milk products*, World Health Organization; Food and Agriculture Organization of the United Nations, Rome, 2011.
75. Barukčić, I. and Tudor Kalit, M., Origin, production and specificities of croatian traditional cheeses, in: *Cheeses Around the World - Types, Production, Properties and Cultural and Nutritional Relevance*, P.F.R. Guine, M.R.P. Correia, A.C. Ferrao, (Eds.), pp. 153–182, Nova Science Publishers, New York, USA, 2019.
76. Lisak Jakopović, K., Barukčić, I., Božanić, R., Bioactive components derived from bovine milk. *Mljekarstvo*, 69, 3, 151, 2019.
77. Mann, B., Athira, S., Sharma, R., Kumar, R., Sakar, P., Bioactive peptides from whey proteins, in: *Whey Proteins. From Milk to Medicine*, Academic Press Elsevier, Cambridge, MA, USA, 2019.
78. Petrotos, K., Tsakali, E., Goulas, P., D'Alessandro, A.G., Casein and whey proteins in human health, in: *Milk and Dairy Products as Functional Foods*, Wiley-Blackwell, 2014.
79. Krolczyk, J.B., Dawidziuk, T., Janiszewska-Turak, E., Solowiej, B., Use of whey and whey preparations in the food industry – A review. *Pol. J. Food Nutr. Sci.*, 66, 3, 157, 2016.
80. Moskovitz, J., Yim, M.B., Chock, P.B., Free radicals and disease. *Arch. Biochem. Biophys.*, 397, 354, 2002.
81. Fang, Y.Z., Yang, S., Wu, G., Free radicals, antioxidants, and nutrition. *Nutrition*, 18, 872, 2002.
82. Guo, L., Harnedy, P.A., Li, B., Hou, H., Zhang, Z., Zhao, X., FitzGerald, R.J., Food protein derived chelating peptides: Biofunctional ingredients for

dietary mineral bioavailability enhancement. *Trends Food Sci. Technol.*, 37, 92, 2014.

83. Chaud, M.V., Izumi, C., Nahaal, Z., Shuhama, T., Bianchi, M.D.L.P., Freitas, O.D., Iron derivatives from casein hydrolysates as a potential source in the treatment of iron deficiency. *J. Agric. Food Chem.*, 50, 871, 2002.
84. Wang, P.F., Huang, G.R., Jiang, J.X., Optimization of hydrolysis conditions for the production of iron-binding peptides from mackerel processing by-products. *Adv. J. Food Sci. Technol.*, 5, 921, 2013.
85. Teschemacher, H., Koch, G., Brantl, V., Milk protein derived opioid receptor ligands. *Pept. Sci.*, 43, 99, 1997.
86. Clare, D.A., Catignani, G.L., Swaisgood, H.E., Biodefense properties of milk: The role of antimicrobial proteins and peptides. *Curr. Pharm. Des.*, 9, 1239, 2003.
87. Austad, N. and Layman, D.K., Dairy bioactive proteins and peptides: A narrative review. *Nutr. Rev.*, 79, S2, 36, 2021.
88. Kell, D.B., Heyden, E.L., Pretorius, E., The biology of lactoferrin, an iron-binding protein that can help defend against viruses and bacteria. *Front. Immunol.*, 11, 1221, 2020.
89. Tsermoula, P., Khakimov, B., Holm Nielsen, J., Balling Engelsen, S., WHEY - The waste-stream that became more valuable than the food product. *Trends Food Sci. Technol.*, 118, A, 230, 2021.
90. Tratnik, L. and Božanić, R., *Milk and milk products*, Croatian Dairy Union, Zagreb, 2012.
91. Narvhus, J.A. and Abrahamsen, R.K., *Buttermilk products*, pp. 409–416, Encyclopedia of Dairy Sciences, Academic Press, USA, 2022.
92. Mudgil, D., Barak, S., Darji, P., Development and characterization of functional cultured buttermilk utilizing Aloe vera juice. *Food Biosci.*, 15, 105–109, 2016.
93. Morin, P., Pouliot, Y., Jiménez-Flores, R., A comparative study of the fractionation of regular buttermilk and whey buttermilk by microfiltration. *J. Food Eng.*, 77, 3, 521–528, 2016.
94. Lee, K., Kim, S., Kim, A., Suh, K.J., Hong, K.B., Sphingolipid identification and skin barrier recovery capacity of a milk sphingolipid-enriched fraction (MSEF) from buttermilk powder. *Int. J. Cosmet. Sci.*, 42, 270–276, 2020.
95. Breij, L.M., Abrahamse-Berkeveld, M., Vandenplas, Y., Jespers, S.N., de Mol, A.C., Khoo, P.C., An infant formula with large, milk phospholipid–coated lipid droplets containing a mixture of dairy and vegetable lipids supports adequate growth and is well tolerated in healthy, term infants. *Am. J. Clin. Nutr.*, 109, 3, 586–596, 2019.
96. Iung, J., Pouliot, Y., Remondetto, G., Jiménez-Flores, R., Brisson, G., On the use on hydroxyapatite suspensions for the separation of milk fat globule membrane components from buttermilk. *Food Chem.*, 404, 134535, 2023.

97. Kiełczewska, K., Ambroziak, K., Krzykowska, D., Aljewicz, M., The effect of high-pressure homogenisation on the size of milk fat globules and MFGM composition in sweet buttermilk and milk. *Int. Dairy J.*, 113, 104898, 2021.
98. Fuller, K.L., Kuhlenschmidt, T.B., Kuhlenschmidt, M.S., Jiménez-Flores, R., Donovan, S.M., Milk fat globule membrane isolated from buttermilk or whey cream and their lipid components inhibit infectivity of rotavirus *in vitro*. *J. Dairy Sci.*, 96, 3488–3497, 2022.
99. Ali, F. and Wang, Z.X., Effect of pasteurization on the enzymatic cross-linking of milk proteins by microbial transglutaminase in view of milk fat globule membrane isolation. *Food Biosci.*, 43, 101100, 2021.
100. Ferreiro, T., Martínez, S., Gayoso, L., Rodríguez-Otero, J.L., Evolution of phospholipid contents during the production of quark cheese from buttermilk. *J. Dairy Sci.*, 99, 4154–4159, 2016.
101. Ubeyitogullari, A. and Rizvi, S.S.H., Production of high-purity phospholipid concentrate from buttermilk powder using ethanol-modified supercritical carbon dioxide. *J. Dairy Sci.*, 103, 8796–8807, 2020.
102. Huang, Z., Zheng, H., Brennan, S.C., Mohan, S.M., Stipkovits, L., Li, L., Kulasiri, D., Production of milk phospholipid-enriched dairy ingredients. *Foods*, 9, 263, 2020.
103. Rotta, E.M., Giroux, H.J., Lamothe, S., Bélanger, D., Sabik, H., Visentainer, V.J., Britten, M., Use of passion fruit seed extract (Passiflora edulis Sims) to prevent lipid oxidation in dairy beverages during storage and simulated digestion. *LWT*, 123, 109088, 2020.
104. Meshram, B.D., Buttermilk based fruit juice beverages. *Asian J. Dairy Food Res.*, 34, 4, 297–299, 2015.
105. Mudgil, D., Barak, S., Darji, P., Development and characterization of functional cultured buttermilk utilizing Aloe vera juice. *Food Biosci.*, 15, 105–109, 2016.
106. Mudgil, D. and Barak, S., Development of functional buttermilk by soluble fiber fortification. *Agro Food Ind. Hi Tech.*, 27, 2, 44–47, 2016.
107. Turkmen, N., Akal, C., Özer, B., Probiotic dairy-based beverages: A review. *J. Funct. Foods*, 53, 62–75, 2019.
108. Burhanuddin Shaikh, M.F. and Dagdulal Rathi, S., Utilisation of buttermilk for the preparation of carbonated fruit-flavored beverages. *Int. J. Dairy Technol.*, 62, 564–570, 2009.

9

Valorization of Wastes and By-Products From the Meat Industry

Senthilkumar Rathnasamy[1]*, Vivek Rangarajan[2], Ramya Munisamy[1], Harish babu Balaraman[1], Subhranshu Samal[2], Vinoth Kannan S.R.[2], Sampatrao Dagu Manjare[2] and Umile Gianfranco Spizzirri[3]

[1]Green Separation Engineering Laboratory, SASTRA Deemed to be University, Thanjavur, Tamil Nadu, India
[2]Department of Chemical Engineering, Birla Institute of Technology and Science - Pilani, K.K. Birla Goa Campus, Zuarinagar, Goa, India
[3]Department of Pharmacy, Health and Nutritional Sciences, University of Calabria, Rende, Italy

Abstract
Sustainable valorization of meat waste has been increasing exponentially in recent decades. The ever-increasing demand for high-value nutraceuticals could be mitigated by exploring their availability in these potential wastes. Besides, the effective reutilization of these waste materials aid in generating a cleaner environment. This chapter primarily focuses on the elucidation of utilizing meat wastes in various methods to generate high-value bioactive compounds. The augmentation of various production and enrichment methods to obtain these bioactive molecules from these potential wastes has been illustrated. It also emphasizes the role of various sustainable solvent systems employed in isolating these bioactive molecules. Further, concluding remarks on the effect of valorization on the environment are also discussed in detail.

Keywords: Valorization, meat waste, nutraceuticals, enrichment, functional foods

**Corresponding author*: senthilrathna@sastra.ac.in

Umile Gianfranco Spizzirri (ed.) Nutraceutics from Agri-Food By-Products, (285–318) © 2023 Scrivener Publishing LLC

9.1 Introduction: Meat Wastes—Sources and Economic Importance

Meat is the second largest food source for primates including human beings. It is considered a staple food in various parts of the world [1]. Globally, the annual consumption of livestock-based meat is estimated to be around 300 million metric tons by 2025 which is expected to generate an estimated amount of 15 million tons of unprocessed and processed meat waste respectively. Meat processing industries (slaughterhouses) generate a huge amount of unprocessed solid wastes (bones, cartilage, skin and intestines) and liquid wastes (washing water, blood & fat). On the other hand, processed meat wastes like offal and undercooked meats equally contribute to the by-products of the meat processing industries. Currently, the solid wastes are being mismanaged ending up in landfills creating a health hazard or anaerobically digested to produce methane, increasing the greenhouse effect [2, 3]. On the other hand, liquid wastes are being deposited in the municipal sewerage system increasing the possibility of groundwater contamination and resulting in various health hazards [4]. With the increase in the global population, consumption of meat products and subsequent generation of meat by-products are estimated to increase exponentially. On the other hand, the absence of organized disposal methods for these wastes has posed a serious issue in various developing and developed countries across the world. This has instigated the need for a standard centralized waste management system applicable across all countries and governed by uniform international laws.

Sustainable utilization of meat by-products has been in existence for a long period of time. Pre-historic events confirm the usage of animal skins as clothing materials and bones for making ornaments [5]. On the other hand, feather residues have been prevalently stuffed inside animal skins for making pillows or winter jackets [6]. With the development of civilization, the effective use of meat processing wastes has evolved meticulously. The fat extracted from raw meat is used as cooking oil or additive in pharmaceutical and bioprocess industries [7, 8]. Further, the unused meat residues and liquid wastes were processed to act as additives in the production of microbial media. With the rise of awareness in sustainability and circular economy over recent decades, the altitude of meat waste management has taken a revolution. Nowadays, peer research individuals concentrate more on effective complete utilization of the generated waste rather than dealing with parts and bits. Additionally, these wastes are exposed to a sequence of valorization approaches which deliver commercially valuable products

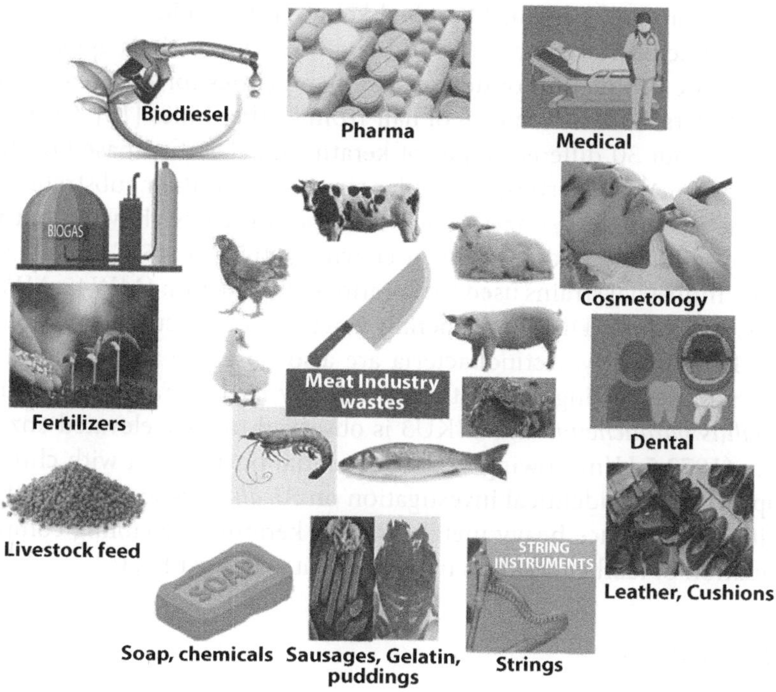

Figure 9.1 Various valuable resources and value-added products extracted from meat wastes.

(collagen, keratin & nutraceuticals) at each stage of the cycle [9]. With the rapid development and modernization, the remains of the sequential approach are being used for the development of value-added products (feed, fuels and fertilizers). The following are some of the most reported applications of valuable products obtained from the valorization of meat wastes (Figure 9.1).

9.2 Bioconversion of Meat Wastes Into Microbial Enzymes

9.2.1 Keratinase

Keratinase is an industrially important enzyme that completely degrades complex proteins. It is the most thermostable serine metalloprotease found in existence which digests insoluble proteinaceous substrates like wool, feathers and epidermis) that could not be digested by other proteases like trypsin, pepsin and papain. Microbial keratinases produced by digesting the feather keratin are harvested and employed for various applications, including the production of detergents, therapeutics for the treatment of acne and scars, and the removal of hair from leather hides [10]. Currently, there are about 30 different types of keratinases classified based on their physicochemical properties, molecular structure and their substrate specificity. Generally, these enzymes are alkaline proteases showing optimal enzyme production and subsequent enzyme activity at 45°C and alkaline pH [11]. Industrial strains used for keratinase production (MW45, MW48, and *Bacillus subtilis* DP1) are generally isolated from accumulated poultry waste [12]. Similarly, Actinobacteria are also observed to produce alkaline keratinases having higher thermostability [13]. Keratinase produced by *Bacillus paralicheniformis* MKU3 is observed to have elevated enzyme activity (1872.5 U/ml) owing to the sequential purification with chromatography [14]. An identical investigation on *Alcaligenes* sp. (AQ05-001) is identified to produce heavy metal-resistant keratinase enabling complete degradation of feather residues in 36 hours at alkaline pH [15].

9.2.2 Protease

Proteases are an important class of enzymes having wide industrial and therapeutic uses. According to recent market research, proteases constitute about 7% to 8% of the CAGR value of total enzyme imports and exports amounting to 6.3 billion USD. This exponentially increasing demand is driven by the versatile properties of proteases including fibrin clot digestion to detergent base for soap manufacturing [16]. Apparently, proteases are also classified based on their physical nature and their specificity into serine, cysteine, and metalloproteases. Liquid meat waste like effluent wastewater, meat residue hydrolysate, coagulated blood, and other liquid discharges obtained by processing meat are enriched with nitrogen sources and thus could potentially be employed as supplements for protease production [17]. Production of protease by *Bacillus mojavensis* SA strain is observed to be enhanced by supplementing the growth media with meat waste hydrolysate [18]. The effluent wastewater from the slaughterhouse is employed as a substrate for *Serratia rubidaea* in the production of serine protease with better enzyme activity [19]. A similar investigation has employed poultry waste for the production of alkaline protease using Streptomyces sp. MAB18 strain. The enzyme showed a high specific activity of 2398.36 U/mg and is stable with antioxidant properties [20]. Further, the production of thermostable protease has employed *Paenibacillus elgii* TKU051 with squid wastes as the production media [21]. Additionally growing *Streptomyces_speibonae* TKU048 with tuna waste as a production medium yields thermostable protease having optimal enzyme activity under acidic conditions [22].

9.2.3 Other Important Enzymes

Chitin is used as a substrate to produce chitinase by microbial fermentation route. The growth factors were optimized using the polynomial equation obtained from artificial neural network optimization method. The highest chitinase activity of 124.78 U/L was obtained [23]. In a novel method to produce cellulase, the deinking paper sludge and pasteurized blood are combinedly and used as cellulose and nitrogen source to culture *Trichoderma reesei*. At optimal conditions, cellulase is harvested from the culture broth [24]. *Staphylococcus xylosus* is commonly grown on meat waste to obtain lipolytic and proteolytic enzymes. The culture broth is treated with ultrasound to enhance the production with reduced fermentation time [25]. Amino acid along with glycerol was used as the substrate for growing *Cryptococcus curvatus* ATCC 20509 under two-stage culture conditions. The substrate combination showed improved lipid production. The lipids showed potential for biodiesel production [26].

9.3 Extraction and Valorization of Meat Wastes and Its Applications

9.3.1 Proteins and Peptides

Proteins are intricate molecules that play a variety of vital functions in the body. The meat business produces a lot of by-products like bones, cartilage, and intestines, which are rich sources of proteins and bioactive peptides. Approximately, 100 g of waste meat is reported to carry 20 g of essential proteins, which can be valorized into functional molecules. The sector can maintain its financial viability and promote waste materials as a potential protein source that is competitively better than plant-derived proteins [27]. With technical advancements, inedible by-products of meat processing industries are being transformed into edible supplements using acidic, alkaline, and enzymatic hydrolysis. Meat wastes from industries like fish, poultry, and chicken are transformed into protein hydrolysates, which find potential application as the seasoning ingredients in soups and beverages [28]. They have also been employed in textile industries (dyeing and paper coating), added as an ingredient in paint, used in animal

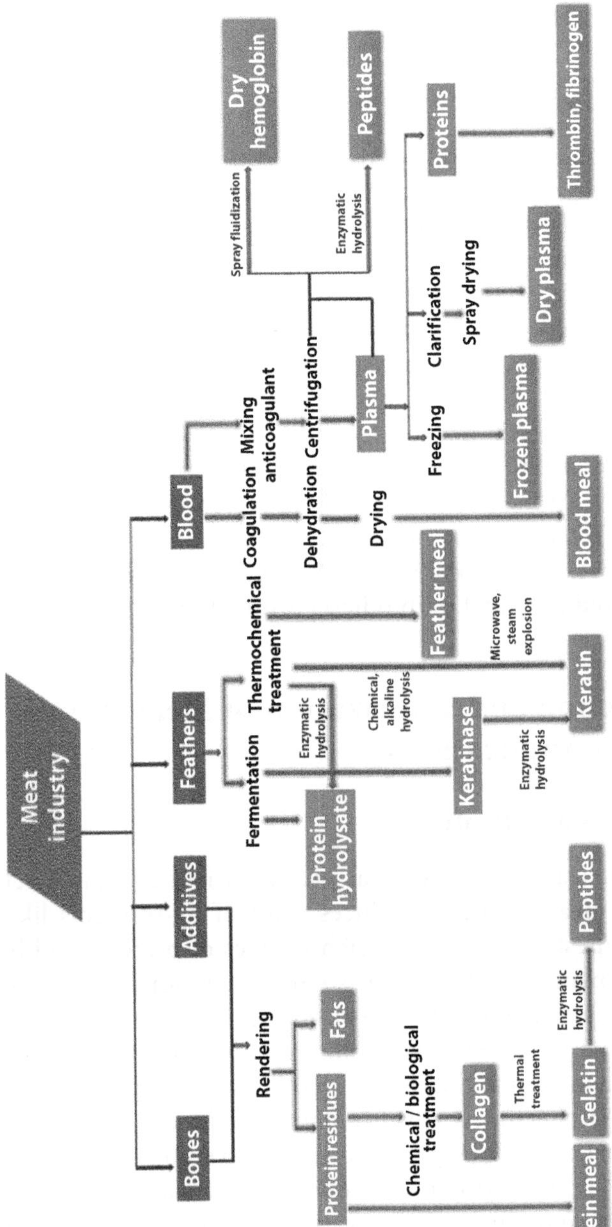

Figure 9.2 Various protein products extracted from meat waste through versatile valorization approaches.

feed, detergents, cosmetics, formulation of media, etc., [29]. Some of the non-edible by-products like wool, hair, skin, fish scales, teeth, trims, and hatchery by-products are either disposed to landfills or are used in clothing, housing components, fuel, and other life-supporting activities that complemented the use of the animal as a basic component for survival and nutrition [30] (Figure 9.2). In contrast to meat, meat waste is a dormant source of valuable proteins that carry essential functions and hence has a considerable market value. However, extraction of these functional proteins is tedious due to their abundance only in connective tissues like lungs (15-20%), liver, and bone cartilage (80-90%) (Table 9.1). Therefore, innovative and sustainable extraction techniques with reduced adverse effects on the nativity of the protein need to be devised.

Table 9.1 Various meat by-products and their applications.

Meat waste	Meat products/ proteins	Applications
Cattle blood, blood plasma	Serum albumin	Blood meal, protein hydrolysates, bio peptides
Porcine blood	Serum albumin	Blood meal, bioactive proteins, bio peptides
Blood	Plasma and Red Blood	Blood meal, animal feed, Adhesives in sausages, plywood, pharmaceutical industry, blood sausages, additives in feed, leather industry, fertilizers
Connective tissue	Collagen	Gelatin type A and B, peptides
Feather and feather meal	Keratin	Meal, hydrolysates, or peptides
Heads and feet (chicken)	Collagen	Gelatin
Bone (chicken)	Collagen	Gelatin
Bone	Collagen, ossein	Gelatin type B, bone collagen hydrolysate in stocks, broths, edible bone phosphate, edible fat, bone glue

(Continued)

Table 9.1 Various meat by-products and their applications. (*Continued*)

Meat waste	Meat products/ proteins	Applications
	Edible fat pieces	bone gelatin, bone meal, livestock and poultry feed
	Inedible fat pieces	Lubricant gelatin: photographic, pharmaceutical, adhesive, abrasive
Collagen - rich tissue	Collagen	Concentrated collagen
Hides	Hair and wool	Leather product, cushioning (furniture)
	Collagen	Cosmetics, facial creams, shampoo, detergents
	Edible	Used in various pulverized meat up to 7-10%. Leather products
Intestines	Sausage casings Intestines	Sausages, sutures, sports guts, musical strings, prosthetic materials collagen sheets, burn dressing, dialyzing membrane, animal feed, protein meal
Hoofs	Foot oil	Lubricant
	Gelatin extracted protein	Jelly-type foods, foaming-in fire extinguishers, felt, furniture cushioning, brushes
Hair, bristles & wool	Meal	Livestock feed or fertilizers Pelting, fabrics and pillows, brushes
Stomach	Pharmaceuticals	Leather industry (blankets, carpets), fertilizer Rennin, the dairy industry
Lungs	Edible products	Human consumption, sausages
Brain	Cholesterol, sphingomyelins	Pharmaceuticals

(*Continued*)

Table 9.1 Various meat by-products and their applications. (*Continued*)

Meat waste	Meat products/ proteins	Applications
Bile/gallstone	Edible products	Bile paste in detergents, acids in pharmaceuticals
Pancreas	Edible products	Trypsin, insulin pharmaceuticals
Liver	Edible products	Human consumption, sausages

9.3.1.1 *Bone and Blood Proteins*

Bone and blood wastes make up the major part (up to 60%) of the waste materials generated by the meat processing industries. The availability of usable bone fraction (11-16%) is restricted and varies with the type of meat sources like pork, beef, and lamb. In general, these wastes have been directly employed as a flavoring and texturing additive in the food processing industries [31]. Given its excellent foaming capacity, processed blood plasma has also been employed as a replacement for egg white proteins in the baking industries. Further, coagulated blood waste is processed and provided as a blood meal additive in cattle feed due to its richness in vitamins and minerals. The clarified blood plasma has also been employed as nutrient supplements in tissue cultures and as agar media in microbiology labs. Other biomedical applications of bone and cartilage residues include their usage as implants in dental and facial surgeries. The marrow of healthy bones is extracted, and the residue is employed as a functional scaffold for promoting bone regeneration.

Modernization has led to the establishment of several separation techniques that are effective in the separation of essential proteins from these wastes, leading to their effective utilization. Extracts from these waste materials (glands, organs, and tissues) are in great demand in several sectors like the pharmaceutical, food, and cosmetic industries. One among them is the extraction of morphogenetic proteins from meat bone marrow [32]. Bone marrow proteins make up 4% to 6% of the total amount of proteins and are an opulent resource of high-value medicinal biomolecules. As these extracts are devoid of antigenic properties, they are readily applied in cosmetic, dental, and dermatological surgeries to ease faster recovery. The liquid plasma (63%) obtained from processed blood waste is rich in therapeutic proteins like albumin, globulin and fibrinogen. While enzymatic proteins like fibrinogen, serotonin, thrombin, and globulins are engaged in clinical applications, bovine serum albumin has a wide market value due to its multifunctional properties [28]. It serves as a stabilizer for vaccines, is used to test Antibiotic sensitivity and is employed for screening the blood type and Rh factor of human beings. Similarly, pork skin rich in proteins is ingested in treating skin ulcers and burns in humans.

9.3.1.2 Collagen

Collagen is the most abundantly available essential protein accounting for over 30% of the whole-body protein. It is an insoluble fibrous protein that anchors all body tissues and remains an integral part of the body. Collagen is the most essential protein extracted from the bone cartilage are widely employed in wound healing and leprosy [33]. Although it is the main constituent of skin, bones, and hides, it has very low nutritional value owing to a lack of necessary amino acids. Recently, therapeutic applications of collagen have elevated from being a wound-healing substrate to structural support for biomedical devices and cementing material for dental implants. Beyond its cosmetic and clinical importance, collagen is also employed in the food industry as a clarifying agent, stabilizer, and corrosion protection material [34]. The unique triple helical structure of the macromolecule constructed with repeating units of glycine, proline and hydroxyproline makes it the most sought protein with a global demand of USD 6.6 billion. Given its omnipresence as a functional macromolecule, the industrial importance of collagen is skyrocketing over recent decades instigating the exploration of these non-edible sources for its extraction [35]. The collagen present in meat waste can be majorly categorized under three different types based on their structural and functional diversity. Type I collagen is elastic in nature (constitutes about 90% of collagen) and acts as building blocks of skin, hair and ligaments in the body. Non-edible sources of type I collagen include eggshells and beef bones. Type II collagen is majorly found to be occurring in mobile joints, cartilage tissues and gut linings. Major sources of type II collagen include chicken and pork meat waste. Type III collagen is found to occur in vital organs like kidneys, the heart and the inner lining of arteries. Therefore, type III collagen is enriched in marine meat waste including fish and prawns [36]. Fish collagen makes up less than 1.5% of all gelatin produced today, with the three most prevalent sources being pig skin (46%), bovine hide (29.4%), and pork and cow bones (23.1%). Nearly 100 million metric tons of bone-laden fish waste land worldwide each year which can be turned into a valuable source for the extraction of collagen.

The structural and functional properties of collagen are observed to be greatly influenced by both the type of waste and the extraction conditions. Therefore, the optimization of the extraction method and the corresponding enzyme employed remains critical in devising the collagen for its desired applications. Pretreatment of the collected meat waste is essential before collagen extraction. Pretreatment with a mild acid (0.1 N HCl) or alkali (0.1 N NaOH) results in the removal of pathogenic microorganisms like spongiform [37]. Various non-thermal methods have been widely reported for the extraction of collagen from pretreated meat waste. Acid-soluble collagen (ASC) fraction from the pretreated waste is extracted using strong acid solutions of more than 0.5 M concentration. It is observed from investigations that employing acetic acid for collagen extraction enhances its yield by 2 folds in comparison with other inorganic acids. This effect is attributed to the lower pKa value of the acid maintaining an optimal pH condition required for the effective extraction of ASC [38]. On the other hand, the application of ultrasound in extraction ASC has also enhanced the yield significantly [39]. Enzymatic digestion of bovine Achilles tendon with (0.5%) pepsin has resulted in a higher yield of type I pepsin soluble collagen (PSC) fraction [40]. A synergetic effect between ultrasound and pepsin enzyme was utilized for the effective extraction of collagen (500 mg/ml) [41]. An identical non-thermal extraction method of pulsating alternating currents of high and low voltages through waste chicken meat is also observed to influence collagen yield positively [42]. However, increased exposure of collagen protein to the ultrasound environment at higher temperatures is observed to induce insignificant structural changes in the macromolecule. Therefore, careful optimization of the ultrasound extraction process is necessary to avoid any structural damage to the extracted collagen.

9.3.1.3 Keratin

Keratin is a natural proteinaceous layer present in hair, feathers, quills, horns, hooves, and nails. It is present in the outermost layer of the epidermis and makes up around 80% of the total protein content of epithelium. The keratin protein found in the skin, hair, and wool of primate organisms is made up of either an α-helical structure or a β-pleated polypeptide. While α-keratin is more prevalent in birds and reptiles, α–keratin is predominant in mammals. Keratin is rich in cysteine residues (7–13%) which are covalently cross-linked via disulfide bonds making it a strong biomolecule available in nature [43]. The amount of sulfur in keratin determines its hardness, owing to its absolute impermeability to water and susceptibility to certain classes of proteases. Among all the meat by-products, feathers remain to be the major source of keratin (25–35%) of which a negligible percentage is used to create low-quality goods like fertilizers and feather meals for animals [44]. The majority of the feathers are discarded as waste, either in landfills or through incineration. The deliberate breakdown of feathers in landfills and subsequent contamination of groundwater make landfills hazardous. On the other hand, incineration and the air pollution caused by the latter, both disposal methods are unfavorable to the environment [45].

Over the recent decades, the industrial demand for keratin is skyrocketing owing to its widened applications extending to tissue engineering, biopharmaceuticals, wound healing, cosmetology, and environmental remediation sectors. Keratin hydrolysates are rich in short-chain peptides that exhibit a stronger antioxidant activity than collagen hydrolysates. Therefore, a sustainable approach to extracting keratin from these feather residues has been in great demand. The disruption of inter and intramolecular bonds linked to keratin has been the main goal of current extraction strategies [46]. Several techniques, like chemical, enzymatic, and microbial hydrolysis employing green solvents like ionic liquids under sustainable energy systems like microwave and ultrasound are carried out for the extraction of keratin.

9.3.1.3.1 Chemical Methods

The conventional method of extracting keratin employs alkali or acid hydrolysis. Treatment of chicken feathers with 2% NaOH at 80°C for 3 hours resulted in a 25% yield of keratin biomolecules. On the other hand, acidic hydrolysis is reported to be more effective with greater yield (up to 40%). However, the acid hydrolysis method impacts the composition and activity of amino acids resulting in a significant loss of essential amino acids including tryptophan. Combining chemical hydrolysis with heating increases the frequency of disulfide bond breakdown subsequently enhancing the keratin yield at low acid/alkali concentrations resulting in a lesser impact on the structure of the protein. However, high temperatures with a minimal reagent concentration tend to damage the amino acids reducing the keratin yield. Therefore, essential factors like the concentration of acid/base, pH, temperature, and reaction duration need to be optimized for native-state keratin extraction. Alternatively, employing reducing agents like 2-mercaptoethanol, bisulfite, and sulfites has reduced keratin to yield free cysteine residues and their derivatives [47]. Extraction of keratin by oxidation technique using a mixture of hydrochloric acid and peracetic acid (2%) or sodium percarbonate with mild ammonia (4.5%) has resulted in enhanced amino acid recovery from wool and hair. Further, various oxidation agents like hydrogen peroxide and sodium thioglycolate and reducing agents like sodium bisulphite, urea and sodium lauryl sulfate have also been widely used for the extraction of keratin [48, 49]. Reports have concluded that the use of reducing agents has resulted in quicker removal of keratin from chicken feathers than the oxidizing counterparts. Denaturing agents like 2-mercaptoethanol or sodium metabisulfite have significantly enhanced the recovery of keratin in its native state. These chemicals break down the non-covalent interactions and the disulfide bonds holding the integrity of keratin fibers resulting in effective extraction of the native protein. Beyond its advantages, employing these methods for large-scale extraction of keratin is restricted due to the environmental hazards caused by the disposal of residual reagents. This lacuna is rectified by the use of thermal methods like a steam explosion or microwave hydrolysis for keratin extraction. Additionally, thermochemical processing is regarded to be the most effective approach in view of the protease resistance exhibited by keratin [50].

9.3.1.3.2 Thermal Methods

Steam explosion process

Hydrothermal steam explosion involves exposing feather residues to a high-pressure supersaturated steam with a temperature range of 180 and 230 °C for brief periods of 1 to 10 minutes to extract keratin. The rapid reduction of pressure results in explosive decompression and consequential rupturing of the stiff keratin fibers extracting them from the feather residues. However, high-temperature exposure for a longer period (over 5 minutes) has been observed to alter the physical structure of the protein [51]. Alternatively, pretreatment of residues with mild alkali (0.4 M NaOH, 1 hour) and subsequent steam explosion (1.6 MPa, 1 minute) enhances the extraction rate of duck keratins (65.78%) keeping the majority of the protein backbone intact in the isolated keratin. Therefore, alternative thermal treatment techniques that enable the extraction of keratin at lower temperature intervals are being investigated [52].

Microwave irradiation process

A modern extraction technique involving microwave-assisted hydrolysis called the "green hydrolysis" approach guarantees sustainable extraction of keratin. During microwave treatment, the keratinous substance receives consistent heating which significantly reduces the activation energy needed for keratin extraction. Due to the rapid homogenous heating taking place in the residue, the keratin extraction efficiency increases in multiple folds. An effective yield (60–70%) of liquid keratin was obtained from wool and chicken feather residues with microwave treatment at 150°C to 180°C for an hour [53]. On the other hand, the presence of hydrolytic reagents like acetic acid and superheated water has significantly improved the performance of green hydrolysis by reducing the treatment time. The hydrolysis process subjects the biomass residue to the acidified water at various pressure and temperature intervals for effective extraction of keratin oligopeptides. Despite various advantages, keratin extraction by green hydrolysis remains dormant due to the use of toxic reagents like detergents and organic acids. This disadvantage has been addressed by the use of ionic liquids. Ionic liquids are eco-friendly solvents widely used to dissolve and rejuvenate wool keratin. For the rejuvenation and dissolving of wool keratin fibers, 1-butyl-3-methylimidazolium chloride ionic liquid has been utilized to prepare wool keratin materials yielding 57% when incubating at 120°C for 30 minutes [54]. The best solvent for dissolving keratin was chloride-containing ionic liquid than bromine owing to its potent nucleophilic behavior and chloride penetration, which are in charge of rupturing the hydrogen bonds.

9.3.1.3.3 Biological Methods

Extraction of keratin by enzymatic hydrolysis consumes little energy than the chemical method. Generally, three major types of enzymatic hydrolysis (deamination, sulfitolysis, and proteolysis) are prevalently used in the extraction of keratin from feather wastes. Proteolytic enzymes like keratinase (serine endopeptidase) and savinase have been widely used for the removal of keratin from feathers and wool [55]. On the other hand, certain microbial enzymes readily hydrolyze the disulfide bonds between cysteine and S-selenocysteine (sulfitolysis) resulting in the disruption of protein. A similar process of keratin denaturation into corresponding free amino acids is called deamination [56]. Enzymatic hydrolysis of keratin residues under an ultrasound environment is observed to enhance the yield. The synergetic effect of ultrasound pretreatment along with enzymatic hydrolysis on chicken feathers yields more keratin peptides with higher bioactivity [57]. In view of its potential advantages over other extraction methods (chemical and thermal), this method is widely employed in meat processing industries where a large amount of feather and epidermal residues are generated [58].

9.3.1.4 Peptides

Bioactive peptides have been of great interest over recent decades owing to their versatile therapeutic benefits. Beyond being rich in antioxidant properties, peptides having specific sequence act as enzyme inhibitors (ACE inhibition) or cofactors (dehydrogenase) in altering the enzyme specificity. In view of these advantages, intense research on peptides with unexploited properties has been in limelight over recent years [59]. On the other hand, meat processing industries generate a substantial amount of protein-rich wastes that are enriched with important amino acids like methyl histidine and hydroxymethyl lysine, which are typically absent in plant proteins. This makes meat proteins the most sought substrate to explore peptides with new properties. Reports describe that enzymatic hydrolysis is the most effective approach for the extraction of bioactive peptides from underutilized meat residues [60]. ACE-inhibitory peptides with a smaller size were obtained by the digestion of bovine myofibril waste with commercial microbial proteases [61]. On the other hand, the action of troponin enzyme on pig skeletal muscles has yielded ACE-inhibitory peptides with identical properties [62]. Employing alcalase (commercial serine protease) followed by flavourzyme over spent hen meat has resulted in short-chain peptides rich in antioxidant activity [63]. Similarly, peptides with higher antioxidant activity were obtained by the action of commercial protease on duck processing residues [64].

Modernization of the peptide extraction process has encouraged employing sustainable hydrolysis methods like microwave and ultrasound for the effective exploitation of meat residues. Extraction of carnosine and anserine (two well-known metal-chelating antioxidant peptides) have been accomplished by employing the pulsed electric field on chicken meat wastes [65]. Microwave-assisted hydrolysis of non-edible sea cucumber meat has resulted in the extraction of collagen peptides with antioxidant properties [66]. Employing ultrasound along with proteolytic enzymes on protein-rich fish by-products has resulted in an enhanced yield of antimicrobial peptides [67]. Additionally, a novel collagen peptide with high antioxidant activity was obtained from pig brain residue by employing ultrasound-assisted extraction [68]. While numerous peptides have been extracted from meat residues and their functional advantages have been investigated, their large-scale application is restricted due to the lack of information relating to the toxicity, dosage level and metabolism profile of these bioactive peptides [69].

9.3.2 Other Valuable Products From Meat Waste

9.3.2.1 Chitosan

Over recent years, the demand for marine-based foods has been exponentially increasing. Crustaceans like shrimps, lobsters and crabs have a hard exoskeleton made of chitin and calcium carbonate composites. These exoskeleton by-products make up 70% of the waste generated from marine meat processing industries [70]. Recent investigations have identified the possibility of extracting chitin from these underutilized resources. Chitin is a long-chain structural polymer made with repeating units of N-acetylglucosamine residues. It is claimed to be the second-largest abundant polysaccharide occurring in nature available in its native form. Chitin makes up to 10% to 20% dry weight of exoskeletons of all marine crustaceans. While crab, oyster, and lobster shells are low in chitin concentration, shrimp shells are reported to harbor high concentrations of chitin [71]. Extraction of chitin from these by-products is accomplished by two methods. The conventional method involves treatment with chemical reagents and is accomplished by a two-step process. The primary step involves the removal of mineral complexes from the powdered shell and is termed demineralization (DM). This process is carried out by incubating the shell powder in a dilute solution of hydrochloric acid (1.0 M) for about 2 hours (time of incubation is reported to vary based on the species) [72]. Some investigations have replaced hydrochloric acid with organic acids like lactic acid to reduce the toxicity of the processed shell residue. Determination of ash content before and after the demineralization process provides the approximate amount of minerals being reduced in the shell powder. The next step in the extraction of chitin involves the removal of residual proteins in the shell powder (deproteination (DP)). This could be accomplished by incubating the source in a dilute alkaline (NaOH) solution (0.5 M) with the replacement of the solution at regular intervals [73]. The process is carried out until the change in the amount of protein extracted becomes insignificant.

Modernization of the chitin extraction process with sustainable energy systems like microwaves and ultrasound has been observed to reduce the reaction time for both DM and DP. Reports have suggested using the microwave environment for the two-step chemical refining of chitin reduces the incubation time significantly (8-15 minutes for DM and 5-10 minutes for DP respectively) [74]. On the other hand, employing ultrasound for DP of shell residue significantly increases the protein removal rate over a short period (40 minutes) [75]. The replacement of alkali chemicals with proteolytic enzymes for the DP of shell residue has also increased the sustainability of chitin extraction. Proteolytic enzymes like trypsin, peptone and ficin are being widely employed for the extraction of residual proteins from shell waste [76]. The efficiency of the DP process is observed to enhance in multiple folds with the introduction of proteases. Over recent decades, *in-situ* pretreatment of shell samples by microbial fermentation with lactic acid-producing microorganisms has also been reported widely. The shell powder with residual proteins intact is added to the sterilized media and lactic acid-producing bacteria like Lactobacillus and Penaeus were inoculated and incubated for the production of lactic acid [77]. The produced lactic acid acts on the shell residues for the removal of mineral composites resulting in simultaneous deproteination and demineralization of the residue.

Alternative to chitin, chitosan has wide applications in the pharmaceutical, biomedical and food industries. The cationic nature of the polymer along with its biocompatible and biodegradable nature makes it the most sought resource with extensive applications including wastewater treatment. In view of its market potential, the formation of chitosan from chitin is of great interest to all pharmaceutical and biomedical industries. The formation of chitosan from chitin involves a chemical modification reaction called deacetylation (DA) [78]. The purity of formed chitosan is the deterministic factor of its application fate. This purity index can be graded by determining the degree of deacetylation (%DD) determined by a simple potentiometric titration method. A known quantity of chitosan is dissolved in a measured volume of 0.1 M HCl. The solution was then titrated with 0.1 M NaOH and the pH of the solution was denoted as a function of NaOH concentration. The volumetric difference between the two infliction points occurring in the resulting plot denotes the %DD of the chitosan employed [79]. Chitosan with high purity and rigidity is employed in biomedical industries while moderately pure chitosan is destined for food applications. Chitosan with the least purity is employed as a flocculant in wastewater treatment plants. Recent investigations have reported the use of ultrasound for enhancing the degree of deacetylation of chitin.

9.3.2.2 *Essential Fats and Nutraceuticals*

Certain residues from the marine meat industries (shrimp, crab and lobster wastes) are rich in nutraceuticals like astaxanthin and omega-3 fatty acids. These nutraceuticals are of high market value due to their wide therapeutic importance. Carotenoids like astaxanthin are fat-soluble antioxidants enriched in the lipid bilayer of the cell membrane and prevent the cells from damage occurring through oxidative stress. Further, purified astaxanthin fractions have been provided as dietary supplements to enhance the antioxidant and anti-inflammatory activities of the metabolic cycle [80]. Polyunsaturated fatty acids like omega-3 acids aid in cognitive development and repair tissue damage. It also prevents heart damage due to the deposition of plagues and amyloidosis. Extraction of these high-value resources from meat industry by-products guarantees an effective valorization approach meeting the sustainability demands of meat industries [81]. Amidst various meat industry wastes, crustacean meat wastes (crabs, lobsters and shrimps) are reported to be affluent in these essential metabolites. Earlier methods have employed organic solvents like ethyl acetate and methanol for the extraction of omega fatty acids and astaxanthin esters from shrimp wastes [82, 83]. Later, bio-derived solvents like limonene and palm oil were employed for the effective extraction of these nutraceuticals [84–86].

The recent development of sustainable solvents has shed the limelight on natural deep eutectic solvents for the extraction of nutraceuticals [87]. Advancements in the existing extraction techniques have evolved the process of purifying these nutraceuticals from meat waste. Among all sustainable extraction methods, supercritical fluid extraction has been prevalently employed for the extraction of nutraceuticals from meat wastes [88, 89]. On the other hand, microwave-assisted solid phase extraction has been reported to have a positive effect on the extraction of astaxanthin [90]. Additionally, ultrasound-assisted extraction has also been observed to effectively assimilate the omega-3 fatty acids and astaxanthin esters from shrimp and crab wastes [91, 92]. Even though various extraction methods have been proposed and validated for sustainable extraction of these functional metabolites, extensive investigation of their mode of action on meat wastes has not been elucidated in detail.

9.4 Utilization of Meat By-Products and Its Wastes Into Value-Added Products

Developing nations like India lacks a proper disposal strategy for abattoir wastes having low optimism for direct valorization. These wastes are disposed of in landfills which raises health hazards for humans and animals due to environmental degradation. Disposal of these wastes into the drainage obstructs the sewers leading to their overflow contaminating surface and groundwater reserves. Alternatively, the valorization of these residual solids into substrates for value-added deliverables like fertilizer and energy is a valuable strategy to reduce pollution. However, there are a variety of strategies to use animal waste effectively. This part of the chapter summarizes the various approaches attempted or employed in the conversion of meat wastes into value-added products. The two best strategies for processing slaughterhouse waste into energy are the generation of biogas and biodiesel.

9.4.1 Animal Feed

Generally, meat wastes generated from meat industries are categorized into three types (bone meal (BM), meat meal (MM), and bone and meat meal (MBM)) based on their composition. However, MBM is observed to constitute the major volume of the residue and remains the most important by-product of the cattle, swine, and poultry industries. MBM contains 4% to 7% water, 8% to 12% fat, 33% to 35% ash, and 48% to 52% proteins along with some minerals and trace elements whose composition varies based on the source industry [93]. According to US regulatory authority, MBM should be free from blood, hair, dung, hide, and, rumen contents with a minimum of 2% to 4% phosphorous and 3% to 5% calcium [94]. On the other hand, by-products with lesser bone residue hence a lower amount of phosphorous and calcium are described as MM. Further, residue rich in calcium and phosphorous with a limited amount of fats and proteins is termed as BM. Earlier, the European Union (EU) used animal proteins processed from meat residues as high-protein animal supplements to enhance the profile of amino acids in the feed. However, its usage was banned due to the prevalence of Bovine Spongiform Encephalopathy (BES) infection, popularly known as mad cow disease. Despite this, certain monogastric animals are still fed with it thus popularly employed as a low-cost animal protein for dogs and cats in the USA and Europe. Whereas a huge amount of MAB is used as a fossil fuel substitute for energy production, as fuel for cement kilns, landfilling, or incineration.

The European Union regulations have classified meat by-products into three categories based on their usability. Meat residues of animals affected by viral infections like BES are classified as category 1 wastes and are discarded by incineration. Recovery of thermal energy from this incineration process can be employed to generate energy. Category 2 wastes contain residual meat from animals that are affected by non-communicable diseases, which can be discarded by incineration or by sterilization at above 33°C and 3 bars for more than 20 minutes. The pressure sterilization process realizes high calorific volatile substances in the process which can be employed as biogas. Meat residues arising from slaughterhouses that are devoid of any diseases are segregated into category 3 waste. These by-products are pasteurized at > 70°C for more than 1 hour and processed into animal feed. Additionally, collagen-rich meat waste from category 2 meat industries is treated with various strong hydrolysis methods to convert them into animal protein supplements. Employing microbial (yeast) consortia to treat the meat waste from category 3 industries would hydrolyze the proteins converting them into feed for broiler chickens [95].

9.4.2 Fertilizers

Farmers are obligated to search for alternate sources of organic materials to act as biofertilizers This could be readily addressed by the utilization of meat wastes unconventional organic fertilizers, which are made from animal by-products. Animal by-products, such as non-edible meat and bones, are being converted into safe nutrient products for the production of organic fertilizer using a new thermal fractionation process. Due to the balanced availability of nutrients MBM, it has been regarded as a potential organic fertilizer. Animal MBM contains high amounts of nitrogen, calcium, and organic matter with less amount of potassium, magnesium, iron, manganese, and zinc compared to inorganic manure [96]. The kind of raw material, however, affects the composition of MBM; for example, MBM derived from soft tissue are rich in nitrogen (C/N ratio of 3.2:4.8), while those generated from bone are rich in phosphorous and calcium. MBM with low carbon content can be readily utilized as nitrogen fertilizer with its efficiency reaching up to 80% of inorganic supplement. The physiochemical metabolism in soil converts MBM into minerals which can then be utilized by plants. Thus, processed MBM is regarded as a slow-release fertilizer that renourishes soil nutrients organically [97]. MBM obtained from category 3 meat industries are pasteurized and defatted by the screw press method. The pressed cake is then exposed to size reduction (50mm) and employed directly as fertilizers. Various thermal treatment like gasification, combustion, or pyrolysis has been applied based on utilization as the BM and MM heating ranges between 13,000 and 30,000 MJ tons. 100 to 300 kg of ash is produced per ton of MBM during combustion is employed in cement kilns. Sintering, melting, and pyrolysis are effective treatments for meat bone meal and bone meal, either alone or in conjunction with other biodegradable waste like dried sewage sludge [98]. Process control in kilns and furnaces allows for the separation of phosphorous from many other elements and the modification of the crystal structure of phosphorous-containing slags that affects the final products' reactivity and phosphorous availability. Both MBM and BM were heated under controlled conditions to temperatures more than 400°C to produce animal bone char with high phosphorous content. Alkaline hydrolysis becomes the replacement for treating and disposing of the wastes through incineration or autoclaving. Chemicals like sodium or potassium hydroxide are used in alkaline hydrolysis to catalyze the breakdown of organic substances into a sterile aqueous environment that is composed of tiny peptides, amino acids, sugars, and soaps. The by-products obtained during this process (bones and teeth) are completely made up of minerals. This leftover is cleaned and ground into a powder that can be used as fertilizer.

9.4.3 Biofuels

Various wastes can be generated in slaughterhouses, including tallows from mutton, beef, pork lard, duck, fleshing oil, mutton fat, lamb meat, feather meal, chicken fat, and animal waste. These wastes have the promising potential to be turned into energy sources. Although these wastes, in general, are made of proteins, fats and carbohydrates, the fat content of these wastes holds significant commercial potential. Thus, they are the natural choice for the generation of energy sources. However, various researches have indicated the importance of the pre-treatment of animal waste fats for pollution abatement and for the recovery of energy [99].

Some of the pretreatment techniques structurally convert these fats of different wastes to the form which renders them usable for various applications. For instance, beef fat may be generated by rendering cow fatty tissue. Beef tallow has been tested to be a good source for the production of second-generation biofuels such as biodiesel and biogas. Similarly, duck tallow with high oleic-acid content—a primary factor for biodiesel- has been reported to be a suitable substrate for biodiesel and biogas production. Additionally, biodiesel may be made using the fleshing oil produced from chicken fat. It is important to note that slaughterhouses are significant suppliers of pork fat. A rendered form of swine fat or suet is lard. The flavor and texture are typically moderate and smooth. Compared to tallow, pork lard may be distinguished by its fat source.

9.4.3.1 Biogas Production

Anaerobic digestion (AD), which is a microbial-based process under anaerobic conditions, is a widely used method for converting organic wastes to biogas. As animal slaughterhouse wastes are rich in different forms of organic carbon that are the major contributors to biological oxygen demand (BOD) is a natural choice for the production of biogas through the anaerobic digestion process. In a two-step conversion process, fatty acid-rich animal wastes are first hydrolyzed into glycerol and long-chain fatty acids before being converted into hydrogen and acetate with the help of acetogenic bacteria [100]. Finally, methanogenic bacteria in the digesters help transform them into methane. Since anaerobic digestion involves numerous genera of methanogenic as well as other types of bacteria, the substrates are needed to be fed in the proper form to obtain good yield and quality of biogas. Solid recirculation and pretreatment of substrates by saponification augmented chemical oxygen demand (COD) degradation and in turn the methane generation through the anaerobic digestion process. The concept of solid recirculation is to reduce the washout of the biomass contents of the slaughterhouse, and also helps retain good sludge of bacteria, which improves the overall bioavailability of substrates and the emulsification of fatty wastes in solid form.

The presence of unwanted microbes is sometimes detrimental to the anaerobic digestion process. Pasteurization is a conventional food preservation process by which undesirable microbes are heat-killed or inactivated so that the shelf life of food products is enhanced. For animal wastes, simple pasteurization at 60°C can destroy unwanted microbes from the surface of the meat. However, caution has been exercised while doing pasteurization as different type of wastes behave differently when subjected to heat treatment, and may sometimes turn the meat wastes to undesirable forms [101]. For instance, thermal pretreatment accelerated the formation of complex nitrogen molecules, which reduced the methane synthesis rate in chicken wastes. On the other hand, heat treatment elevated the number of labile chemicals in pig wastes, which increased the rate of degradation of organic particles thereby facilitating the increased generation of methane [102]. In general, enzymatic pretreatment of wastes is beneficial for the improved generation of methane gas. Enzymes such as cellulase, protease and lipases have proven to accelerate biomethane production through the conversion of complex substrates to the simple and easily digestible form. A cocktail of these enzymes can be added to meat wastes to increase biogas generation. More specifically, the more rigid fat molecules can be hydrolyzed to simply fatty acids by adding lipase enzymes. Usually, the prevailing thermophilic conditions of 50°C to 55°C in most of the anaerobic digestors would augment this lipase-mediated fat hydrolysis. For instance, lipase-mediated hydrolysis increased biogas production by approximately 26.50% when fatty material was used in the process [103].

Chicken feathers, made of keratin, are considered to be challenging substrates in the digestion process. However, supplementation of protein-digesting enzymes such as keratinases helps improve the digestibility of these wastes. Hydrolysis of keratin feathers using alkaline endopeptidase has been used as an effective pretreatment method that increased the overall production of methane by significant folds [104]. Despite the increased efficiency of such enzymatic-based hydrolysis approaches, these enzyme-based hydrolysis methods are considered to be cost-intensive, So, the current research endeavors focus on reducing the cost of production of such enzymes, when they are used for low-end applications such as waste treatment. Another well-explored pretreatment strategy is hydrothermal pretreatment (HTP), which improved the solubilization potential and the over-physical characteristics of the waste that make them more amenable to the AD process. Based on a lab-scale study, a temperature of 190°C has been suggested to be optimal for this HTP process [105]. Biochemical methane potential (BMP) tests carried out in bottles revealed that autoclave pretreatment (thermal) was efficient in the case of meat wastewater, which enhanced methane production by 15 folds as compared to the untreated waste [106].

9.4.3.2 Biodiesel Production

In general, the transesterification process for biodiesel generation is carried out at an optimal temperature of 60°C to 65°C. This process requires the substrate to be fed in the proper form for the transesterification process to be very effective. A methyl ester of fatty acid generated from animal fats and wastes can be used directly as a biodiesel source or as a substitute for traditional diesel fuel. Transesterification typically transforms triglycerides derived from animal fats and waste products into methyl esters of fatty acids. Animal fats and wastes are incorporated with alcohol (methanol/ethanol) to complete the alcoholysis process. However, ethanol dissolves animal fats better than methanol.

Various pretreatment procedures such as extraction, extraction using supercritical CO_2, and thermal pretreatments are used to remove unwanted materials that can influence the transesterification process. The pretreatment should ensure that the product contains at least 3% free fatty acids [107]. Extraction using supercritical carbon dioxide removes contaminants and moisture levels in animal wastes. However, the requirement of supercritical CO_2 varies with respect to the type of animal waste under consideration [99, 108]. In the heat treatment process, also called the melting process, animal fats are subjected to heating up to 110°C, which ensures the removal of 70% moisture in them. But for the complete separation of liquids from the solids, the mixture should be subjected to a filtration process to exclude impurities such as protein remnants, particles, and gums [109]. Low-pressure operations are often desirable to remove wax and other contaminants from animal fats and wastes when thermal pretreatment is carried out. In general, the fatty acid composition in animal fats and wastes dictates the temperature requirement for the pretreatment processes [110].

On the other hand, the supercritical CO_2 extraction approach for animal fat preparation is an effective method to recover high-fat content in animal wastes produced in slaughterhouses [111]. The choice of suitable alcohol is important for a transesterification process, although methanol is a commonly used alcohol. As stated, animal fat dissolves relatively better in ethanol than in methanol, it is important to determine the solubility of fats in methanol. In order to turn beef tallow into biodiesel, its solubility in ethanol or methanol must be established. In a typical transesterification process involving beef tallow, a blend of methanol and ethanol was added to the beef tallow in a 6:1 molar ratio for the improved synthesis of biodiesel [112]. Figure 9.3 depicts the optimal temperature and molar ratio for several sources of animal wastes and fats that can be transesterified into methyl/ethyl ester for use as biodiesel. Similarly, other animal fats and wastes, such as chicken fat, pork lard, mutton fat, and so on, are transesterified into methyl/ethyl ester, which may be utilized as biodiesel or as an addition to diesel fuel. However, the molar ratio and reaction temperature may differ depending on the type of animal fats and wastes employed.

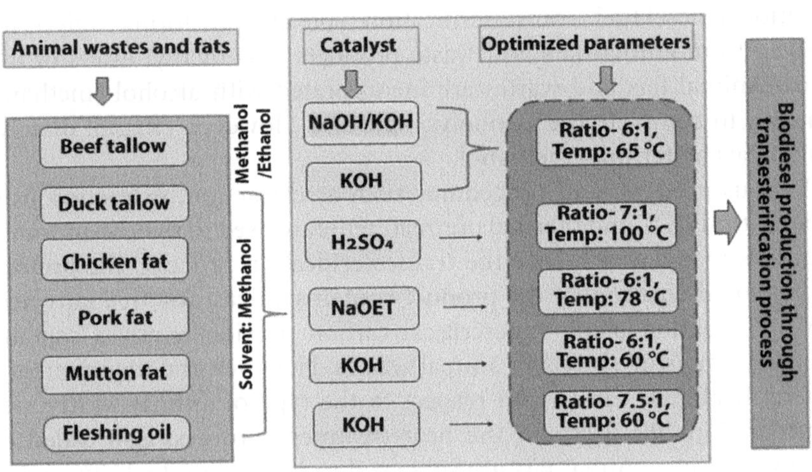

Figure 9.3 Optimum parameters for biodiesel generation from various animal fats and wastes.

9.5 Conclusion

As the saying goes "One man's waste is another one's treasure," by-products and residues generated from meat industries have always been a treasure trove for the isolation and production of various commercially important products. As discussed throughout this chapter, all solid and liquid waste materials from the meat industries have been converted into usable products meeting the global sustainability metrics. Further, the valorization approach also delivers a circular bioeconomy where the industries could generate revenue by reducing the waste to be discarded. Even though this chapter covers all critical aspects of meat by-products, the possibility of valorizing the aforementioned residues is not limited only to the scope discussed in here. Further, this chapter provides the essence of the versatility of meat waste valorization from an aerial view. With the intention of kindling more peers into circular bioeconomy and sustainability, this chapter leaves an open end to the readers to employ these ideologies including but not limited to those observed here to the benefit of society.

References

1. Al-Gheethi, A., Ma, N.L., Rupani, P.F., Sultana, N., Yaakob, M.A., Mohamed, R.M., Soon, C.F., Biowastes of slaughterhouses and wet markets: An overview of waste management for disease prevention. *Environ. Sci. Pollut. Res.*, 28, 1–4, 2021.
2. Fearon, J., Mensah, S.B., Vivian, B., Abattoir operations, waste generation and management in the Tamale metropolis: Case study of the Tamale slaughterhouse. *J. Public Health Epidemiol.*, 6, 1, 14–19, 2014.
3. Ali, A.M., Nawaz, A.M., Al-Turaif, H.A., Shahzad, K., The economic and environmental analysis of energy production from slaughterhouse waste in Saudi Arabia. *Environ. Dev. Sustain.*, 23, 3, 4252–4269, 2021.
4. Rashid, F., Hoque, M.E., Aziz, M., Sakib, T.N., Islam, M.T., Robin, R.M., Investigation of optimal hybrid energy systems using available energy sources in a rural area of Bangladesh. *Energies*, 14, 18, 5794, 2021.
5. Sultana, R., Dwivedi, A., Moktadir, M.A., Investigating the role of consumers, producers, and policymakers: A case of leather supply chain towards sustainable chemistry. *Curr. Opin. Green Sustain. Chem.*, 39, 100724, 2022.
6. Provin, A.P., de Aguiar Dutra, A.R., Machado, M.M., Cubas, A.L., New materials for clothing: Rethinking possibilities through a sustainability approach-A review. *J. Cleaner Prod.*, 282, 124444, 2021.
7. Solomando, J.C., Antequera, T., Perez-Palacios, T., Lipid digestion and oxidative stability in ω-3-enriched meat model systems: Effect of fish oil

microcapsules and processing or culinary cooking. *Food Chem.*, 328, 127125, 2020.

8. Gao, Y., Li, M., Zhang, L., Wang, Z., Yu, Q., Han, L., Preparation of rapeseed oil oleogels based on beeswax and its application in beef heart patties to replace animal fat. *LWT*, 149, 111986, 2021.
9. Toldrá, F., Reig, M., Mora, L., Management of meat by- and co-products for an improved meat processing sustainability. *Meat Sci.*, 181, 108608, 2021.
10. Vidmar, B. and Vodovnik, M., Microbial keratinases: Enzymes with promising biotechnological applications. *Food Technol. Biotechnol.*, 56, 3, 312–328, 2018.
11. Ashraf, M.M., Asad, W., Kiran, T., Isolation of keratinolytic bacterial strains from poultry farm soil samples; production, optimization and thermal characterization of the keratinase enzyme (s). *J. Chem. Soc. Pak.*, 40, 05, 953–960, 2018.
12. Sanghvi, G., Patel, H., Vaishnav, D., Oza, T., Dave, G., Kunjadia, P., Sheth, N., A novel alkaline keratinase from Bacillus subtilis DP1 with potential utility in cosmetic formulation. *Int. J. Biol. Macromol.*, 87, 256–262, 2016.
13. Saha, S. and Dhanasekaran, D., Isolation and screening of keratinolytic actinobacteria form keratin waste dumped soil in Tiruchirappalli and Nammakkal, Tamil Nadu, India. *Curr. Res. J. Biol. Sci.*, 2, 2, 124–131, 2010.
14. Kalaikumari, S.S., Vennila, T., Monika, V., Chandraraj, K., Gunasekaran, P., Rajendhran, J., Bioutilization of poultry feather for keratinase production and its application in leather industry. *J. Cleaner Prod.*, 208, 44–53, 2019.
15. Yusuf, I., Ahmad, S.A., Phang, L.Y., Syed, M.A., Shamaan, N.A., Khalil, K.A., Dahalan, F.A., Shukor, M.Y., Keratinase production and biodegradation of polluted secondary chicken feather wastes by a newly isolated multi heavy metal tolerant bacterium-Alcaligenes sp. AQ05-001. *J. Environ. Manage.*, 183, 182–195, 2016.
16. Sharma, K.M., Kumar, R., Panwar, S., Kumar, A., Microbial alkaline proteases: Optimization of production parameters and their properties. *J. Genet. Eng. Biotechnol.*, 15, 1, 115–126, 2017.
17. Ramakodi, M.P., Santhosh, N., Pragadeesh, T., Mohan, S.V., Basha, S., Production of protease enzyme from slaughterhouse effluent: An approach to generate value-added products from waste. *Bioresour. Technol. Rep.*, 12, 100552, 2020.
18. Hammami, A., Bayoudh, A., Abdelhedi, O., Nasri, M., Low-cost culture medium for the production of proteases by Bacillus mojavensis SA and their potential use for the preparation of antioxidant protein hydrolysate from meat sausage by-products. *Ann. Microbiol.*, 68, 8, 473–484, 2018.
19. Doddapaneni, K.K., Tatineni, R., Vellanki, R.N., Gandu, B., Panyala, N.R., Chakali, B., Mangamoori, L.N., Purification and characterization of two novel extra cellular proteases from Serratia rubidaea. *Process Biochem.*, 42, 8, 1229–1236, 2007.

20. Manivasagan, P., Venkatesan, J., Sivakumar, K., Kim, S.K., Production, characterization and antioxidant potential of protease from Streptomyces sp. MAB18 using poultry wastes. *BioMed. Res. Int.*, 2013, 12, 2013.
21. Lee, D.H., Doan, C.T., Tran, T.N., Nguyen, V.B., Nguyen, A.D., Wang, C.L., Wang, S.L., Proteases production and chitin preparation from the liquid fermentation of chitinous fishery by-products by Paenibacillus elgii. *Mar. Drugs*, 19, 9, 477, 2021.
22. Tran, T.N., Doan, C.T., Nguyen, V.B., Nguyen, A.D., Wang, S.L., Conversion of fishery waste to proteases by streptomyces speibonae and their application in antioxidant preparation. *Fishes*, 7, 3, 140, 2022.
23. Suryawanshi, N. and Eswari, J.S., Chitin from seafood waste: Particle swarm optimization and neural network study for the improved chitinase production. *J. Chem. Technol. Biotechnol.*, 97, 2, 509–519, 2022.
24. Weiss, R., Eischer, A., Tadic, T., Gritsch, S.M., Ortner, M., Prall, K., Neunteufel, E., Putz, R.F., Guebitz, G.M., Nyanhongo, G.S., Valorisation of slaughter house and deinking paper waste streams for the production of enzyme by Trichoderma reesei. *J. Cleaner Prod.*, 275, 122882, 2020.
25. Nehring, P., Lorenzo, J.M., Santos, S.P., Wagner, R., de Menezes, C.R., Dos Santos, B.A., Barin, J.S., Campagnol, P.C., Cichoski, A.J., Effect of ultrasound application on the growth of S. xylosus inoculated in by-products from the poultry industry. *Curr. Res. Food Sci.*, 5, 345–350, 2022.
26. Kamal, R., Huang, Q., Wang, Q., Yu, X., Song, J., Limtong, S., Zhao, Z.K., Co-utilization of amino acid-rich wastes and glycerol for microbial lipid production. *Biofuel Bioprod. Biorefin.*, 16, 1, 142–154, 2022.
27. Osner, R., Food wastage. *Nutr. Food Sci.*, 82, 4, 13–16, 1982.
28. Jayathilakan, K., Sultana, K., Radhakrishna, K., Bawa, A.S., Utilization of by-products and waste materials from meat, poultry and fish processing industries: A review. *J. Food Sci. Technol.*, 49, 3, 278–293, 2012.
29. Lynch, S.A., Mullen, A.M., O'Neill, E., Drummond, L., Álvarez, C., Opportunities and perspectives for utilisation of co-products in the meat industry. *Meat Sci.*, 144, 62–73, 2018.
30. Ramirez, J., McCabe, B., Jensen, P.D., Speight, R., Harrison, M., Van Den Berg, L., O'Hara, I., Wastes to profit: A circular economy approach to value-addition in livestock industries. *Anim. Prod. Sci.*, 61, 6, 541–550, 2021.
31. Hicks, T.M. and Verbeek, C.J., Meat industry protein by-products: Sources and characteristics, in: *Protein By-products*, pp. 37–61, Academic Press, Cambridge, Massachusetts, 2016.
32. Irshad, A. and Sharma, B.D., Abattoir by-product utilization for sustainable meat industry: A review. *J. Anim. Prod. Adv.*, 5, 6, 681–696, 2015.
33. Kulkarni, V.V. and Devatkal, S.K., Utilization of by-products and waste materials from meat and poultry processing industry: A review. *J. Meat Sci.*, 11, 1, 1–10, 2015.

34. Ahmed, M., Verma, A.K., Patel, R., Collagen extraction and recent biological activities of collagen peptides derived from sea-food waste: A review. *Sustain. Chem. Pharm.*, 18, 100315, 2020.
35. Atef, M., Ojagh, S.M., Latifi, A.M., Esmaeili, M., Udenigwe, C.C., Biochemical and structural characterization of sturgeon fish skin collagen (Huso huso). *J. Food Biochem.*, 44, 8, e13256, 2020.
36. Coppola, D., Oliviero, M., Vitale, G.A., Lauritano, C., D'Ambra, I., Iannace, S., de Pascale, D., Marine collagen from alternative and sustainable sources: Extraction, processing and applications. *Mar. Drugs*, 18, 4, 214, 2020.
37. Toniasso, D.P., da Silva, C.G., Junior, B.D., Somacal, S., Emanuelli, T., Kubota, E.H., Dornelles, R.C., Mello, R., Collagen extracted from rabbit: Meat and by-products: Isolation and physicochemical assessment. *Food Res. Int.*, 162, 111967, 2022.
38. Sotelo, C.G., Comesaña, M.B., Ariza, P.R., Pérez-Martín, R.I., Characterization of collagen from different discarded fish species of the west coast of the Iberian Peninsula. *J. Aquat. Food Prod. Technol.*, 25, 3, 388–399, 2016.
39. Shaik, M.I., Effendi, N.A., Sarbon, N.M., Functional properties of sharpnose stingray (Dasyatis zugei) skin collagen by ultrasonication extraction as influenced by organic and inorganic acids. *Biocatal. Agric. Biotechnol.*, 35, 102103, 2021.
40. Fujimoto, D., Akiba, K.Y., Nakamura, N., Isolation and characterization of a fluorescent material in bovine achilles tendon collagen. *Biochem. Biophys. Res. Commun.*, 76, 4, 1124–1129, 1977.
41. Manimegalai, N.P., Ramanathan, G., Gunasekaran, D., Jeyakumar, G.F., Sivagnanam, U.T., Cardinal acuity on the extraction and characterization of soluble collagen from the underutilized abattoir junks for clinical demands. *Process Biochem.*, 122, 29–37, 2022.
42. Ghosh, S., Gillis, A., Sheviryov, J., Levkov, K., Golberg, A., Towards waste meat biorefinery: Extraction of proteins from waste chicken meat with non-thermal pulsed electric fields and mechanical pressing. *J. Cleaner Prod.*, 208, 220–231, 2019.
43. Senthilkumar, N., Chowdhury, S., Sanpui, P., Extraction of keratin from keratinous wastes: Current status and future directions. *J. Mater. Cycles Waste Manage.*, 6, 1–6, 2022.
44. Bragulla, H.H. and Homberger, D.G., Structure and functions of keratin proteins in simple, stratified, keratinized and cornified epithelia. *J. Anat.*, 214, 4, 516–559, 2019.
45. Blumenberg, M. and Younis, S., Introductory chapter: Keratins-what to do with too much? What to do with too little?, in: *Keratin*, IntechOpen, London, UK, 2018.
46. Shavandi, A., Silva, T.H., Bekhit, A.A., Bekhit, A.E., Keratin: Dissolution, extraction and biomedical application. *Biomater. Sci.*, 5, 9, 1699–1735, 2017.

48. Cassoni, A.C., Freixo, R., Pintado, A.I., Amorim, M., Pereira, C.D., Madureira, A.R., Pintado, M.M., Novel eco-friendly method to extract keratin from hair. *ACS Sustain. Chem. Eng.*, 6, 9, 12268–12274, 2018.
49. Pourjavaheri, F., Pour, S.O., Jones, O.A., Smooker, P.M., Brkljača, R., Sherkat, F., Blanch, E.W., Gupta, A., Shanks, R.A., Extraction of keratin from waste chicken feathers using sodium sulfide and l-cysteine. *Process Biochem.*, 82, 205–214, 2019.
50. Alahyaribeik, S. and Ullah, A., Methods of keratin extraction from poultry feathers and their effects on antioxidant activity of extracted keratin. *Int. J. Biol. Macromol.*, 148, 449–456, 2020.
51. Bertini, F., Canetti, M., Patrucco, A., Zoccola, M., Wool keratin-polypropylene composites: Properties and thermal degradation. *Polym. Degrad. Stab.*, 98, 5, 980–987, 2013.
52. Zhang, Y., Zhao, W., Yang, R., Steam flash explosion assisted dissolution of keratin from feathers. *ACS Sustain. Chem. Eng.*, 3, 9, 2036–2042, 2015.
53. Pulidori, E., Micalizzi, S., Bramanti, E., Bernazzani, L., Duce, C., De Maria, C., Montemurro, F., Pelosi, C., De Acutis, A., Vozzi, G., Tinè, M.R., One-pot process: Microwave-assisted keratin extraction and direct electrospinning to obtain keratin-based bioplastic. *Int. J. Mol. Sci.*, 22, 17, 9597, 2021.
54. Zoccola, M., Aluigi, A., Patrucco, A., Vineis, C., Forlini, F., Locatelli, P., Sacchi, M.C., Tonin, C., Microwave-assisted chemical-free hydrolysis of wool keratin. *Text. Res. J.*, 82, 19, 2006–2018, 2012.
55. Khumalo, M., Sithole, B., Tesfaye, T., Valorisation of waste chicken feathers: Optimisation of keratin extraction from waste chicken feathers by sodium bisulphite, sodium dodecyl sulphate and urea. *J. Environ. Manage.*, 262, 110329, 2020.
56. Tasaki, K., A novel thermal hydrolysis process for extraction of keratin from hog hair for commercial applications. *Waste Manage.*, 104, 33–41, 2020.
57. Alahyaribeik, S. and Ullah, A., Effects of ultrasound-assisted alkaline extraction on antioxidant activity and functional characteristics of chicken feather keratin peptides. *ChemistrySelect*, 5, 44, 13788–13794, 2020.
58. Ningthoujam, D.S., Tamreihao, K., Mukherjee, S., Khunjamayum, R., Devi, L.J., Asem, R.S., Keratinaceous wastes and their valorization through keratinolytic microorganisms, in: *Keratin*, IntechOpen, London, UK, 2018.
59. Zhang, Y., Wang, Y., Wang, S., Fang, B., Engineering bi-functional enzyme complex of formate dehydrogenase and leucine dehydrogenase by peptide linker mediated fusion for accelerating cofactor regeneration. *Eng. Life Sci.*, 17, 9, 989–996, 2017.
60. Bechaux, J., Gatellier, P., Le Page, J.F., Drillet, Y., Sante-Lhoutellier, V., A comprehensive review of bioactive peptides obtained from animal by-products and their applications. *Food Funct.*, 10, 10, 6244–6266, 2019.
61. Ryder, K., Bekhit, A.E., McConnell, M., Carne, A., Towards generation of bioactive peptides from meat industry waste proteins: Generation of peptides using commercial microbial proteases. *Food Chem.*, 208, 42–50, 2016.

62. Katayama, K., Tomatsu, M., Fuchu, H., Sugiyama, M., Kawahara, S., Yamauchi, K., Kawamura, Y., Muguruma, M., Purification and characterization of an angiotensin I-converting enzyme inhibitory peptide derived from porcine troponin C. *Anim. Sci. J.*, 74, 1, 53–58, 2003.
63. Kumar, D., Jyoti, A., Tarafdar, A., Kumar, A., Badgujar, P.C., Comparative functional and spectroscopic analysis of spent hen meat hydrolysate by individual and combined treatment of microbial proteases. *Prep. Biochem. Biotechnol.*, 51, 6, 618–627, 2021.
64. Lee, S.J., Kim, E.K., Hwang, J.W., Oh, H.J., Cheong, S.H., Moon, S.H., Jeon, B.T., Lee, S.M., Park, P.J., Purification and characterisation of an antioxidative peptide from enzymatic hydrolysates of duck processing by-products. *Food Chem.*, 123, 2, 216–20, 2010.
65. Robin, A., Ghosh, S., Gabay, B., Levkov, K., Golberg, A., Identifying critical parameters for extraction of carnosine and anserine from chicken meat with high voltage pulsed electric fields and water. *Innov. Food Sci. Emerg. Technol.*, 76, 102937, 2022.
66. Li, Y., Li, J., Lin, S.J., Yang, Z.S., Jin, H.X., Preparation of antioxidant peptide by microwave-assisted hydrolysis of collagen and its protective effect against H2O2-induced damage of RAW264. 7 cells. *Mar. Drugs*, 17, 11, 642, 2019.
67. Siewe, F.B., Kudre, T.G., Narayan, B., Optimisation of ultrasound-assisted enzymatic extraction conditions of umami compounds from fish by-products using the combination of fractional factorial design and central composite design. *Food Chem.*, 334, 127498, 2021.
68. Zou, Y., Wang, W., Li, Q., Chen, Y., Zheng, D., Zou, Y., Zhang, M., Zhao, T., Mao, G., Feng, W., Wu, X., Physicochemical, functional properties and antioxidant activities of porcine cerebral hydrolysate peptides produced by ultrasound processing. *Process Biochem.*, 51, 3, 431–443, 2016.
69. Lafarga, T. and Hayes, M., Bioactive peptides from meat muscle and by-products: Generation, functionality and application as functional ingredients. *Meat Sci.*, 98, 2, 227–239, 2014.
70. Mohan, K., Ganesan, A.R., Ezhilarasi, P.N., Kondamareddy, K.K., Rajan, D.K., Sathishkumar, P., Rajarajeswaran, J., Conterno, L., Green and eco-friendly approaches for the extraction of chitin and chitosan: A review. *Carbohydr. Polym.*, 287, 119349, 2022.
71. Yadav, M., Goswami, P., Paritosh, K., Kumar, M., Pareek, N., Vivekanand, V., Seafood waste: A source for preparation of commercially employable chitin/chitosan materials. *Bioresour. Bioprocess.*, 6, 1, 1–20, 2019.
72. Kumari, S., Rath, P., Kumar, A.S., Tiwari, T.N., Extraction and characterization of chitin and chitosan from fishery waste by chemical method. *Environ. Technol. Innovation*, 1, 77–85, 2015.
73. Omar, B.A., Elmasry, R., Eita, A., Soliman, M.M., El-Tahan, A.M., Sitohy, M., Upgrading the preparation of high-quality chitosan from Procambarus clarkii wastes over the traditional isolation of shrimp chitosan. *Saudi J. Biol. Sci.*, 29, 2, 911–919, 2022.

74. El Knidri, H., El Khalfaouy, R., Laajeb, A., Addaou, A., Lahsini, A., Eco-friendly extraction and characterization of chitin and chitosan from the shrimp shell waste via microwave irradiation. *Process Saf. Environ. Prot.*, 104, 395–405, 2016.
75. Vallejo-Domínguez, D., Rubio-Rosas, E., Aguila-Almanza, E., Hernández-Cocoletzi, H., Ramos-Cassellis, M.E., Luna-Guevara, M.L., Rambabu, K., Manickam, S., Munawaroh, H.S., Show, P.L., Ultrasound in the deproteinization process for chitin and chitosan production. *Ultrason. Sonochem.*, 72, 105417, 2021.
76. Marzieh, M.N., Zahra, F., Tahereh, E., Sara, K.N., Comparison of the physicochemical and structural characteristics of enzymatic produced chitin and commercial chitin. *Int. J. Biol. Macromol.*, 139, 270–276, 2019.
77. Sedaghat, F., Yousefzadi, M., Toiserkani, H., Najafipour, S., Bioconversion of shrimp waste Penaeus merguiensis using lactic acid fermentation: An alternative procedure for chemical extraction of chitin and chitosan. *Int. J. Biol. Macromol.*, 104, 883–888, 2017.
78. Bajaj, M., Winter, J., Gallert, C., Effect of deproteination and deacetylation conditions on viscosity of chitin and chitosan extracted from Crangon crangon shrimp waste. *Biochem. Eng. J.*, 56, 51–62, 2011.
79. Ngo, T.H. and Ngo, D.N., Effects of low–frequency ultrasound on heterogenous deacetylation of chitin. *Int. J. Biol. Macromol.*, 104, 1604–1610, 2017.
80. Prameela, K., Venkatesh, K., Immandi, S.B., Kasturi, A.P., Krishna, C.R., Mohan, C.M., Next generation nutraceutical from shrimp waste: The convergence of applications with extraction methods. *Food Chem.*, 237, 121–132, 2017.
81. Dave, D. and Routray, W., Current scenario of Canadian fishery and corresponding underutilized species and fishery by-products: A potential source of omega-3 fatty acids. *J. Cleaner Prod.*, 180, 617–641, 2018.
82. Gómez-Estaca, J., Calvo, M.M., Álvarez-Acero, I., Montero, P., Gómez-Guillén, M.C., Characterization and storage stability of astaxanthin esters, fatty acid profile and α-tocopherol of lipid extract from shrimp (L. vannamei) waste with potential applications as food ingredient. *Food Chem.*, 216, 37–44, 2017.
83. López-Cervantes, J., Sánchez-Machado, D.I., Gutiérrez-Coronado, M.A., Ríos-Vázquez, N.J., Quantification of astaxanthin in shrimp waste hydrolysate by HPLC. *Biomed. Chromatogr.*, 20, 10, 981–984, 2006.
84. Handayani, A.D., Indraswati, N., Ismadji, S., Extraction of astaxanthin from giant tiger (Panaeus monodon) shrimp waste using palm oil: Studies of extraction kinetics and thermodynamic. *Bioresour. Technol.*, 99, 10, 4414–4149, 2008.
85. Scurria, A., Fabiano Tixier, A.S., Lino, C., Pagliaro, M., D'agostino, F., Avellone, G., Chemat, F., Ciriminna, R., High yields of shrimp oil rich in omega-3 and natural astaxanthin from shrimp waste. *ACS Omega*, 5, 28, 17500–17505, 2020.

86. Ciriminna, R., Scurria, A., Fabiano-Tixier, A.S., Lino, C., Avellone, G., Chemat, F., Pagliaro, M., Omega-3 extraction from anchovy fillet leftovers with limonene: Chemical, economic, and technical aspects. *ACS Omega*, 4, 13, 15359–15363, 2019.
87. Roy, V.C., Ho, T.C., Lee, H.J., Park, J.S., Nam, S.Y., Lee, H., Getachew, A.T., Chun, B.S., Extraction of astaxanthin using ultrasound-assisted natural deep eutectic solvents from shrimp wastes and its application in bioactive films. *J. Cleaner Prod.*, 284, 125417, 2012.
88. Sánchez-Camargo, A.P., Meireles, M.Â., Lopes, B.L., Cabral, F.A., Proximate composition and extraction of carotenoids and lipids from Brazilian redspotted shrimp waste (Farfantepenaeus paulensis). *J. Food Eng.*, 102, 1, 87–93, 2011.
89. Sánchez-Camargo, A.P., Martinez-Correa, H.A., Paviani, L.C., Cabral, F.A., Supercritical CO2 extraction of lipids and astaxanthin from Brazilian redspotted shrimp waste (Farfantepenaeus paulensis). *J. Supercrit. Fluids*, 56, 2, 164–173, 2011.
90. Nunes, A.N., Roda, A., Gouveia, L.F., Fernández, N., Bronze, M.R., Matias, A.A., Astaxanthin extraction from marine crustacean waste streams: An integrate approach between microwaves and supercritical fluids. *ACS Sustain. Chem. Eng.*, 9, 8, 3050–3059, 2021.
91. Sharayei, P., Azarpazhooh, E., Zomorodi, S., Einafshar, S., Ramaswamy, H.S., Optimization of ultrasonic-assisted extraction of astaxanthin from green tiger (Penaeus semisulcatus) shrimp shell. *Ultrason. Sonochem.*, 76, 105666, 2021.
92. Gulzar, S. and Benjakul, S., Impact of pretreatment and atmosphere on quality of lipids extracted from cephalothorax of Pacific white shrimp by ultrasonic assisted process. *Food Chem.*, 309, 125732, 2020.
93. Darch, T., Dunn, R.M., Guy, A., Hawkins, J.M., Ash, M., Frimpong, K.A., Blackwell, M.S., Fertilizer produced from abattoir waste can contribute to phosphorus sustainability, and biofortify crops with minerals. *PLoS One*, 14, 9, e0221647, 2019.
94. Chen, S., Ai, X., Dong, T., Li, B., Luo, R., Ai, Y., Chen, Z., Li, C., The physico-chemical properties and structural characteristics of artificial soil for cut slope restoration in Southwestern China. *Sci. Rep.*, 6, 1, 20565, 2016.
95. Truong, L., Morash, D., Liu, Y. *et al.*, Food waste in animal feed with a focus on use for broilers. *Int. J. Recycl. Org. Waste Agric.*, 8, 417–429, 2019.
96. Stępień, A., Wojtkowiak, K., Grzywińska-Rąpca, M., Pawluczuk, J., Meat processing waste as a source of nutrients and its effect on the physicochemical properties of soil. *Sustainability*, 14, 3, 1341, 2022.
97. Stępień, A., Wojtkowiak, K., Kolankowska, E., Use of meat industry waste in the form of meat-and-bone meal in fertilising maize (Zea mays L.) for grain. *Sustainability*, 13, 5, 2857, 2021.

98. Chew, K.W., Chia, S.R., Yen, H.W., Nomanbhay, S., Ho, Y.C., Show, P.L., Transformation of biomass waste into sustainable organic fertilizers. *Sustainability*, 11, 8, 2266, 2019.
99. Chakraborty, R., Gupta, A.K., Chowdhury, R., Conversion of slaughterhouse and poultry farm animal fats and wastes to biodiesel: Parametric sensitivity and fuel quality assessment. *Renewable Sustain. Energy Rev.*, 29, 120–134, 2014.
100. Ning, Z., Ji, J., He, Y., Huang, Y., Liu, G., Chen, C., Effect of lipase hydrolysis on biomethane production from swine slaughterhouse waste in China. *Energy Fuels*, 30, 9, 7326–7330, 2016.
101. Affes, R., Palatsi, J., Flotats, X., Carrère, H., Steyer, J.P., Battimelli, A., Saponification pretreatment and solids recirculation as a new anaerobic process for the treatment of slaughterhouse waste. *Bioresour. Technol.*, 131, 460–467, 2013.
102. Rodríguez-Abalde, Á., Gómez, X., Blanco, D., Cuetos, M.J., Fernández, B., Flotats, X., Study of thermal pre-treatment on anaerobic digestion of slaughterhouse waste by TGA-MS and FTIR spectroscopy. *Waste Manage. Res.*, 12, 1195–1202, 2013.
103. Meng, Y., Li, S., Yuan, H., Zou, D., Liu, Y., Zhu, B., Li, X., Effect of lipase addition on hydrolysis and biomethane production of Chinese food waste. *Bioresour. Technol.*, 179, 452–459, 2015.
104. Forgács, G., Lundin, M., Taherzadeh, M.J., Sárvári Horváth, I., Pretreatment of chicken feather waste for improved biogas production. *Appl. Biochem. Biotechnol.*, 169, 2016–2028, 2013.
105. Park, S., Yoon, Y.M., Han, S.K., Kim, D., Kim, H., Effect of hydrothermal pre-treatment (HTP) on poultry slaughterhouse waste (PSW) sludge for the enhancement of the solubilization, physical properties, and biogas production through anaerobic digestion. *J. Waste Manage.*, 64, 327–332, 2017.
106. Hernández-Fydrych, V.C., Benítez-Olivares, G., Meraz-Rodríguez, M.A., Salazar-Peláez, M.L., Fajardo-Ortiz, M.C., Methane production kinetics of pretreated slaughterhouse wastewater. *Biomass Bioenergy*, 130, 105385, 2019.
107. Seffati, K., Honarvar, B., Esmaeili, H., Esfandiari, N., Enhanced biodiesel production from chicken fat using CaO/CuFe2O4 nanocatalyst and its combination with diesel to improve fuel properties. *Fuel*, 235, 1238–1244, 2019.
108. Sai Akhil, U. and Alagumalai, A., A short review on valorization of slaughterhouse wastes for biodiesel production. *ChemistrySelect*, 4, 45, 13356–13362, 2009.
109. Mata, T.M., Cardoso, N., Ornelas, M., Neves, S., Caetano, N.S., Evaluation of two purification methods of biodiesel from beef tallow, pork lard, and chicken fat. *Energy Fuels*, 25, 10, 4756–4762, 2011.
110. Bhatti, H.N., Hanif, M.A., Qasim, M., Biodiesel production from waste tallow. *Fuel*, 87, 2961–2966, 2008.

111. Abd-Alla, M. and El-Enany, A.W., Production of acetone-butanol-ethanol from spoilage date palm (Phoenix dactylifera L.) fruits by mixed culture of Clostridium acetobutylicum and Bacillus subtilis. *Biomass Bioenergy*, 42, 172–178, 2012.
112. Toldrá-Reig, F., Mora, L., Toldrá, F., Trends in biodiesel production from animal fat waste. *Appl. Sci.*, 10, 10, 3644, 2020.

10

Valorization of Seafood Processing By-Products for Bioactive Compounds

Puthanveetil V. Suresh[1,2*], A.R. Brundha[1,2], Tanaji G. Kudre[1,2] and S.K. Sandesh[1,2]

[1]*Meat and Marine Sciences Department, CSIR-Central Food Technological Research Institute, Mysuru, India*
[2]*Academy of Scientific and Innovative Research (AcSIR), Ghaziabad, India*

Abstract
Seafood is one of the major commodities contributing to global food security. Global seafood production was about 177.8 million tons in 2019, with a total first sale value equivalent to US$ 406 billion, out of which about 158 million tons (88%) were used for direct human consumption. Commercial seafood processing generates a huge volume of non-edible portions as by-products that are either underutilized or discarded as waste in many parts of the world. These seafood processing by-products are rich sources of various bioactive compounds, such as peptides, lipids, omega-3-fatty acid, pigment, enzymes, vitamins, minerals, chitin, collagen, and so on, with promising beneficial effects for human health. However, recovery of these bioactive compounds from the seafood processing by-products is a challenging task owing to the extremely complex and highly perishable nature of the by-products from the different seafood processing plant. Identifying the bioactive compounds and their promising applications in areas like biomedicine, nutraceutical, and functional food has led to a new perspective on the sustainable valorization of an enormous quantity of different by-products from seafood processing plant.

Keywords: Seafood, seafood processing, processing by-products, valorization, bioactive compounds

**Corresponding author*: drsureshpv@cftri.res.in; drsureshpv@hotmail.com

Umile Gianfranco Spizzirri (ed.) Nutraceutics from Agri-Food By-Products, (319–360) © 2023 Scrivener Publishing LLC

10.1 Introduction

Fish and other aquatic foods (seafood) play a pivotal role in healthy diets in many countries, which provide the animal protein required to consume the least expensive healthy diet [1]. Aquatic foods are increasingly being recognized for their crucial role in food security and nutrition, not only as a source of protein but also as a unique and highly diverse source of essential omega-3 fatty acids and bioavailable micronutrients [2, 3]. According to the Food and Agriculture Organization (FAO) of the United Nations, the total global seafood (finfish and shellfish) production (by capture fisheries and aquaculture) was about 177.8 million tons (MT) in 2019, excluding aquatic plants and other aquatic animals [4]. Out of total production in 2019 (177.8 MT), capture production was estimated at 92.5 MT, and aquaculture production was estimated at 85.3 MT. The first sale price of the total seafood production was calculated at US$ 406 billion in 2019. Of the total seafood (fishery and aquaculture) production in 2019, over 158 MT (89%) were used for direct human consumption and the remaining (11%, 20 MT) were destined for non-food products, mainly for the production of fishmeal and fish oil [4].

In 2019 of the total fishery and aquaculture production, 44% were destined in live and fresh form, while the remaining 56% (88.5 MT) was processed for human consumption [3]. This commercial processing of fisheries and aquaculture production has resulted in an enormous volume of non-edible by-products, which may constitute up to 70% of processed raw material, depending on the size, kind and type of handling methods used [3]. Although no detailed estimation for non-edible by-products is available from the processing of fisheries and aquaculture production, a calculation (based on the quantity processed for human consumption, 88.48 MT) revealed that 61.9 MT (equivalent to 50-70% of processed for human consumption) with an average of 53.1 MT globally. In the current decade, the seafood industry, particularly inland water aquaculture, has been growing fast as the food production industries.

The seafood supply chain generates a variety of by-products, commencement with the collection of seafood and ending with its consumption [5]. The seafood is processed by several approaches as the raw material for the manufacture of processed products and as a method of preservation, immediately after harvesting, either in onboard handling facilities of the fishing fleets or in offshore handling factories and during transport, storage, retailers, consumers and so on [6]. Additionally, enormous amounts of overall harvest are discarded annually while capturing the wild fishes as "by-catch" or "accidentally draped" species and also from aquaculture farms. It was reported that about 17.9-39.5 MT of whole seafood is annually discarding worldwide by industrial wild catcher fisheries operations [5]. Moreover, to the term 'by-product', many other terms such as 'co-product', 'fish waste', 'fish offal', 'fish visceral mass', 'fish discards', etc. have been used in the literature to describe the non-edible portion generating from seafood processing [7–9].

It should be emphasized that 'by-product' is the most commonly used and appropriate phrase to describe the non-edible portion resulting from seafood processing. Hence, we intended to use the word 'by-product' throughout this chapter to represent only the non-edible parts generated from commercial seafood (finfish and shellfish) processing.

Since seafood includes varied types of aquatic animals, the non-edible part produced during their processing varies significantly in composition and quantity. The variety of seafood and the processing method have a significant impact on the composition and percentage of different non-edible portions [10]. It was reported that processing of finfish generates 10% to 50% of the total quantity of the seafood processed as non-edible by-products parts, which are generally composed of heads (9%-12%), gut (viscera) (12%-18%), skin (1%-3%), bones (9%-15%) and scales (~5%) [4, 9, 11, 12]. Some of the major processing of finfish and the by-products produced are summarized in Figure 10.1 and Table 10.1. Similarly, shellfishes, particularly crustaceans, produce up to 85% of the raw material processed as non-edible portions, which are mainly composed of head/cephalothorax, shell/carapace, viscera, appendages, etc. The main crustacean species processed by the seafood industries are shrimp/prawns, crabs and lobsters. Shrimp/prawn processing by-products account for up to 55%-60% of the original weight of raw materials. Crab processing generates 60% to 70% of actual weight (wet basis) as non-edible discard comprising back shells, viscera, gills, and other shell wastes. The lobster processing by-products may account for up to 60% of the original weight of raw materials [13]. In addition, Molluscan shellfish such as Squids, Octopus, etc., are also commercially processed in large quantities in different parts of the world. Some of the major processing of shellfish and by-products produced are summarized in Figure 10.2 and Table 10.1.

The seafood industries have attempted several physical/mechanical, chemical, and biological approaches during the last several decades to treat or reduce the disposal of seafood processing by-products while minimizing the environmental impact [8, 11, 12, 14, 15]. In addition, in recent years, green techniques such as ultrasound-assisted extraction and supercritical fluid extraction have also been applied by various research teams in the

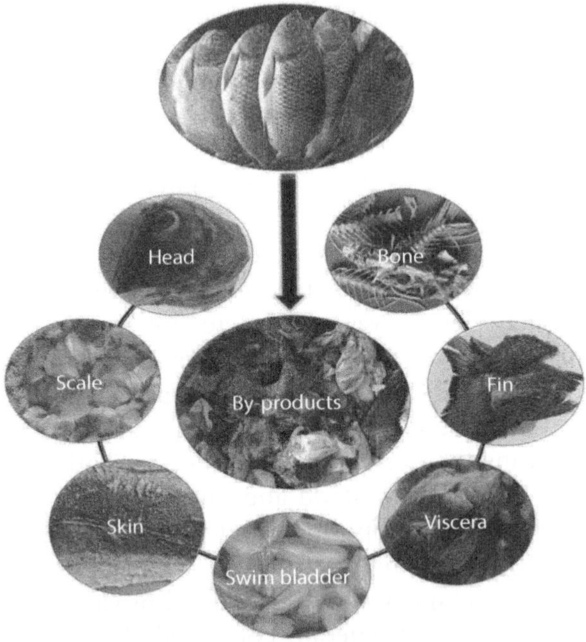

Figure 10.1 Finfish processing by-products.

Table 10.1 Major seafood (finfish and shellfish) processing by-products (values shown in the table are compiled from various authors mentioned somewhere else in the text).

Seafood type	By-products	% (w/w) by-product
Finfish		
❖ Boney fish	➢ Viscera/gut/internal organ	15-20
❖ Cartilaginous fish	➢ Head	15-20
❖ Jawless fish (Hagfish and Lampreys)	➢ Skin with scale	1-3
	➢ Bone/frame	10-15
	➢ Trimming	2-5

(*Continued*)

Table 10.1 Major seafood (finfish and shellfish) processing by-products (values shown in the table are compiled from various authors mentioned somewhere else in the text). (*Continued*)

Seafood type	By-products	% (w/w) by-product
Shellfish		
Crustacean		
• Shrimp	Cephalothorax/head, carapace/shell and tail	65-85
• Lobster	,,	up to 60
• Crayfish	,,	up to 85
• Krill	,,	70-75
• Crab	Back shell, appendages, gills, viscera, etc.	60-70
Mollusca		
○ Scallop	Outer shell	60-80
○ Mussel	Non-edible portion	8-12
○ Oyster		
○ Clam		
Cephalopods		
▪ Squid	Gladius/pen, ink bag, liver and other organs	25-42
▪ Cuttlefish	,,	--
▪ Octopi	Intestine, mouth apparatus, eye, etc.	11-20
Coelenterates and echinoderms		
▪ Sea urchin	Shell and viscera	--
▪ Sea cucumber	Viscera	--
▪ Jellyfish	--	--

-- Not available

Figure 10.2 Shellfish (crab and shrimp) processing by-products.

treatment and sustainable valorization of seafood processing by-products [16, 17]. In general, these seafood processing by-products are rich and least exploited sources of several bioactive compounds with significant commercial applications in different fields. This chapter explores the possibility for valorizing the enormous volume of seafood processing by-products for their potent bioactive compounds.

10.2 Bioactive Compounds From Seafood Processing By-Products

The word "bioactive molecules/compounds" is defined as the biologically derived molecules/compounds that have the capacity to interact with one or more molecules/compounds of a living cell and deploy a beneficial impact on human health [19]. Because of exceptional diversity in living organisms and distinctive habitats, aquatic biosphere is one of the most important natural reservoirs of various physiologically active molecules/compounds with a beneficial effect on human health. Substances formed from aquatic organisms, especially from salt water, are exciting since the organisms of the salty atmosphere, with changing intensity of salt, temperature, and light. Overall, seafood by-products generated from finfish and shellfish processing are rich sources of several bioactive compounds with beneficial effects on human health. These bioactive compounds from the seafood processing by-products can be grouped as (i) bioactive proteins such as enzymes, collagen, bioactive peptides, etc., (ii) lipids, fatty acids, etc. (iii) chitin, chitosan, chitin/chitosan oligosaccharides/oligomers, *N*-Acetyl-D-glucosamine, D-glucosamine, etc. (iv) vitamin, pigment (carotenoids), etc., (v) minerals, trace elements, etc., (vi) squalene, (vii) chondroitin sulphate.

Table 10.2 Major finfish processing by-products and associated bioactive compounds and their bioactivities.

Seafood type	By-products	Bioactive compounds	Bioactivities	References
Finfish	Viscera	• Fatty acids • Enzymes • Bioactive peptide • Vitamins	Antioxidant ACE inhibitory Anticancer Antidiabetic	[191–195]
	Head	• Fatty acids • Collagen • Bioactive peptides • Minerals	Antioxidant	[195–197]
	Skin	• Collagen • Bioactive peptide	Antioxidant ACE inhibitory Iron-binding Neurobehavirol Anticoagulant Antihypertensive	[99, 196, 198–202]
	Scale and fin	• Collagen • Bioactive peptides	ACE inhibitory Antioxidant	[200, 203–205]
	Bone/frame	• Collagen • Minerals • Bioactive peptide • Squalene	Antioxidant ACE inhibitory Ca-binding Neuroprotective Anticoagulant Antihypertensive Antiproliferative	[206–214]
	Swim bladder	• Collagen		[95, 205]

Table 10.3 Major shellfish processing by-products and associated bioactive compounds and their bioactivities.

Seafood type	By-products	Bioactive compounds	Bioactivities	References
Shellfish				
Crustacean	Head/shell/tail/ appendages	▪ Chitin ▪ Chitosan ▪ Chitin/ chitosan oligomer ▪ GlcNAC GlcN ▪ Carotenoid Bioactive peptide ▪ Fatty acids	Antioxidant Antimicrobail Antihypertensive Antidiabetic Antiinflamatory	[85, 146, 215–217]
Mollusca	Gladius/chiton (Endoskeleton)	▪ Chitin	Antioxidant Antimicrobial Antihypertensive Antiproliferative	[215, 218, 219]
	External shell	▪ Minerals (Calcium, iron, zinc and phosphorus)		[220, 221]
	Viscera	▪ Fatty acids ▪ Enzymes ▪ Bioactive peptide ▪ Lecithin	Antioxidant Antihypertensive Antidiabetic Antiinflamatory	[148, 152]
	Non-edible body parts such as skin and tunics	▪ Collagen ▪ Gelatin ▪ Bioactive peptide ▪ Lecithin ▪ Fatty acids	Antioxidant Immunostimulator Antiinflamatory Antihypertensive Antidiabetic	[122, 123, 222]
Jellyfish	Body parts	▪ Collagen ▪ Bioactive peptide	Antioxidant Immunostimulator	[100, 223–225]

Different bioactive compounds from finfish and shellfish processing by-products are summarized and presented in Table 10.2 and Table 10.3, respectively.

10.3 Extraction Methods of Bioactive Compounds From Seafood Processing By-Products

The treatment and disposal of an enormous quantity of highly perishable seafood processing by-products is not easy compared to other agro-food processing by-products. Further, disposal of seafood processing by-products involves various strict concerns, including those related to safety, hygiene, ecology, etc. and is frequently regulated by government agencies [5, 12]. Traditionally, seafood processing by-products have been disposed of by landfill or ocean dumping, resulting in severe environmental issues and the loss of various extractable high-value components found in these natural resources.

Furthermore, to reduce the dumping of the enormous amount of seafood processing by-products produced annually, seafood industries across the globe adopted various approaches for treating their by-products and extracting valuable components to enhance the marketable value of these resources. These treating and extraction procedures of seafood processing by-products to recover the high-value components can be broadly classified as physical/mechanical, chemical, and biological processing [12]. Generally, seafood or other allied industries are applied in combination of two or more of these techniques to maximize the recovery yield of the valuable components as well as to recover these components with better quality/property.

Among the various methods, the most frequently used extraction method for valuable components from seafood processing by-products is the chemical method, including solvent extraction procedures. In most cases, the chemical extraction methods involve some sort of mechanical pretreatment processing to enhance the recovery percentage of valuable components. The chemical extraction methods have been applied extensively for the extraction of bioactive compounds from seafood processing by-products such as bioactive peptides, lipids, fatty acids, chitin and chitosan and their bioactive derivatives, vitamin, pro-vitamins, pigment (carotenoids), minerals and so on.

Currently, owing to the progress in biotechnological techniques, biological process has gained much significance as sustainable processes and is at the edge of substituting some of the chemical/mechanical processing for treating the seafood processing by-products and recovery of bioactive compounds. In comparison to conventional processes, the sustainable processes and recovery systems are more lucrative since they require less energy use of resources and power, which results in lesser releases of contaminants, and permit superior and more effective usage of renewable resources, to reduce the requirement on non-renewable materials [12, 20].

Furthermore, this biological processing would aid in the recovery of new bioactive compounds from the seafood processing by-products without losing the bioactivity of the extracted compounds. Hence, in recent decades biological methods have gained more importance as preferred methods for the recovery of bioactive compounds from various seafood processing by-products. The various biological approaches for retrieving bioactive compounds from seafood processing by-products can be divided into 'enzymatic/bio-catalytic methods' and 'fermentation methods'. Enzymes have been used in the process of enzymatic methods in either soluble free form or immobilized form. Various commercial proteases from microbial sources, such as alcalase, neutrase, flavourzyme, umamizyme, and protamex; papain, bromelain and ficin from plant sources and pepsin, trypsin; and chymotrypsin from animal sources, have been widely used by various seafood industry for the recovery of bioactive compounds from their by-products [8, 11, 12, 21]. Furthermore, direct fermentation using specific live microorganisms with the ability to synthesize one or more preferred extracellular enzymes has been used in the enzymatic processes. In addition, whole cells (microorganism) that are metabolically inactive but consist of one or more desirable enzyme activities are also exploited in the recovery of bioactive compounds from seafood processing by-products. In the fermentation processes of extraction, in addition to the enzymatic activity, exploiting several other metabolic processes of the growing microorganisms, such as lactic acid fermentation. The biological extraction methods have been widely used for the recovery of bioactive compounds from different seafood processing by-products such as bioactive peptides, lipids, fatty acids, chitin and chitosan and their bioactive derivatives, vitamin, pro-vitamins, pigment (carotenoids), minerals, and so on.

Recently, emerging/innovative green technologies have been attempted [18, 22] as an attractive substitute to the approaches traditionally applied for the valorization of seafood processing by-products and recovery of bioactive compounds [23]. These innovative green approaches have been reported to have several advantages over the conventional practices of extraction, such as improving the quality and the recovery efficacy, as well as reducing the losses of biological properties of the bioactive compounds recovered from the seafood processing by-products [18, 22]. In order to isolate bioactive compounds, consequently valorizing seafood processing by-products, from various techniques can be used, such as pulsed electric field (PEF) extraction, ultrasound-assisted extraction (UAE), microwave-assisted extraction (MAE), and supercritical fluid extraction (SFE) [16–18]. Among these emerging alternative processing approaches, UAE and SFE have received much importance for their application in the extraction of different bioactive compounds from seafood processing by-products. UAE is well documented as an effective, quick, green, repeatable substitute for non-thermal extraction systems as compared to traditional extraction approaches in the food industry [24]. SFE is an unconventional green extraction technique that has received more attention in the food industries for several applications in the last decade [22].

SFE is reviewed as a green process owing to the use of non-harmful solvents that make it a much more sustainable approach and minimal power usage and ecological issues. It has been studied widely in the extraction of fish oil and concentration of ω-3 long-chain polyunsaturated fatty acids (eicosapentaenoic acid (EPA) and docosahexaenoic acid (DHA) from different seafood processing by-products [25]. Application of UAE extraction [26] and SFE [27] for the recovery of collagen from different seafood processing by-products has been reported. Supercritical fluid extraction has been defined as an effective method to get collagen in a highly purified form to be used in high-end applications such as in the biomedical field [27]. It has been well documented that innovative approaches permit for the concurrent recovery of different compounds such as lipids and proteins by means of green edible-grade solvents, a very encouraging method that could transform the multi-step mining of bioactive compounds from the enormous volume of biomaterial like seafood processing by-products [23, 28].

10.4 Recovery of Enzymes From Seafood Processing By-Products

Enzymes can be used as a biotechnological tool to transform other compounds/molecules into important products in the form of nutraceutical, functional food and other industries. [11, 21, 28–30]. Commercially available enzymes are derived from various living organisms, including bacteria, fungi, plants and animals from terrestrial and aquatic sources, including marine. It is well known that enzymes from marine organisms differ from those of terrestrial equivalents in their characteristics in terms of superior catalytic, physical, and chemical properties. Marine-derived enzymes exhibit distinct features, such as high tolerance to salt, resistance at high pressure and temperature, superior catalytic activity at minimal concentration, lesser sensitivity to substrate rate and a wide pH level [31–33]. Consequently, enzymes from marine bio-sources have gained great interest for their potent commercial uses in a different field.

Enzymes remain a major share of bioactive compounds in the seafood processing by-products and extracted commercial scale [23]. Enzymes that can be extractable from seafood by-products comprise digestive proteinases, extracellular gastric proteinases, chitinases, lipases, phospholipases, transglutaminases, polyphenol oxidase and others [11, 32]. It is well recognized that among the different components of seafood processing by-products, viscera (fish internal organ/gut) are the most important source of marine enzymes, particularly digestive proteases, due to the high content of enzymes and abundant availability. The viscera comprise ~5% of the total finfish weight and represent one of the main components of by-products from the seafood processing industry. Besides the fish digestive system, diverse extracellular enzymes are liberated, constituting viscera, a promising natural and inexpensive source of marine derived enzymes [34, 35]. However, a number of elements, such as growth stage, period, habitats, etc., can create a usual difference in the quantity and composition of different enzymes found in different seafood species [36]. Among various enzymes from seafood processing by-products, particularly from finfish viscera, proteinases (including the gastric protease pepsin; intestinal proteases trypsin and chymotrypsin; collagenase, etc.) forms the most significant group with commercial importance [37]. The digestive proteases from finfish viscera include pepsin (an aspartic protease) and serine proteases (trypsin, chymotrypsin, collagenase, and elastase) [23, 38–45]. These proteases can be endogenous or secreted extracellularly by the fish microbiota. These marine proteases from finfish viscera are one of the widely used enzymes in various industries due to their unusual catalytic activities (such as higher catalytic efficacy at low temperatures, lesser sensitivity to substrate rate, and better resistance at boarder pH range); however, reports on the potential uses of these enzymes are limited to date [23, 38, 39, 46–50].

Among different digestive proteases, trypsin (EC 3.4.21.4) extracted from the finfish viscera by-products represent the most important, and its application has increased considerably in the recent year owing to its unique property of high stability and activity at high temperature and pH [23, 51]. Trypsin and trypsin-like proteases have been extracted and characterized from different fish species found in different niches like cold water chinook salmon [52] and warm water Monterey sardine [38], true sardine [53], sardine [54], striped seabream [55], bogue [56], sardinella [57], and yellowfin tuna [58]. In addition to digestive proteases, various other enzymes, which are capable of hydrolyzing the polypeptide backbone of native collagen protein, have been recovered and characterized from different seafood processing by-products [59]. The extraction of proteases from seafood processing by-products might be advance to develop novel inexpensive proteases for commercial applications and concomitantly answer the concern of seafood processing by-products management [23, 51].

In addition, to the digestive proteases from finfish viscera, a number of other enzymes like lipases, chitinases, alkaline phosphatase, hyaluronidase, and *N*-Acetyl-D-glucosaminidase have been extracted and characterized from seafood processing by-products [9, 11, 33]. Lipases are commercial enzymes that have a number of uses in detergent, food, dairy, paper, chemical, etc. industries [33]. Lipases have been recovered and characterized from processing by-products of cod, salmon, rohu, oil sardine, mullet, Indian mackerel, and red sea bream. Owing to their high catalytic activity at mild pH (6.0 to 9.0) and temperature (30-45 °C) conditions, the marine lipases are very suitable for different food applications [11, 33, 60]. Mardina *et al.* extracted and characterized lipase from the pancreas of little tuna (*Euthynnus affinis*) with an enzyme activity of 11.96 U/ml at the optimum pH 9 [61].

Chitinases catalyze the hydrolysis of β-(1,4)-*N*-acetylglucosaminide bonds in the chitin polymer. Chitin is considered the second largest biopolymer in the world, only after cellulose. The chitinolytic enzymes are broadly distributed in nature and found in many microorganisms, plants, and animals. In animals, chitinases play a significant physiological function in nutrition and morphogenesis [62, 63]. Endo-chitinases (EC 3.2.1.14) and β-*N*-acetylhexosaminidases (EC 3.2.1.52) are two groups of enzymes of chitinase complex, and their actions are known to be synergistic and consecutive.

The combined action of these enzymes (endo-chitinases and β-*N*-acetylhexosaminidases) results in the digestion of insoluble chitin polymer into its water-soluble low molecular weight derivatives [11, 64, 65]. Chitinases have gained much attention in recent years due to their promising application in the bioconversion of seafood processing chitinous by-products to bioactive molecules [29, 66, 67]. Chitinases have been recovered and characterized from processing by-products of shellfish and finfish such as squid, cuttlefish, clam, shrimp, red sea bream, cod, Japanese eel, silver croaker, etc. [68–74]. Crude chitinases from Japanese common squid and cuttlefish were used to produce *N*-Acetyl-D-glucosamine from chitin polymer [72].

In addition, the enzymes like transglutaminases, alginate lyase and chlamysin have been recovered and analyzed from different seafood processing by-products [75–77]. Chlamysin, a lysozyme-like enzyme with potential inhibitory effect on both gram^{+ve} and gram^{-ve} bacteria, have isolated from the viscera of marine bivalve *Chlamys islandica* [76]. Transglutaminases have been recovered and analyzed from walleye pollock by-products [75]. Transglutaminases enzyme (meat glue) has wide commercial applications, particularly in surimi and meat industries. Alginate lyase has been extracted and characterized from viscera, gland and heptopancrease of marine shellfish molluscs [77]. It has huge applications, specifically in the extraction of biomolecules from seaweeds and in algal biotechnology to prepare algal protoplast.

Localization of the targeted enzymes in the tissue, such as intracellular or extracellular, is one of the key factors affecting the extraction of enzymes from seafood by-products. The extraction and purification of enzymes seafood processing by-products are commonly performed by various separation methods such as membrane technology, ohmic heating, salt and organic precipitation, chromatography, phase separation by aqueous two-phase systems, etc. [12, 33, 78]. However, most of the enzyme recovery methods are laborious, highly expensive, and require labor with high technical skills [23, 51].

10.5 Extraction of Bioactive Pigment (Carotenoids) From Seafood Processing By-Products

The bioactive pigment present in the by-products of seafood processing is the carotenoids. Carotenoids are accountable for the beautiful coloration of the flesh and skin of salmonid fish (e.g., salmon and trout) and the crustaceans (e.g., shrimp, lobster, crab, crayfish, krill, etc.) exoskeleton [79–81]. Processing by-products of crustaceans shellfish (shrimp, lobster, crabs, crayfish, krill, etc.) are the main raw material for the preparation of natural carotenoids, astaxanthin [33, 82–84]. Carotenoids are also recovered from Salmons processing by-products [84]. The carotenoids found in the free and complex caroteno-protein forms in crustaceans and the total quantity of it have been found differ from species to species as well as different body parts such as the head (35.8 to 185.3 μg/g) and carapace (59.8 to 117.4, μg/g) of marine shrimp [83]. Generally, the free form of carotenoids is highly unstable, while its complex caroteno-protein form exhibits comparatively better stability. In order to enhance the stability of carotenoid (astaxanthin) extracted from shrimp (*Litopenaeus vannamei*) processing shells, by-product immobilization in alginate-chitosan beads was reported [85]. Carotenoids are exciting bioactive compounds that have the potential to alleviate the destructive power of oxidative stress [86]. The antioxidant potential of carotenoids in a different system and their positive effect on human health are well documented [83, 86, 87]. Furthermore, the antioxidant action of carotenoids (astaxanthin) controls the physiological activity linked to lipid peroxidation, having a beneficial effect on chronic ailments like cardiovascular diseases, muscular deterioration, and tumors [80, 81, 84, 88].

Several research groups have reported the recovery and characterization of carotenoids from different shellfish processing by-products [33, 40, 82, 83, 90–93, 98]. Chemical and biological (enzymatic or fermentative) methods have been used to recover and purify carotenoids from shellfish processing by-products. It is well documented that biological methods (enzymatic or fermentative) are more preferable in the extraction of carotenoids as caroteino-protein from crustacean by-products since these methods result in a higher yield of carotenoids with better quality for their possible applications [33].

Sachindra *et al.* studied the extraction of carotenoids from shrimp (*Penaeus indicus*) by-products by lactic acid fermentation using *Lactobacillus plantarum* B 4496 followed by extraction with a solvent mixture (hexane and isopropanol, 60:40 v/v) and refined sunflower oil [83]. Authors observed that lactic acid fermentation was found to be comparatively better to acid (15% propionic acid) ensilaging with respect to recovery yield and stability of carotenoids from the shrimp by-products ensilage, both in inorganic solvent and oil extraction.

Recovery of stable carotenoprotein from different shellfish processing by-products using enzymatic approaches has been reported [40, 91–93]. An enzymatic method for recovery of carotenoprotein from shrimp processing by-products using a chelating agent (EDTA) and the proteolytic enzyme trypsin has developed [91]. The authors extracted astaxanthin in the form of a carotenoprotein complex with 65% protein and 49% astaxanthin. Cano-Lopez *et al.* used a trypsin preparation from Atlantic cod at 4 °C for extraction of carotenoprotein with 81% astaxanthin and 64% protein from shrimp processing by-products [92]. In a study, Chakrabarti comparatively evaluated the yield of carotenoprotein from shrimp processing by-products using different proteases [93]. The author reported that trypsin exhibited higher carotenoid yield (55%), followed by pepsin and papain (50%) during the same period. Klomklao *et al.* used bluefish (*Pomatomus saltatrix*) for the trypsin preparation and the commercial bovine trypsin for its isolation of carotenoprotein from black tiger shrimp (*P. monodon*) processing shell by-products and comparatively evaluated their yields [40]. They observed that the hydrolytic activity of bluefish trypsin and the yield of carotenoprotein were similar to that obtained using commercial bovine trypsin. A semi solid-state fermentation (SSF) process using *L. paracasei* strain A3 was reported for the recovery of carotenoprotein from crawfish processing by-products [89]. Bhaskar *et al.* extracted carotenoprotein from shrimp (*P. monodon*) by-products by lactic acid fermentation using *Pediococcus acidolactici* CFR2182 [90]. They used a statistical design (Response surface methodology, RSM) procedure to optimize the fermentation and succeeded >75% recovery of carotenoids.

10.6 Extraction of Collagen and Its Derivatives From Seafood Processing By-Products

Collagen is the structural protein found in vertebrates and invertebrates and comprises ~25% of the total quantity of protein [94–97]. In contrast, gelatin does not exist in nature but is formed by the denaturation of triple-helix polymer collagen [11, 33]. The major amino acids found in the collagen or gelatin are glycine (Gly), alanine (Ala), proline (Pro, X) and hydroxyproline (Y), with principally a Gly-X-Y triplet as the replication unit [94, 98]. Collagen and gelatin find many applications in various fields, including in the food, cosmetic, biomedical, and pharmaceutical industries [94]. Currently, collagens are widely used in health and functional food products due to their unique functional and structural features and potential beneficial effects on human beings [94, 97]. Bioactive property, particularly the antioxidant activity of seafood-derived collagen, has been reported [95, 96, 99]. These studies established the antioxidant capability of native collagens from diverse seafood species. In addition, Morishige *et al.* have reported immunomodulation activity of collagen isolated from giant edible jellyfish (*Nemopilema nomurai*) by a cell-line study [100].

Currently, collagens are produced commercially from slaughterhouse by-products of land-based animals, mainly bovine and porcine [94, 97]. However, the utilization of collagens and collagen-derived products from these land-based animals have made issues concerning consumer safety due to the outburst of bovine spongiform encephalopathy, foot-and-mouth disease, and other prions ailments, along with certain ethical concern among some religious thought [94, 97, 101]. Seafood processing by-products have been well reported as a potentially sustainable resource of marine-derived alternative collagens [94]. Collagens can be extracted from processing by-products of finfish and shellfish [94–96, 98, 102]. Among 29 forms of collagen (Type I -Type XXIX), the Type I collagen is the most commercially significant form of collagen, and it is effectively extracted, purified, and characterized from by-products of several species of finfish and shellfish [94]. The collagen can be grouped as salt-solubilized collagen, acid-solubilized collagen, and pepsin-solubilized collagens based on their extraction approaches. Different methods coupled with mechanical, chemical, and enzymatic processing have been attempted to extract collagens from seafood processing by-products.

Extensive studies have been reported on collagen isolation and characterizations from processing by-products of different seafood species [94–96, 101–110]. In a comparative study carried out by Nagai *et al.* reported that the yield of pepsin-soluble collagen was found to be higher (44.7%) than acid-soluble collagen (10.7%) from Ocellate Puffer fish (*Takifugu rubripes*) skin [111]. Liu *et al.* extracted and characterized pepsin-soluble collagen from Giant Red Sea Cucumbers (*Parastichopus sp*) [112]. The application of finfish pepsin for the extraction of collagen from the skin of various fish species and increased collagen yield has been reported [113]. Pal and Suresh have extensively studied extraction and characterization of acid-solubilized and pepsin-solubilized collagens from processing by-products like skin, scale and swim bladder of Indian major carp Catla (*Catla catla*) and Rohu (*Labeo rohita*) [95, 96, 114, 115]. Lidiya *et al.* recently isolated and characterized Type I collagen from swim bladder by-products of freshwater Red-bellied Pacu [116] and skin of marine Malabar sole [99].

As mentioned above application of emerging green technologies USA extraction and SFE has been attempted for the recovery of collagen from various seafood processing by-products [18, 26, 27, 109]. USA extraction of collagen from the skin of the sea bass *Lateolabrax japonicus* has been developed [26]. SFE has been described as a successful method to get collagen in a highly purified form to be used in high-end applications such as in the biomedical and pharmaceutical fields [27].

Gelatin from seafood processing by-products can be prepared by hydrolyzing the collagen [117, 118] present in the raw material through a pre-treatment step (acid treatment for type A gelatin or alkaline treatment for type B gelatin) process followed by an extraction step [119]. The properties of raw material (collagen) and the extraction methods strongly determined the characteristics of the gelatin prepared [120]. Gelatin is largely used as a functional component in foods and cosmetics, but it can produce bioactive peptides with different bioactivities such as antioxidant, ACE-Inhibitory, antimicrobial, etc., by enzyme (protease) hydrolysis [117, 121]. Production of bioactive gelatin peptides with antioxidant activity from tuna processing by-products and giant squid tunics by enzyme (trypsin, α-chymotrypsin, etc.) hydrolysis has been reported [122]. Inner and outer tunics of the jumbo flying squid (*D. gigas*), tuna (*Thunnus sp*) and halibut (*Hypoglossus sp*) skins were used for the enzymatic production of bioactive gelatin peptides [123]. Gelatin peptides with free radical scavenging activity and Angiotensin-I converting enzyme (ACE) inhibitory potential were produced from Pacific cod (*Gadus macrocephalus*) skin by-products [124].

10.7 Extraction of Bioactive Lipids and Fatty Acids From Seafood Processing By-Products

Lipids belong to the essential group of nutrients for human beings and play a vital role in maintaining a healthy body [125, 126]. Fatty acids are the main component of lipids and are classified into various groups, such as saturated fatty acids, unsaturated fatty acids, monosaturated fatty acids and poly unsaturated fatty acids, based on their chain length and presence or absence of double bonds [127]. Fatty acids are also classified as essential and non-essential fatty acids based on their effect on human health. Essential fatty acids are considered as functional foods and nutraceuticals due to their several beneficial effects on human health. The potential beneficial effects of essential fatty acids on human nutrition and health are well-documented [128, 129].

The beneficial effects of seafood lipids, particularly from finfish, on human health are well recognized [130, 131.] The presence of polyunsaturated fatty acids (n-PUFAs), particularly omega-3 fatty acids (n-3 fatty acids, ω-3 fatty acids) such as eicosapentaenoic acid (EPA, 20:5 n-3) and docosahexaenoic acid (DHA, 22:6 n-3) in fish oils/lipids are accountable for the human health benefits [132, 133]. Seafood processing by-products like head, viscera and liver are a rich source of several lipid-based compounds [7]. Bioactive lipid-based compounds that can be extracted from seafood processing by-products include n-PUFAs, particularly ω-3 fatty acids, phospholipids, squalene, vitamins, etc. Lipids enriched with n-PUFAs can be extracted from seafood processing by-products by different approaches. These lipid extraction methods include the conventional chemical methods using Soxhlet or Goldfisch, Folch, Bligh, Dryer technique, and acid digestion process using different organic solvents such as diethyl ether, petroleum ether, chloroform/methanol and hexane [134–138]. The biological (enzymatic) methods are also widely applied for the extraction of lipids enriched with n-PUFAs from different seafood processing by-products. These biological processes of lipid extraction can be divided into two groups (i) protease-catalyzed digestion and (ii) lipase-catalyzed digestion and esterification [8, 12, 139].

Generally, protease-catalyzed digestion methods are applied to recover the n-PUFAs from the initial raw material (seafood processing by-products). Commercial proteases from plants (papain), animals (chymotrypsin) and microorganisms (protamex, flavourzyme, neutrase and alcalase) have been widely used for the recovery of lipids from seafood processing by-products from protease-catalyzed digestion method [140–143]. Laset *et al.* achieved a recovery of 77% of total lipids with high ω-3 fatty acids content from salmon frames using protamex at 55 °C and pH 6.5 for 60 min hydrolysis [140]. Generally, the quantity of ω-3 fatty acids in the natural sources is minimal and may not offer a necessary amount of fatty acids in therapeutic applications [132]. Lipase-catalyzed digestion and esterification can be applied to concentrate the ω-3 fatty acids present in the lipids previously recovered from seafood processing by-products [8, 32, 144]. The concentration of ω-3 fatty acids by lipase-catalyzed digestion and esterification using the lipase from salmon by-products have been reported [145].

Supercritical carbon dioxide extraction has been reported as the emerging technology for the enhanced recovery of lipids enriched with n-PUFAs from seafood processing by-products [146–148]. Nguyen *et al.* have reported the preparation of lipids rich in n-PUFAs from the liver of the Australian lobster (*Jasus edwardsii*) using the supercritical carbon dioxide extraction method [148]. Lipids enriched with n-PUFAs, particularly EPA and DHA, have been prepared by supercritical carbon dioxide extraction from different by-products such as cephalothorax, carapace and tail of the Northern shrimp (*Pandalus borealis*) [146] and skin of the Indian mackerel (*Rastrelliger kanagurta*) [149]. Ferdosh *et al.* reported a higher recovery of n-PUFAs, particularly DHA, from the head by-product of longtail tuna (*Thunnus tonggol*) using ethyl alcohol as a co-solvent in supercritical carbon dioxide extraction [147].

Lecithin is a bioactive fatty compound mainly composed of phospholipid mixes [121] and has found applications in the nutraceutical and pharmaceutical industries [150, 151]). Shellfish squid visceral by-products are the main source of marine lecithin, mainly composed of phosphatidylcholine and phosphatidylethanolamine [152, 153]. Lecithin was extracted and analyzed from squid (*Todarodes pacificus*) visceral by-product after de-oiling by supercritical carbon dioxide treatment [152, 153].

Squalene is a bioactive natural lipid ($C_{30}H_{50}$) belonging to the polyunsaturated terpenoid group that is partially cratered from endogenous cholesterol synthesis and partly from dietary sources [154, 155]. Naturally, it is a precursor of a number of bioactive molecules, bile acid, cholesterol, hopanoids, steroids, etc., in living organisms [156]. Due to its exceptional biological, functional and chemical characteristics, squalene has several uses in the nutraceutical, pharmaceutical, cosmetics, and food sectors [155, 157]. The bioactive potentials of squalene, such as antioxidant, cardio-protective, anti-tumor, antimicrobial, and anti-cholesterol properties, have been well recognized [157–161]. It is widely used as a dietary supplement in human nutrition [162] and as a moisturizer, antioxidant (to quench singlet oxygen, $^1O2$) and ant-itching agent in cosmetics and dermatology [163]. Recently its potential as an adjuvant in vaccines for seasonal influenza and COVID-19 has been reported [164]. Deep sea shark liver oil is the richest natural and commercial source of squalene [165, 166]. In addition to sharks, various species of finfish from fresh water and marine have been observed to present squalene in considerable quantities (70-1803 mg/kg on a wet weight basis), mainly from viscera by-product [167, 168]. Ali *et al.* reported an ultrasound-assisted direct *in situ* saponification (UDS) process to extract squalene from the liver by-products of different fish (sea bass, skipjack tuna, gray bamboo shark, and spot-tail shark) species [169].

10.8 Extraction of Essential and Bioactive Minerals From Seafood Processing By-Products

These compounds are classified into macro-minerals and micro-minerals. They are essential elements necessary for several physiological functions, nerve functions and many enzymatic reactions in the human body. Thus, these essential elements can be considered as bioactive compounds. Macro-minerals are often stated as major minerals, are found at a higher amount in the body or requisite in higher quantity in food. It includes calcium, chlorine, magnesium, phosphorus, potassium, sodium, and sulfur. Micro-minerals, also known as trace minerals, are found in small amounts in the body or are required in lower quantities in food. Micro-minerals include chromium, cobalt, copper, fluorine, iodine, iron, manganese, molybdenum, selenium, and zinc. Both essential minerals (macro-minerals and micro-minerals) occur in seafood because of their capacity to obtain inorganic atoms from marine or fresh water. Both macro-minerals and micro-minerals are found in a higher quantity in seafood than in terrestrial foods [170]. Thus, seafood-processing by-products are rich sources of some essential bioactive minerals.

Among the various essential elements, calcium is one of the most important. It is essential for many physiological activities in the human body, including the development and maintenance of teeth and bones, nerve activity, many enzymatic processes, and biochemical reaction [171]. Fishbone includes a substantial quantity of seafood processing by-products and is a potential source of various minerals. Fishbone by-products contain 69.46% (on a dry basis) inorganic minerals and mainly constitute calcium (59.69%) and phosphorus (35.81%) with a mole ratio of Ca/P of 1.67 [171]. Thus, skeleton by-products with higher mineral content from big fish processing can be a sustainable natural bioresource for isolating calcium with potential application in food supplements and nutraceuticals. Beak bone stones of different finfish species, such as Ghol (*Protonibea diacanthus*) and Koth (*Otolithes biauratusa*), could be potential raw materials for bioactive calcium powder [172]. Generally, the calcium powders prepared from the finfish skeleton also contain a minor amount of other essential elements such as Cu, Mn, Zn, Se and Fe besides calcium and phosphorus. The finfish skeleton is also one of the rich sources of functionally active phosphorous. Phosphorous is the second major constituent of bone and teeth, a major controller of energy metabolisms, and a critical element that facilitates in the development of nucleic acids [171]. However, the human body cannot absorb calcium in a precipitated form such as calcium phosphate. Various studies demonstrated that in order to use fish bones as a calcium supplement, the bone would be transformed into its actual adsorption type by solubilization and ionization processes [121].

Physiologically active mineral compounds from finfish processing skeletal by-products can be extracted by enzymatic processes [171]. Jung *et al.* reported that skeleton by-product from commercial processing of Hoki (*Johnius belengerii*) and its hydrolyzed products using a proteolytic enzyme from bluefin tuna (*Thunnus thynnus*) could be used as a mineral nutraceutical with high bioavailability of calcium [173]. Kim *et al.* prepared a bone powder for food supplements from hoki (*J. belengerii*) frame using a protease from mackerel viscera [174]. Malde *et al.* reported that papain treated salmon bone could be used as a natural source of calcium and phosphorus for animal feed [175]. The main stage in enzymatic production of soluble calcium powder from fish skeletal includes: Raw material (Fish frame) → Treatments to get rid of non-bone parts → Fish bone → Crushing →

Mixing of enzyme and buffer to bone powder → Enzyme hydrolysis → Enzyme deactivation → Filtration → Concentration of filtrate → Calcium powder.

10.9 Recovery of Bioactive Peptides From Seafood Processing By-Products

Bioactive peptides are organic compounds made by amino acids joined by amide/peptide links. Though few bioactive peptides occur free in their natural source, the wide variety of known bioactive peptides are encoded in the polypeptide chain of the parent polymer proteins and are liberated mostly by hydrolytic processes of the enzyme [176]. Thus, bioactive peptides can be described as specific protein fragments that positively influence body physiology or metabolisms and may affect human health [178]. Bioactive peptides play an important role in human health by controlling the digestion, endocrine and cardiovascular systems, as well as immune and nerve function. Bioactive peptides are considered a new class of biologically active compounds and are widely used in foods system to avert oxidation and microbial deterioration and the treatment of various chronic diseases [172, 177, 178].

Seafood proteins have received great consideration as a sustainable bioresource of bioactive peptides because of the abundant availability of raw materials in the form of processing by-products [8, 177, 179]. Chemical (acid and alkali) and biological (enzymatic and fermentative) approaches are the most usually applied for the digestion of seafood proteins. Enzymatic digestion using exogenous enzymes has been the most frequently used technique for making bioactive peptides from seafood proteins [179]. Bioactive peptides from seafood resources displayed several bioactivities, including antihypertensive, antioxidant, immunomodulatory, neuroactive (opioid activity), hormone-regulating, antimicrobial and mineral-binding properties [178].

Bioactive peptides with calcium-binding properties from pepsinolytic hydrolysates of Alaska pollock backbone [180] and Hoki frame [181] have been prepared and analyzed. Bordenave *et al.* reported the neuroactive (opioid) activities of bioactive peptides in the protein hydrolysates prepared from cod and salmon heads [182]. Bioactive peptides with immune-stimulating activity from Atlantic cod stomach hydrolysate have been prepared and characterized [183]. Antihypertensive peptides like angiotensin-converting enzyme inhibitory peptides have been identified from protein hydrolysates formed from numerous finfish and shellfish processing by-products [184–186]. Antimicrobial activities of protamines (a natural cationic polypeptide), isolated from various finfish species such as Pacific herring, chum salmon and rainbow trout, have been reported [179, 187].

10.10 Recovery of Vitamins From Seafood Processing By-Products

Vitamins are bioactive compounds, and their crucial beneficial roles in human physiology and health are well known. Seafood processing by-products, mainly viscera, are one of the rich and sustainable sources of several vitamins, mainly vitamin A, vitamin D, and vitamin E [11]. Oils extracted from the fish liver are a potent source of these vitamins (A, D, and E) [11]. Vitamin A occurs chiefly in fish liver oils [172]. Oils recovered from Halibut and cod liver by-products are rich sources of vitamins like A and D. It was recorded that Sardine oil is a potent source of vitamins A and D with a quantity of 125 μg/g oil [11, 172].

10.11 Recovery of Chondroitin Sulphate From Seafood Processing By-Products

Chondroitin sulphate is a glycosaminoglycan, and its bioactive potential and health benefits are well-documented [188, 189]. It has much commercial significance as a bioactive and therapeutic compound. Using chondroitin sulphate along with Glucosamine alleviates pain and inflammation from osteoarthritis and has a favorable effect on degenerated joints [133]. It is also used in food as a dietary supplement along with GlcN for people who suffer from osteoarthritis and others such as athletes and dancers. It is the main bioactive constituent of shark cartilage and is preferably prepared from shark processing skeleton by-products. In processed shark skeletons, chondroitin sulphate content was reported to be about 21.3% [190]. It is also obtained from other processing by-products of sharks, such as fin and head [11, 133].

10.12 Extraction of Chitin and Its Derivatives From Seafood Processing By-Products

Chitin and its derivatives are amazing molecules with distinctive characteristics that present a variety of applications in various fields. Chitin and its derivatives find widespread applications in biomedicine, pharmaceuticals, textile, cosmetics, agriculture, food, nutraceuticals, and environmental filed owing their exceptional chemical properties as well as their biodegradable, nontoxic, biocompatible, etc. nature [9, 11, 12, 29, 67, 131].

Chitin, is a nitrogenous polysaccharide composed of β-1,4-linked *N*-Acetyl-D-glucosamine (GlcNAc). It is measured as the second most copious and renewable biomaterial on the world, subsequent only to cellulose [67]. Chitin is widely distributed in the environment, particularly as the major structural and protective component in the exoskeleton of insects and crustaceans and in the cell wall of most fungi [9]. Even though, chitin distributed widely in various group of organisms, crustacean such as shrimp, crab, lobster, krill, etc. processing by-product are presently the main source for commercial chitin available in the market [9]. Crustacean chitin commercially produced from shrimp or crab processing by-product by chemical methods includes the removal of minerals (step 1) using a strong acid (demineralization), proteins (step 2) using a strong alkali (deproteinization) and pigments (step 3) using a solvent (decolouration). In addition, chitin production from crustacean by-products by biological procedures like enzymatic and fermentation have been reported [29, 67, 90]. Bhaskar et al. reported that the biological method of chitin extraction using milder procedure resulted in chitin with reliable quality besides letting simultaneous retrieval of other valuable constituent present in the crustacean by-products [90]. A number of enzymatic processes for the removal of proteins from the decalcified crustacean by-products have been reported [9]. Fermentation of crustacean by-products using lactic acid bacteria with proteolytic activity has been found to be very effective in protein and mineral removal [90].

Chitosan is a nitrogenous polysaccharide, a copolymer of β-1,4-linked *N*-Acetyl-D-glucosamine and D-glucosamine (Glc) [9, 67]. It is not occurring naturally in any of the seafood products or by-products. However, it can be derived from crustacean chitin by N-deacetylation process using concentrated alkali. Hence, most industries produce chitosan, instead of chitin from crustacean by-products, due to its wide variety of biological and commercial applications than chitin [9, 11, 12, 29, 67, 131]. At present commercially a huge quantity of chitosan produced by *N*-deacetylation of chitin prepared from seafood processing crustacean (shrimp, crab, lobster, etc.) by-products by harsh thermo-chemical treatment with a concentrated alkaline solution (up to 40%) and at very high temperature (up to 140 °C).

N-acetyl chitooligosaccharides and chitooligosaccharides are the partially digested derivative of chitin and chitosan, respectively. These oligosaccharides have more advantageous than parent polymer chitin and chitosan [29, 67]. They are widely applied in food, biomedicine, pharmaceutical, functional food, etc., due to their high solubility in water and better bio-functional properties [29]. They are manufactured by partial hydrolysis of chitin and chitosan or chitinous by-products using acid or enzyme treatment [29, 67]. *N*-Acetyl-D-glucosamine and D-glucosamine are the primary components of chitin and chitosan. These molecules have attracted much attention due to their potential therapeutic and health benefits [131, 216]. Commonly, *N*-Acetyl-D-glucosamine is produced by acid hydrolysis of crustacean chitin with concentrated HCl at high temperature (>80 °C). Biological methods (enzymatic and fermentative) have attempted for the production of *N*-Acetyl-D-glucosamine from crustacean by-product/discard [72, 216].

10.13 Conclusion

Seafood is one of the major commodities contributing to global food security. A large portion of the seafood landed for direct human consumption is processed through the supply chain. This seafood processing generates massive amounts of non-edible by-products, which are underutilized or discarded as waste in many parts of the world, leading to a negative impact on the environment, the economy, and society. These seafood processing by-products are one of the rich natural resources for some high-value materials with wide applications in various fields. Even though some of these valuable components are already commercially extracted, a much more sustainable valorization of these natural bio-resources is advantageous. As a whole, these seafood processing by-products are promising sources of several commercially important bioactive compounds such as a bioactive peptide, omega-3-fatty acid, carotenoid, enzyme, vitamins, minerals, chitin, collagen, etc. Many of these bioactive compounds from seafood processing by-products have exhibited promising beneficial effects on human health. Nonetheless, the enormous quantity of seafood processing by-products generated annually continues to exist as a major unexploited bio-resources for many of these biologically active compounds. Though numerous methods have been reported so far in the literature, it is feasible to recover various bioactive compounds from the same by-product.

Extraction of bioactive compounds from seafood processing by-products is a tedious job due to the complex and highly perishable nature of the raw materials. However, significant development has been made in the last two decades to identify several novel bioactive compounds from seafood processing by-products and their sustainable recovery. This chapter debates the inherent advantage of secondary handling of seafood by-products for the recovery of numerous bioactive compounds. Identifying these bioactive compounds and their promising applications in biomedicine, nutraceutical, and functional food has led to a new perception of the valorization of by-products from the seafood processing industry.

References

1. FAO, IFAD, UNICEF, WFP and WHO, *The state of food security and nutrition in the world 2022. Repurposing food and agricultural policies to make healthy diets more affordable*, Food and Agriculture Organization of the United Nations, Rome, 2022, https://www.fao.org/3/cc0639en/cc0639en.pdf.
2. Suresh, P.V., Nutraceutical potential of seaweeds. *Indian Food Ind. Mag.*, 3, 5, 23–37, 2021.
3. FAO, *The state of world fisheries and aquaculture 2022, towards blue transformation*, Food and Agriculture Organization of the United Nations, Rome, 2022, https://doi.org/10.4060/cc0461en.
4. FAO, *FAO year book, fishery and aquaculture statistics 2019*, Food and Agriculture Organization of the United Nations, Rome, 2021, https://www.fao.org/3/cb7874t/cb7874t.pdf.
5. Ghosh, P.R., Fawcett, D., Sharma, S.B., Poinern, G.E.J., Progress towards sustainable utilization and management of food wastes in the global economy. *Int. J. Food Sci.*, 2016, 3563478, 2016.
6. Sharp, M. and Mariojouls, C.C., Waste not, want not: Better utilization of fish waste in the Pacific. *SPC Fish. Newsl.*, 138, 44–48, 2012.
7. Rustad, T., Utilization of marine by-products. *EJEAF Che.*, 2, 458–463, 2003.
8. Guerard, F., Enzymatic methods for marine by-products recovery, in: *Maximizing the Value of Marine By-products*, F. Shahidi (Ed.), pp. 107–136, Woodhead Publishing Ltd., Cambridge, UK, 2007.
9. Suresh, P.V., Kudre, T.G., Johny, L.C., Sustainable valorization of seafood processing by-product/discard, in: *Waste to Wealth. Energy, Environment, and Sustainability*, R.R. Singhania, R.A. Agarwal, R.P. Kumar, R.K. Sukumaran (Eds.), pp. 111–139, Springer, Singapore, 2018.
10. Lee, C.M., Fish mince: Cryostabilization and product formulation, in: *Handbook of Seafood Quality, Safety and Health Application*, C. Alasalvar, F. Shahidi, K. Miyashita, U. Wanasundara (Eds.), pp. 156–170, Wiley-Blackell, London, UK, 2011.
11. Suresh, P.V. and Prabhu, G.N., Seafood, in: *Valorization of Food Processing By-products*, M. Chandrasekaran (Ed.), pp. 685–736, CRC Press, Boca Raton, FL, 2013.
12. Venugopal, V., Valorization of seafood processing discards: Bioconversion and bio-refinery approaches. *Front. Sustain. Food Syst.*, 5, 611835, 2021.
13. Lee, C.M., Seafood flavour from processing by-products, in: *Maximizing the Value of Marine By-products*, F. Shahidi (Ed.), pp. 304–327, Woodhead Publishing, Cambridge, UK, 2007.

14. Bhaskar, N., Sachindra, N.M., Suresh, P.V., Mahendrakar, N.S., Microbial reclamation of fish industry by-products, in: *Aquaculture Microbiology and Biotechnology*, D. Montet and R.C. Ray (Eds.), pp. 249–276, Science Publishers, Eneld, New Hampshire, USA, 2011.
15. Sachindra, N.M. and Mahendrakar, N.S., Fishery by-products-an overview, in: *Fish Processing By-products: Quality Assessment and Applications*, N.M. Sachindra and N.S. Mahendraker (Eds.), pp. 299–345, Studium Press LLC, Houston, USA, 2015.
16. Zhao, Y.M., de Alba, M., Sun, D.-W., Tiwari, B., Principles and recent applications of novel non-thermal processing technologies for the fish industry-a review. *Crit. Rev. Food Sci. Nutr.*, 59, 728–742, 2019.
17. Gómez, B., Munekata, P.E.S., Gavahian, M., Barba, F.J., Martí-Quijal, F.J., Bolumar, T., Campagnol, P.C.B., Tomasevic, I., Lorenzo, J.M., Application of pulsed electric fields in meat and fish processing industries: An overview. *Food Res. Int.*, 123, 95–105, 2019.
18. Khawli, F., Pateiro, M., Domínguez, R., Lorenzo, J.M., Gullon, P., Kousoulaki, K., Ferrer, E., Berrada, H., Barba, F.J., Innovative green technologies of intensification for valorization of seafood and their by-products. *Mar. Drugs*, 17, 689, 2019.
19. Durazzo, A., Di Lena, G., Gabrielli, P., Santini, A., Lombardi-Boccia, G., Lucarini, M., Nutrients and bioactive compounds in seafood: Quantitative literature research analysis. *Fishes*, 7, 132, 2022.
20. Kim, J.S. and Park, J.W., Mince from seafood processing by-products and surimi as food ingredients, in: *Maximizing the Value of Marine By-products*, F. Shahidi (Ed.), pp. 246–255, Woodhead Publishing, Cambridge, UK, 2007.
21. Suresh, S.K., Anushma, V., Suresh, P.V., Tanaji, G.K., Enzymes in meat, fish and poultry product processing and preservation-II, in: *Novel Food Grade Enzymes*, D.A. Tripathi, K.K. Darani, S.K. Srivastava (Eds.), pp. 193–217, Springer, Singapore, 2022.
22. Bruno, S.F., Ekorong, F.J.A.A., Karkal, S.S., Cathrine, M.S.B., Kudre, T.G., Green and innovative techniques for recovery of valuable compounds from seafood by-products and discards: A review. *Trends Food Sci. Technol.*, 85, 10–22, 2019.
23. Caruso, G., Floris, R., Serangeli, C., Paola, L.D., Fishery wastes as a yet undiscovered treasure from the sea: Biomolecules sources, extraction methods and valorization. *Mar. Drugs*, 18, 622, 2020.
24. Gallo, M., Ferrara, L., Naviglio, D., Application of ultrasound in food science and technology: A perspective. *Foods*, 7, 164, 2018.
25. Rubio-Rodríguez, N., De Diego, S.M., Beltran, S., Jaime, I., Sanz, M.T., Rovira, J., Supercritical fluid extraction of the omega-3 rich oil contained in hake (*Merluccius capensis-Merluccius paradoxus*) by-products: Study of the influence of process parameters on the extraction yield and oil quality. *J. Supercrit. Fluids*, 47, 215–226, 2008.

26. Kim, H.K., Kim, Y.H., Park, H.J., Lee, N.H., Application of ultrasonic treatment to extraction of collagen from the skins of sea bass *Lateolabrax japonicas*. *Fish. Sci.*, 79, 849–856, 2013.
27. Silva, J.C., Barros, A.A., Aroso, I.M., Fassini, D., Silva, T.H., Reis, R.L., Duarte, A.R.C., Extraction of collagen/gelatin from the marine demosponge *Chondrosia reniformis* (Nardo, 1847) using water acidified with carbon dioxide-process optimization. *Ind. Eng. Chem. Res.*, 55, 6922–6930, 2016.
28. Di Paola, L., Cicci, A., Bravi, M., Toward an efficient biorefining of microalgae and biomass alike: A unit operating view on how to mimick the optimisation history of the crude oil refining industry. *Chem. Eng. Trans.*, 43, 1321–1326, 2015.
29. Suresh, P.V., Enzymatic production of different types of chitooligosaccharides, in: *Chitooligosaccharides - Prevention and Control of Diseases*, S.K. Kim (Ed.), pp. 27–56, Springer Nature, Switzerland, 2022.
30. Ghosh, S., Sarkar, T., Pati, S., Kari, Z.A., Edinur, H.A., Chakraborty, R., Novel bioactive compounds from marine sources as a tool for functional food development. *Front. Mar. Sci.*, 9, 832957, 2022.
31. Chandrasekaran, M., Industrial enzymes from marine microorganisms: The Indian scenario. *J. Mar. Biotechnol.*, 5, 86–89, 1997.
32. Shahidi, F. and Kamil, J., Enzymes from fish and aquatic invertebrates and their application in the food industry. *Trends Food Sci. Technol.*, 12, 435–464, 2001.
33. Sachindra, N.M., Bhaskar, N., Hosokawa, M., Value addition of seafood processing discards, in: *Handbook of Seafood Quality, Safety and Health Application*, C. Alasalvar, F. Shahidi, K. Miyashita, U. Wanasundara (Eds.), pp. 390–398, Wiley-Blackell, London, UK, 2011.
34. Kim, S.K. and Mendis, E., Bioactive compounds from marine processing by-products-a review. *Food Res. Int.*, 39, 383–393, 2006.
35. Ideia, P., Pinto, J., Ferreira, R., Figueiredo, L., Spínola, V., Castilho, P.C., Fish processing industry residues: A review of valuable products extraction and characterization methods. *Waste Biomass Valorization*, 11, 3, 2020.
36. Gildberg, A., Enzymes and bioactive peptides from fish waste related to fish silage, fish feed and fish sauce production. *J. Aquat. Food Prod. Technol.*, 13, 3–11, 2004.
37. Ferraro, V., Cruz, I.B., Jorge, R.F., Malcata, F.X., Pintado, M.E., Castro, P.M.L., Valorisation of natural extracts from marine source focused on marine by-products: A review. *Food Res. Int.*, 43, 2221–2233, 2010.
38. Castillo-Yañez, F.J., Pacheco-Aguilar, R., Garcia-Carreño, F.L., Toro, M.D.L.A.N.D., Characterization of acidic proteolytic enzymes from Monterey sardine (*Sardinops sagax caerulea*) viscera. *Food Chem.*, 85, 343–350, 2004.
39. Kishimura, H., Klomklao, S., Benjakul, S., Chun, B.S., Characteristics of trypsin from the pyloric ceca of walleye pollock (*Theragra chalcogramma*). *Food Chem.*, 106, 194–199, 2008.

40. Klomklao, S., Benjakul, S., Visessanguan, W., Kishimura, H., Simpson, B.K., Extraction of carotenoprotein from black tiger shrimp shells with the aid of bluefish trypsin. *J. Food Biochem.*, 33, 2, 201–217, 2009.
41. Nalinanon, S., Benjakul, S., Kishimura, H., Biochemical properties of pepsinogen and pepsin from the stomach of albacore tuna (*Thunnus alalunga*). *Food Chem.*, 121, 49–55, 2010.
42. Zhou, L., Budge, S.M., Ghaly, A.E., Brooks, M.S., Dave, D., Extraction, purification and characterization of fish chymotrypsin: A review. *Am. J. Biochem. Biotechnol.*, 7, 104–123, 2011.
43. Zhao, L., Budge, S.M., Ghaly, A.E., Brooks, M.S., Dave, D., Extraction, purification, and characterization of fish pepsin: A critical review. *J. Food Process. Technol.*, 2, 750, 2011.
44. Ghaly, A.E., Ramakrishnan, V.V., Brooks, M.S., Budge, S.M., Dave, D., Fish processing wastes as a potential source of proteins, amino acids and oils: A critical review. *J. Microb. Biochem. Technol.*, 5, 107–129, 2013.
45. Wangkheirakpam, M.R., Mahanand, S.S., Majumdar, R.K., Sharma, S., Hidangmayum, D.D., Netam, S., Fish waste utilization with reference to fish protein hydrolysate-a review. *Fish. Technol.*, 56, 169–178, 2019.
46. Klomklao, S., Kishimura, H., Yabe, M., Benjakul, S., Purification, characterization of two pepsins from the stomach of pectoral rat tail (*Coryphaenoides pectoralis*). *Comp. Biochem. Physiol. B, Biochem. Mol. Biol.*, 147, 682–689, 2007.
47. Ketnawa, S., Benjakul, S., Ling, T.C., Alvarez, O.M., Rawdkuen, S., Enhanced recovery of alkaline protease from fish viscera by phase partitioning and its application. *Chem. Cent. J.*, 7, 79, 2013.
48. Simpson, B.K., Digestive proteinases from marine animals, in: *Seafood Enzymes: Utilization and Influence on Postharvest Seafood Quality*, N.F. Haard and B.K. Simpson (Eds.), pp. 191–124, Marcel Dekker, New York, USA, 2000.
49. Khantaphant, S. and Benjakul, S., Purification, and characterization of trypsin from the pyloric caeca of brown stripe red snapper (*Lutjanus vitta*). *Food Chem.*, 120, 658–664, 2010.
50. Espósito, T.S., Amaral, I.P.G., Buarque, D.S., Oliveira, G.B., Carvalho, L.B., Bezerra, R.S., Fish processing waste as a source of alkaline proteases for laundry detergent. *Food Chem.*, 112, 125–130, 2009.
51. Murthy, L.N., Phadke, G.G., Unnikrishnan, P., Annamalai, J., Joshy, C.G., Zynudheen, A.A., Ravishankar, C.N., Valorization of fish viscera for crude proteases production and its use in bioactive protein hydrolysate preparation. *Waste Biomass Valorization*, 9, 1735–1746, 2017.
52. Kurtovic, I., Marshall, S.N., Simpson, B., Isolation, and characterization of a trypsin fraction from the pyloric ceca of chinook salmon (*Oncorhynchus tshawytscha*). *Comp. Biochem. Physiol. B, Biochem. Mol. Biol.*, 143, 432–440, 2006.
53. Kishimura, H., Hayashi, K., Miyashita, Y., Nonami, Y., Characteristics of trypsins from the viscera of true sardine (*Sardinops melanostictus*) and the

pyloric ceca of arabesque greenling (*Pleuroprammus azonus*). *Food Chem.*, 97, 65–70, 2006.

54. Bougatef, A., Souissi, N., Fakhfakh, N., Ellouz-Triki, Y., Nasri, M., Purification and characterization of trypsin from the viscera of sardine (*Sardina pilchardus*). *Food Chem.*, 102, 343–350, 2007.
55. Ali, N.E.H., Hmidet, N., Bougatef, A., Nasri, R., Nasri, M.A., Laundry detergent-stable alkaline trypsin from striped seabream (*Lithognathus mormyrus*) viscera: Purification and characterization. *J. Agric. Food Chem.*, 57, 10943–10950, 2009.
56. Barkia, A., Bougatef, A., Nasri, R., Fetoui, E., Balti, R., Nasri, M., Trypsin from the viscera of Bogue (*Boops boops*): Isolation and characterization. *Fish Physiol. Biochem.*, 36, 893–902, 2009.
57. Ben Khaled, H., Bougatef, A., Balti, R., Souissi, N., Nasri, M., Triki-Ellouz, Y., Isolation and characterization of trypsin from sardinelle (*Sardinella aurita*) viscera. *J. Sci. Food Agric.*, 88, 2654–2662, 2008.
58. Klomklao, S., Benjakul, S., Visessanguan, W., Kishimura, H., Simpson, B.K., Saeki, H., Trypsins from yellowfin tuna (*Thunnus albacores*) spleen: Purification and characterization. *Comp. Biochem. Physiol. B, Biochem. Mol. Biol.*, 144, 47–56, 2006.
59. Slizyte, R., Mozuraityte, R., Martınez-Alvarez, O., Falch, E., Fouchereau-Peron, M., Rustad, T., Functional, bioactive and antioxidative properties of hydrolysates obtained from cod (*Gadus morhua*) backbones. *Process Biochem.*, 44, 668–677, 2009.
60. Swapna, C.H., Amit, K.R., Sachindra, M.N., Bhaskaran, N., Seafood enzymes and their potential industrial application, in: *Handbook of Seafood Quality, Safety and Health Application*, C. Alasalvar, F. Shahidi, K. Miyashita, U. Wanasundara (Eds.), pp. 522–532, Wiley-Blackell, London, UK, 2011.
61. Mardina, V., Fitriani, F., Harmawan, T., Hildayani, G.M., Valorisation of Tongkol fish (*Eutynnus affinis*) pancreas for enzyme lipase production. *J. Islam. Sci. Technol.*, 4, 89–97, 2018.
62. Gooday, G.W., Aggressive and defensive roles for chitinases, in: *Chitin and Chitinases*, P. Jolles and R.A.A. Muzzarelli (Eds.), pp. 157–169, Birkhauser Publishing, Basel, 1999.
63. Fines, B. and Holt, G., Chitinase and apparent digestibility of chitin in the digestive tract of juvenile cobia *Rachycentron canadum*. *J. Aquac.*, 303, 34–39, 2010.
64. Suresh, P.V. and Chandrasekaran, M., Utilization of prawn waste for chitinase production by the marine fungus *Beauveria bassiana* by solid state fermentation. *World J. Microbiol. Biotechnol.*, 14, 655–660, 1998.
65. Suresh, P.V. and Chandrasekaran, M., Impact of process parameters on chitinase production by an alkalophilic marine *Beauveria bassiana* in solid state fermentation. *Process Biochem.*, 34, 3, 257–267, 1999.
66. Suresh, P.V., Kumar, P.K.A., Sachindra, N.M., Thermoactive β-*N*-acetyl hexosaminidase production by a soil isolate of *Penicillium monoverticillium*

CFR 2 under solid state fermentation: Parameter optimization and application for *N*-acetyl chitooligosaccharides preparation from chitin. *World J. Microbiol. Biotechnol.*, 27, 1435–1447, 2011.

67. Suresh, P.V., Enzymatic technologies of chitin and chitosan, in: *Enzymatic Technologies for Marine Polysaccharides*, T. Antonio (Ed.), pp. 449–466, CRC Press, Boca Raton, FL, 2019.
68. Fange, R., Lundblad, G., Lind, J., Slettengren, K., Chitinolytic enzymes in the digestive system of marine fishes. *Mar. Biol.*, 53, 317–321, 1979.
69. Matsumiya, M. and Mochizuki, A., Distribution of chitinase and *N*-acetylhexosaminidase in the organs of several fishes. *Fish. Sci.*, 62, 150–151, 1996.
70. Matsumiya, M., Miyauchi, K., Mochizuki, A., Distribution of chitinase and *N*-acetylhexosaminidase in the organs of a few squid and a cuttlefish. *Fish. Sci. Res.*, 64, 166–167, 1998.
71. Matsumiya, M., Kouji, M., Atsushi, M., Characterization of 38 kDa and 42 kDa chitinase isozymes from the liver of Japanese common squid *Todarodes pacificus*. *Fish. Sci.*, 68, 3, 603–609, 2002.
72. Matsumiya, M., Enzymatic production of *N*-Acetyl-D-Glucosamine using crude enzyme from the liver of squids. *Food Sci. Technol. Res.*, 10, 3, 296–299, 2004.
73. Gutowska, M.A., Drazen, J.C., Robison, B.H., Digestive chitinolytic activity in marine fishes of Monterey Bay, California. *Comp Biochem Physiol A Mol Integr Physiol.*, 139, 351–358, 2004.
74. Mana, I., Kouji, M., Atsushi, M., Masahiro, M., Purification, and characterization of chitinase from the stomach of silver croaker *Pennahia argentatus*. *Protein Expr. Purif.*, 65, 2, 214–222, 2009.
75. Kumazwa, Y., Nakanishi, K., Yasuda, H., Motoki, M., Purification, and characterization of six transglutaminase walleye pollock liver. *Fish. Sci.*, 62, 959–964, 1996.
76. Nilsen, I.W., Overbo, K., Sandsdalen, E., Sandaker, E., Sletten, K., Myrnes, B., Protein purification and gene isolation of chlamysin, a cold active lysozyme-like enzyme with antibacterial activity. *FEBS Lett.*, 464, 453–158, 1999.
77. Wong, T.Y., Preston, L.A., Schiller, N.L., Alginate lyase: A review of major sources and enzyme characteristic structure-function analysis, biological roles, and applications. *Annu. Rev. Microbiol.*, 54, 289–340, 2000.
78. Morrissey, M.T. and Okada, T., Marine enzymes from seafood by-product, in: *Maximizing the Value of Marine By-products*, F. Shahidi (Ed.), pp. 374–389, Woodhead Publishing, Cambridge, UK, 2007.
79. Simpson, B.K., Pigments from by-products of seafood processing, in: *Maximizing the Value of Marine By-products*, F. Shahidi (Ed.), pp. 413–432, Woodhead Publishing, Cambridge, UK, 2007.
80. Sila, A., Ayed-Ajmi, Y., Sayari, N., Nasri, M., Martinez-Alvarez, O., Bougatef, A., Antioxidant and anti-proliferative activities of astaxanthin extracted

from the shell waste of deep-water pink Shrimp (*Parapenaeus longirostris*). *Nat. Prod. J.*, 3, 82–89, 2013.

81. Bohn, T., Carotenoids, chronic disease prevention and dietary recommendations. *Int. J. Vitam. Nutr. Res.*, 87, 121–130, 2017.
82. Sachindra, N.M., Bhaskar, N., Mahendrakar, N.S., Carotenoids in different component of Indian shrimps. *J. Sci. Food Agric.*, 85, 167–172, 2005.
83. Sachindra, N.M., Bhaskar, N., Siddegowda, G.S., Sathisha, A.D., Suresh, P.V., Recovery of carotenoids from ensilaged shrimp waste. *Bioresour. Technol.*, 98, 8, 1642–1646, 2007.
84. Seabra, L.M.J. and Pedrosa, L.F.C., Astaxanthin: Structural and functional aspects. *Rev. Nutr.*, 23, 1041–1050, 2010.
85. Taksima, T., Limpawattana, M., Klaypradit, W., Astaxanthin encapsulated in beads using ultrasonic atomizer and application in yogurt as evaluated by consumer sensory profile. *LWT - Food Sci. Technol.*, 62, 431–437, 2015.
86. Rao, A.V. and Rao, L.G., Carotenoids and human health. *Pharmacol. Res.*, 55, 207–216, 2007.
87. Eggersdorfer, M. and Wyss, A., Carotenoids in human nutrition and health. *Arch. Biochem. Biophys.*, 652, 18–26, 2018.
88. Kurihara, H., Koda, H., Asami, S., Kiso, Y., Tanaka, T., Contribution of the antioxidative property of astaxanthin to its protective effect on the promotion of cancer metastasis in mice treated with restraint stress. *Life Sci.*, 70, 2509–2520, 2002.
89. Cremades, O., Ponce, E., Corpas, R., Gutierrez, J.F., Jover, M., Alvarez-Ossorio, M.C., Parrado, J., Bautista, J., Processing of crawfish (*Procambarus clarkii*) for the preparation of caroteno-proteins and chitin. *J. Agric. Food Chem.*, 49, 11, 5468–5472, 2001.
90. Bhaskar, N., Suresh, P.V., Sakhare, P.Z., Sachindra, N.M., Shrimp biowaste fermentation with (*Pediococcus acidolactici*) CFR2182: Optimization of fermentation conditions by response surface methodology and effect of optimized conditions on deproteination/demineralization and carotenoid recovery. *Enzyme Microb. Technol.*, 40, 1427–1434, 2007.
91. Simpson, B.K. and Haard, N.F., The use of proteolytic enzymes to extract carotenoproteins from shrimp wastes. *J. Appl. Biochem.*, 7, 212–222, 1985.
92. Cano-Lopez, A., Simpson, B.K., Haard, N.F., Extraction of carotenoprotein from shrimp process waste with the aid of trypsin from Atlantic cod. *J. Food Sci.*, 52, 503–506, 1987.
93. Chakrabarti, R., Carotenoprotein from tropical brown shrimp shell waste by enzymatic process. *Food Biotechnol.*, 16, 1, 81–90, 2002.
94. Pal, G.K. and Suresh, P.V., Sustainable valorisation of seafood by-products: Recovery of collagen and development of collagen-based novel functional food ingredients. *Innov. Food Sci. Emerg. Technol.*, 37, 201–215, 2016.
95. Pal, G.K. and Suresh, P.V., Physico-chemical characteristics and fibril-forming capacity of carp swim bladder collagens and exploration of their

potential bioactive peptides by in silico approaches. *Int. J. Biol. Macromol.*, 101, 304–313, 2017.
96. Pal, G.K. and Suresh, P.V., Comparative assessment of physicochemical characteristics and fibril formation capacity of thermostable carp scales collagen. *Mater. Sci. Eng. C*, 70, 32–40, 2017.
97. Jafari, H., Lista, A., Siekapen, M.M., Ghaffari-Bohlouli, P., Nie, L., Alimoradi, H., Shavandi, A., Fish collagen: Extraction, characterization, and applications for biomaterials engineering. *Polymers*, 12, 2230, 2020.
98. Regenstein, J.M. and Zhou, P., Collagen and gelatin from marine by-products, in: *Maximizing the Value of Marine By-products*, F. Shahidi (Ed.), pp. 279–303, Woodhead Publishing, Cambridge, UK, 2007.
99. Lidiya, C.J., Vijaykumar, M., Kudre, T.G., Suresh, P.V., Malabar sole (*Cynoglossus macrostomus*) skin as promising source of type I acid and pepsin solubilized collagens with potential bioactivity. *J. Food Sci. Technol.*, 59, 1, 157–167, 2022.
100. Morishige, H., Sugahara, T., Nishimoto, S., Muranaka, A., Ohno, F., Shiraishi, R., Doi, M., Immunostimulatory effects of collagen from jellyfish *in vivo*. *Cytotechnology*, 63, 5, 481–492, 2011.
101. Liu, D., Zhou, P., Li, T., Regenstein, J.M., Comparison of acid-soluble collagens from the skins and scales of four carp species. *Food Hydrocoll.*, 41, 290–297, 2014.
102. Suseela, M., Ninan, G., Hema, G.S., Shiny, K., Lakshman, P.T., Fish collagen and gelatin, in: *Fish Processing By-products: Quality Assessment and Applications*, N.M. Sachindra and N.S. Mahendraker (Eds.), pp. 173–235, Studium Press LLC, Houston, USA, 2015.
103. Nagai, T., Yamashita, E., Taniguch, K., Kanamori, N., Suzuki, N., Isolation, and characterization of collagen from the outer skin waste material of Cuttlefish (*Sepia lycidas*). *Food Chem.*, 72, 425–429, 2001.
104. Nagai, T. and Suzuki, N., Partial characterization of collagen from purple sea urchin (*Anthocidaris crassispina*) test. *Int. J. Food Sci. Technol.*, 35, 497501, 2000.
105. Nagai, T. and Suzuki, N., Preparation, and partial characterization of collagen from paper nautilus (*Argonauta argo*, Linnaeus) outer skin. *Food Chem.*, 76, 149–153, 2002.
106. Arsesen, J.A. and Gildberg, A., Preparation and characterization of gelatin from the skin of harp seal (*Phoca groendlandica*). *Bioresour. Technol.*, 82, 191–194, 2002.
107. Arsesen, J.A. and Gildberg, A., Extraction and characterization of gelatin from the Atlantic salmon (*Salmo salar*) skin. *Bioresour. Technol.*, 82, 191–194, 2007.
108. Muyonga, J.H., Cole, C.G.B., Duodu, K.G., Extraction and physico-chemical characterization of Nile perch (*Lates niloticus*) skin and bone gelatin. *Food Hydrocoll.*, 18, 581–592, 2004.

109. Ali, A.M.M., Kishimura, H., Benjakul, S., Extraction efficiency and characteristics of acid and pepsin soluble collagens from the skin of golden carp (*Probarbus jullieni*) as affected by ultrasonication. *Process Biochem.*, 66, 237–244, 2018.
110. Pal, G.K., Nidheesh, T., Suresh, P.V., Comparative study on characteristics and *in vitro* fibril formation ability of acid and pepsin soluble collagen from the skin of catla (*Catla catla*) and rohu (*Labeo rohita*). *Food Res. Int.*, 76, 804–812, 2015.
111. Nagai, T., Araki, Y., Suzuki, N., Collagen of the skin of ocellate puffer fish (*Takifugu rubripes*). *Food Chem.*, 78, 173–17, 2002.
112. Liu, Z., Oliveira, A.C.M., Su, Y.C., Purification and characterization of pepsin-solubilized collagen from skin and connective tissue of giant red sea cucumber (*Parastichopus californicus*). *J. Agric. Food Chem.*, 58, 1270–1274, 2010.
113. Benjakul, S., Thiansilakul, Y., Visessanguan, W., Roytrakul, S., Kishimura, H., Prodpran, T., Meesane, J., Extraction and characterisation of pepsin-solubilised collagens from the skin of bigeye snapper (*Priacanthus tayenus* and *Priacanthus macracanthus*). *J. Sci. Food Agric.*, 90, 1, 132–138, 2010.
114. Pal, G.K., Nidheesh, T., Govindaraju, K., Jyoti, Suresh, P.V., Enzymatic extraction, and characterisation of a thermostable collagen from swim bladder of rohu (*Labeo raita*). *J. Sci. Food Agric.*, 97, 1451–1458, 2017.
115. Pal, G.K., Nidheesh, T., Suresh, P.V., Comparative study on characteristics and *in vitro* fibril formation ability of acid and pepsin soluble collagen from the skin of *catla* (*Catla catla*) and rohu (*Labeo rohita*). *Food Res. Int.*, 76, 804–812, 2015.
116. Lidiya, C.J., Tanaji, G.K., Suresh, P.V., Acid and pepsin soluble collagens from skin by-product of Red-bellied Pacu (*Piaractus brachypomus*): Extraction and comparative characterizations towards finding substitute to bovine and porcine collagen. *J. Aquat. Food Prod. Technol.*, 30, 1–13, 2021.
117. Al-Nimry, S., Dayah, A.A., Hasan, I., Daghmash, R., Cosmetic, biomedical and pharmaceutical applications of fish gelatin/hydrolysates. *Mar. Drugs*, 19, 145, 2021.
118. Benjakul, S., Kittiphattanabawon, P., Regenstein, J.M., Fish gelatin, in: *Food Biochemistry and Food Processing*, 2nd Edition, B. Simson, L.M.L. Nollet, F. Toldra, S. Benjakul, G. Paliyath, Y.H. Hui (Eds.), pp. 388–340, John Wiley and Sons, Iowa, USA, 2012.
119. Zhou, P. and Regenstein, J.M., Optimization of extraction conditions for Pollock skin gelatin. *J. Food. Sci.*, 69, 393–398, 2004.
120. Karim, A. and Bhat, R., Fish gelatin: Properties, challenges, and prospects as an alternative to mammalian gelatins. *Food Hydrocoll.*, 23, 563–576, 2009.
121. Mutalipassi, M., Esposito, R., Ruocco, N., Viel, T., Costantini, M., Zupo, V., Bioactive compounds of nutraceutical value from fishery and aquaculture discards. *Foods*, 10, 1495, 2021.

122. Gómez-Guillén, M., López-Caballero, M.E., López de Lacey, A., Alemán, A., Giménez, B., Montero, P., Antioxidant and antimicrobial peptide fractions from squid and tuna skin gelatin, in: *Sea By-products as a Real Material: New Ways of Application*, E. Le Bihan, and N. Koueta (Eds.), pp. 89–115, Transworld Research Network Signpost, Kerala, India, 2010.
123. Aleman, A., Giménez, B., Montero, P., Gómez-Guillén, M.C., Antioxidant activity of several marine skin gelatins. *LWT - Food Sci. Technol.*, 44, 407–41, 2011.
124. Ngo, D.H., Ryu, B., Vo, T.S., Himaya, S., Wijesekara, I., Kim, S.K., Free radical scavenging and angiotensin-I converting enzyme inhibitory peptides from Pacific cod (*Gadus macrocephalus*) skin gelatin. *Int. J. Biol. Macromol.*, 49, 1110–1116, 2011.
125. Cullis, P.R. and Kruijff, B.D., Lipid polymorphism and the functional roles of lipids in biological membranes. *Biochim. Biophys. Acta Rev. Biomembr.*, 559, 399–420, 1979.
126. Adamovich, Y., Aviram, R., Asher, G., The emerging roles of lipids in circadian control. *Biochim. Biophys. Acta Mol. Cell Biol. Lipids*, 1851, 1017–1025, 2015.
127. Rustan, A.C. and Drevon, C.A., Fatty acids: Structures and properties, in: *Encyclopedia of Life Sciences*, 2001, https://doi.org/10.1038/npg.els.0000715.
128. Calder, P., Very long-chain n-3 fatty acids and human health: Fact, fiction and the future. *Proc. Nutr. Soc.*, 77, 52–72, 2018.
129. Li, X., Bi, X., Wang, S., Zhang, Z., Li, F., Zhao, A.Z., Therapeutic potential of ω-3 polyunsaturated fatty acids in human autoimmune diseases. *Front. Immunol.*, 10, 2241, 2019.
130. Berge, A. and Barnathan, G., Fatty acids from lipids of marine organisms: Molecular biodiversity, roles as biomarkers, biologically active compounds, and economical aspects, in: *Marine Biotech, Advances in Biochemical Engineering/Biotechnology*, vol. 96, Y. Le Gal and R. Ulber (Eds.), pp. 49–126, Springer-Verlag, Berlin, Heidelberg, 2005.
131. Venugopal, V., *Marine products for healthcare: Functional and bioactive nutraceutical compounds from the ocean*, Taylor and Francis, CRC Press, Boca Raton, 2009.
132. Osborn, H.T. and Akok, C.C., Structured lipids-novel fats with medical, nutraceutical, and food applications. *Compr. Rev. Food Sci. Food Saf.*, 1, 93–103, 2002.
133. Shahidi, F. and Alasalvar, C., Marine oils and other marine nutraceuticals, in: *Handbook of Seafood Quality, Safety and Health Application*, C. Alasalvar, F. Shahidi, K. Miyashita, U. Wanasundara (Eds.), pp. 444–454, Wiley-Blackell, London, UK, 2011.
134. Folch, J., Less, M., Stanley, H.S., A simple method for the isolation and purification of total lipids from animal tissues. *J. Biol. Chem.*, 226, 497–509, 1957.
135. Aryee, A.N. and Simpson, B.K., Comparative studies on the yield and quality of solvent-extracted oil from salmon skin. *J. Food Eng.*, 92, 353–358, 2000.

136. Shahidi, F., Extraction and measurement of total lipids. *Curr. Protoc. Food Anal. Chem.*, 7, 1, D1.1.1–D1.1.11, 2003.
137. Zhong, Y., Madhujith, T., Mahfouz, N., Shahidi, F., Compositional characteristics of muscle and visceral oil from Steelhead trout and their oxidative stability. *Food Chem.*, 104, 602–608, 2007.
138. Xiao, L., *Evaluation of extraction methods for recovery of fatty acids from marine products*, Masters Dissertation, University of Bergen, Norway, 2010.
139. Wanasundara, U., Preparative and Industrial-scale isolation, and purification of omega-3 polyunsaturated fatty acids from marine sources, in: *Handbook of Seafood Quality, Safety and Health Application*, C. Alasalvar, F. Shahidi, K. Miyashita, U. Wanasundara (Eds.), pp. 464–475, Wiley-Blackell, London, UK, 2011.
140. Laset, B., Julshamn, K., Espe, M., Chemical composition and theoretical nutritional evaluation of the produced fractions from enzymatic hydrolysis of salmon frames with Protamex. *Process Biochem.*, 38, 1747–1759, 2003.
141. Dumay, J., Barthomeuf, C., Bergea, J.P., How enzymes may be helpful for upgrading fish by-products: Enhancement of fat extraction. *J. Aquat. Food Prod. Technol.*, 13, 2, 69–84, 2004.
142. Amit, K.R., Swapna, H.C., Bhaskar, N., Halami, P.M., Sachindra, N.M., Effect of fermentation ensilaging on recovery of oil from fresh water fish viscera. *Enzyme Microb. Technol.*, 46, 9–13, 2010.
143. Chaurasia, S.P., Bhandaria, K., Sharmaa, A., Dalai, A.K., A review on lipase catalysed synthesis of DHA rich glyceride from fish oils. *Int. J. Res. Sci. Innov.*, 3, 9–19, 2016.
144. Shahidi, F. and Wanasundara, U., Omega-3 fatty acid concentrates: Nutritional aspects and production technologies. *Trends Food Sci. Technol.*, 9, 230–240, 1998.
145. Linder, M., Matouba, E., Fanni, J., Parmentier, M., Enrichment of salmon oil with n-3 PUFA by lipolysis, filtration and enzyme re-esterification. *Eur. J. Lipid Sci. Technol.*, 104, 455–462, 2002.
146. Amiguet, V.T., Kramp, K.L., Mao, J., McRae, C., Goulah, A., Kimpe, L.E., Blais, J., Arnason, J.T., Supercritical carbon dioxide extraction of polyunsaturated fatty acids from Northern shrimp (*Pandalus borealis* Kreyer) processing by-products. *Food Chem.*, 130, 853–858, 2012.
147. Ferdosh, S., Sarker, Z.I., Rahman, N.N.N.A., Akanda, J.H.A., Ghafoor, K., Kadir, M.O.A., Simultaneous extraction and fractionation of fish oil from tuna by-product using supercritical carbon dioxide (SC-CO_2). *J. Aquat. Food Prod. Technol.*, 25, 2, 230–239, 2014.
148. Nguyen, T.T., Zhang, W., Barber, A.R., Su, P., He, S., Significant enrichment of polyunsaturated fatty acids (PUFAs) in the lipids extracted by supercritical CO_2 from the livers of Australian rock lobsters (*Jasus edwardsii*). *J. Agric. Food Chem.*, 63, 4621–4628, 2015.

Application, C. Alasalvar, F. Shahidi, K. Miyashita, U. Wanasundara (Eds.), pp. 485–493, Wiley-Blackell, London, UK, 2011.
180. Jung, K., Karawita, R., Heo, S.J., Lee, B.J., Kim, S.K., Jeon, Y.J., Recovery of novel Ca-binding peptides from Alaska pollok (*Theragra chalcogramma*) backbone by pepsinolytic hydrolysis. *Process Biochem.*, 41, 2097–2100, 2006.
181. Jung, W.K. and Kim, S.K., Calcium binding peptide derived from pepsinolytic hydrolysates of hoki (*Johnius belengerrii*) frame. *Eur. Food Res. Technol.*, 224, 763–767, 2007.
182. Bordenave, S., Fruitier, I., Ballandier, I., Sannier, F., Gildberg, A., Batista, I., Piot, M.J., HPLC preparation of fish waste hydrolysate fractions. Effect on guinea pig ileum and ACE activity. *Prep. Biochem. Biotechnol.*, 32, 1, 65–77, 2002.
183. Gildberg, A., Recovery of proteinases and protein hydrolysates from fish viscera. *Bioresour. Technol.*, 39, 271–276, 1992.
184. Vercruysse, L., John, V.C., Guy, S., ACE Inhibitory peptides derived from enzymatic hydrolysates of animal muscle protein: A Review. *J. Agric. Food Chem.*, 53, 2, 8106–8115, 2005.
185. Murry, B.A. and Fitz-Gerald, R.J., Angiotensin converting enzyme inhibitory peptides derived from food protein: Biochemistry, bioactivity and production. *Curr. Pharm. Des.*, 53, 773–791, 2007.
186. Liu, Z., Chen, D., Su, Y.C., Zeng, M.Y., Optimization of hydrolysis conditions for the production of the angiotensin-I converting enzyme inhibitory peptides from sea cucumber collagen hydrolysates. *J. Aquat. Food Prod. Technol.*, 20, 2, 222–232, 2011.
187. Johansen, C., Verheul, A., Gram, L., Gill, T., Abee, T., Protamine-induced permeabilization of cell envelopes of gram-positive and gram-negative bacteria. *Appl. Environ. Microbiol.*, 63, 1155–115, 1997.
188. Uebelhart, D., Thonar, E.J., Delmas, P.D., Chantraine, A., Vignon, E., Effects of oral chondroitin sulfate on the progression of kneedosteoarthritis: A pilot study. *Osteoarthr. Cartil.*, 6, 1, 39–46, 1998.
189. Vangsness, C.T., Spiker, W., Erickson, J., A review of evidence-based medicine for glucosamine and chondroitin sulfate use in knee osteoarthritis. *Arthroscopy*, 25, 1, 86–94, 2009.
190. Sim, J.S., Rang, I.A., Seung, M.C., Hae, J.J., Jo, J.H., Kim, Y.S., Evaluation of chondroitin sulfate in shark cartilage powder as a dietary supplement: Raw materials and finished products. *Food Chem.*, 101, 20, 532–539, 2007.
191. Bougatef, A., Nedjar-Arroume, N., Manni, L., Ravallec, R., Barkia, A., Guillochon, D., Nasri, M., Purification and identification of novel antioxidant peptides from enzymatic hydrolysates of sardinelle (*Sardinellaaurita*) by-products proteins. *Food Chem.*, 118, 559–565, 2010.
192. Hsu, K.C., Li-Chan, E.C.Y., Jao, C.L., Antiproliferative activity of peptides prepared from enzymatic hydrolysates of tuna dark muscle on human breast cancer cell line MCF-7. *Food Chem.*, 126, 617–622, 2011.

193. Jia, J., Zhou, Y., Lu, J., Chen, A., Li, Y., Zheng, G., Enzymatic hydrolysis of Alaska pollack (*Theragra chalcogramma*) skin and antioxidant activity of the resulting hydrolysate. *J. Sci. Food Agric.*, 90, 635–64, 2010.
194. Kumar, N.S., Nazeer, R., Jaiganesh, R., Purification and biochemical characterization of antioxidant peptide from horse mackerel (*Magalaspis cordyla*) viscera protein. *Peptides*, 32, 1496–1501, 2011.
195. Barkia, A., Bougatef, A., Khaled, H.B., Nasri, M., Antioxidant activities of sardinelle heads and/or viscera protein hydrolysates prepared by enzymatic treatment. *J. Food Biochem.*, 34, 303–320, 2010.
196. Chi, C.F., Wang, B., Wang, Y.M., Zhang, B., Deng, S.G., Isolation, and characterization of three antioxidant peptides from protein hydrolysate of bluefin leatherjacket (*Navodon septentrionalis*) heads. *J. Funct. Foods*, 12, 1–10, 2015.
197. Sathivel, S., Bechtel, P.J., Babbitt, J., Smiley, S., Crapo, C., Reppond, K.D., Prinyawiwatkul, W., Biochemical and functional properties of herring (*Clupea harengus*) by-product hydrolysates. *J. Food Sci.*, 68, 2196–2200, 2003.
198. Nikoo, M., Benjakul, S., Ehsani, A., Li, J., Wu, F., Yang, N., Xu, X., Antioxidant and cryoprotective effects of a tetrapeptide isolated from Amur sturgeon skin gelatin. *J. Funct. Foods*, 7, 609–620, 2014.
199. Xu, L., Dong, W., Zhao, J., Xu, Y., Effect of marine collagen peptides on physiological and neurobehavioral development of male rats with perinatal asphyxia. *Mar. Drugs*, 13, 3653–3671, 2015.
200. Byun, H.G. and Kim, S.K., Purification and characterization of angiotensin I converting enzyme (ACE) inhibitory peptides from Alaska pollack (*Theragra chalcogramma*) skin. *Process Biochem.*, 36, 1155–1162, 2001.
201. Chi, C.F., Wang, B., Hu, F.Y., Wang, Y.M., Zhang, B., Deng, S.G., Wu, C.W., Purification and identification of three novel antioxidant peptides from protein hydrolysate of bluefin leatherjacket (*Navodon septentrionalis*) skin. *Food Res. Int.*, 73, 124–139, 2015.
202. Guo, L., Hou, H., Li, B., Zhang, Z., Wang, S., Zhao, X., Preparation, isolation, and identification of iron-chelating peptides derived from Alaska pollock skin. *Process Biochem.*, 48, 988–993, 2013.
203. Fahmi, A., Morimura, S., Guob, H.C., Shigematsu, T., Kida, K., Uemura, Y., Production of angiotensin I converting enzyme inhibitory peptides from sea bream scale. *Process Biochem.*, 39, 1195–1200, 2004.
204. Ahn, C.B., Kim, J.G., Je, J.Y., Purification, and antioxidant properties of octapeptide from salmon by-product protein hydrolysate by gastrointestinal digestion. *Food Chem.*, 147, 78–83, 2014.
205. Pal, G.K., *Fish collagen and microbial collagenase: Production, characterization and their potential food applications*, Ph.D. Thesis, Academy of Scientific & Innovative Research, Ghaziabad, India, 2018.
206. Ko, J.Y., Kang, N., Lee, J., Kim, J.S., Kim, W.S., Park, S.J., Jeon, Y.J., Angiotensin I-converting enzyme inhibitory peptides from an enzymatic hydrolysate of flounder fish (*Paralichthys olivaceus*) muscle as a potent antihypertensive agent. *Process Biochem.*, 51, 535–541, 2016.

207. Lee, S.H., Qian, Z.J., Kim, S.K., A novel angiotensin I converting enzyme inhibitory peptide from tuna frame protein hydrolysate and its antihypertensive effect in spontaneously hypertensive rats. *Food Chem.*, 118, 96–102, 2010.
208. Chai, H.J., Wu, C.J., Yang, S.H., Li, T.L., Pan, B.S., Peptides from hydrolysate of lantern fish (*Benthosema pterotum*) proved neuroprotective *in vitro* and *in-vivo*. *J. Funct. Foods*, 24, 438–449, 2016.
209. Wu, S., Feng, X., Lan, X., Xu, Y., Liao, D., Purification and identification of Angiotensin-I Converting Enzyme (ACE) inhibitory peptide from lizard fish (*Saurida elongata*) hydrolysate. *J. Funct. Foods*, 13, 295–299, 2015.
210. Rajapakse, N., Mendis, E., Jung, W.K., Je, J.Y., Kim, S.K., Purification of a radical scavenging peptide from fermented mussel sauce and its antioxidant properties. *Food Res. Int.*, 38, 175–182, 2005.
211. Toopcham, T., Roytrakul, S., Yongsawatdigul, J., Characterization and identification of angiotensin I-converting enzyme (ACE) inhibitory peptides derived from tilapia using *Virgibacillus halodenitrificans* SK1-3-7 proteinases. *J. Funct. Foods*, 14, 435–444, 2015.
212. Naqash, S.Y. and Nazeer, R.A., Antioxidant activity of hydrolysates and peptide fractions of *Nemipterus japonicus* and exocoetus volitans muscle. *J. Aquat. Food Prod. Technol.*, 19, 180–192, 2010.
213. Pan, X., Zhao, Y., Hu, F., Wang, B., Preparation and identification of antioxidant peptides from protein hydrolysate of skate (*Raja porosa*) cartilage. *J. Funct. Foods*, 25, 220–230, 2016.
214. Jung, W.K., Park, P.J., Byun, H.G., Moon, S.H., Kim, S.K., Preparation of hoki (*Johnius belengerii*) bone oligophosphopeptide with a high affinity to calcium by carnivorous intestine crude proteinase. *Food Chem.*, 91, 333–340, 2005.
215. Apetroaei, M.R., Zgârian, R.G., Manea, A.M., Rau, I., Tihan, G.T., Schroder, V., New source of chitosan from Black Sea marine organisms identification. *Mol. Cryst. Liq. Cryst.*, 628, 102–109, 2016.
216. Suresh, P.V., Biodegradation of shrimp processing bio-waste and concomitant production of chitinase enzyme and *N*-Acetyl-D-glucosamine by marine bacteria: Production and process optimization. *World J. Microbiol. Biotechnol.*, 28, 2945–2962, 2012.
217. Kumar, P.A. and Suresh, P.V., Biodegradation of shrimp biowaste by marine *Exiguobacterium* sp. CFR26M and concomitant production of extracellular protease and antioxidant materials: Production and process optimization by response surface methodology. *Mar. Biotechnol.*, 16, 202–218, 2014.
218. Rasti, H., Parivar, K., Baharara, J., Iranshahi, M., Namvar, F., Chitin from the Mollusc chiton: Extraction, characterization, and chitosan preparation. *Iran J. Pharm. Res.*, 16, 366–379, 2017.
219. Jothi, N. and Nachiyar, R.K., Identification and isolation of chitin and chitosan from cuttle bone of *Sepia prashadi* Winckworth. *Glob. J. Biotechnol. Biochem.*, 8, 1–7, 2013.

220. Lee, Y.K., Kim, A.Y., Min, S.G., Kwak, H.S., Characteristics of milk tablets supplemented with nano-powdered eggshell or oyster shell. *Int. J. Dairy Technol.*, 69, 337–345, 2016.
221. Alsuhaibani, A.M.A., Rheological and nutritional properties and sensory evaluation of bread fortified with natural sources of calcium. *J. Food Qual.*, 2018, 1–7, 2018.
222. Ktari, N., Fakhfakh, N., Balti, R., Khaled, H.B., Nasri, M., Bougatef, A., Effect of degree of hydrolysis and protease type on the antioxidant activity of protein hydrolysates from cuttlefish (*Sepia officinalis*) by-products. *J. Aquat. Food Prod. Technol.*, 22, 436–448, 2013.
223. Leone, A., Lecci, R.M., Durante, M., Meli, F., Piraino, S., The bright side of gelatinous blooms: Nutraceutical value and antioxidant properties of three Mediterranean Jellyfish (*Scyphozoa*). *Mar. Drugs*, 13, 4654–4681, 2015.
224. Zhang, Q., Song, C., Zhao, J., Shi, X., Sun, M., Liu, J., Fu, Y., Jin, W., Zhu, B., Separation and characterization of antioxidative and angiotensin converting enzyme inhibitory peptide from Jellyfish gonad hydrolysate. *Molecules*, 23, 94, 2018.
225. Zhuang, Y.L., Sun, L.P., Zhao, X., Hou, H., Li, B.F., Investigation of gelatin polypeptides of jellyfish (*Rhopilema esculentum*) for their antioxidant activity *in vitro*. *Food Technol. Biotechnol.*, 48, 222–228, 2010.

11

Recovery of Bioactive Compounds From Agro-Food By-Products by Membrane-Based Operations

Alfredo Cassano[1]*, Carmela Conidi[1] and Esperanza Maria Garcia-Castello[2]

[1]Institute on Membrane Technology, ITM-CNR, Rende (Cosenza), Italy
[2]Institute of Food Engineering for Development, IuIAD, Universitat Politècnica de València, Valencia, Spain

Abstract

Pressure-driven membrane operations today are well-established technologies in food processing industries. These processes, in relation to their intrinsic properties, have demonstrated to meet the requirements for the recovery, separation, and concentration of bioactive molecules also in the treatment of downstream processing streams of the agro-food production.

This chapter provides a comprehensive outlook on the recovery of bioactive compounds from agro-food by-products by membrane-based technologies, in the light of the recent research advancements in this field. Typical applications and sustainable solutions in the treatment of fruit, vegetables, meat, and fish by-products are presented and discussed.

Keywords: Membrane-based processes, pressure-driven membrane operations, integrated membrane systems, circular economy, agro-food by-products, bioactive compounds

11.1 Introduction

The production, handling, and processing of agricultural products is characterized by the production of huge quantities of waste and by-products

**Corresponding author*: a.cassano@itm.cnr.it

Umile Gianfranco Spizzirri (ed.) Nutraceutics from Agri-Food By-Products, (361–390) © 2023 Scrivener Publishing LLC

whose management poses a serious challenge for food industries due to their potential impact on the environment.

It has been estimated that approximately 26% of food wastes are generated from the beverage industry, followed by dairy (21%), fruit and vegetable (14.8%), cereal (12.9%), meat (8%), vegetable and animal oils (3.9%), fish (0.4%), and others (12.7%) processing industries [1].

Agro-food by-products are a precious resource of potentially useful chemical substances. In particular, bioactive compounds isolated from fruits and vegetables by-products mainly include polyphenols, vitamins, essential minerals, fatty acids, volatiles, and pigments [2], whereas, animal- and marine-based ones include mainly bioactive proteins and peptides [3]. Most of these compounds have been recognized for their health benefits and anti-inflammatory, antimicrobial, antioxidant, and anticarcinogenic properties [4, 5]. These compounds can be recovered and converted in products with marketable value such as food additives, functional foods, nutra-/pharmaceuticals, cosmeceuticals, beauty products, and bio-packaging, in agreement with the principles of circular economy, so providing considerable opportunities for the sustainable utilization of agro-food by-products [6].

The development of appropriate methods for the recovery of bioactive compounds from their original sources is a difficult task and is generally accomplished through a multistep process. The processes employed should be environmentally sustainable and efficient, resulting in high yields.

The general flowchart for the recovery of bioactive compounds from agro-food waste and by-products is based on pre-treatment, extraction, and purification steps [7]. Pretreatment methods aim at removing insoluble and macromolecular compounds from the original product in order to facilitate the release of target compounds. They include physical (milling, grinding), chemical (acid and alkaline hydrolysis), biological (use of microorganisms, such as fungi and bacteria) and enzymatic processes.

Conventional extraction methods are based on solid-liquid extraction with various solvents; however, they are characterized by extended extraction times, low yields of recovered target compounds and the use of expensive, toxic organic solvents, which require their further elimination from the extract.

Nonthermal technologies including supercritical and subcritical fluid extraction, microwave-assisted extraction, ultrasound-assisted extraction, enzyme-assisted extraction, pulsed electric field treatment, high pressure homogenization, membrane-based processes, and hurdle technologies provide several advantages over conventional extraction methods including higher efficiency and improved extraction yields, lower or no solvent consumption, reduced process time, better preservation of bioactive compounds (nontoxic procedures), and lower energy consumption [8–15]. Among these methodologies, membrane-based processes are increasingly been used in agro-food industries due to their intrinsic properties, such as selectivity, high surface area per unit volume, simple operating conditions in terms of pressure and temperature, non-use of chemical or biological agents, and consequently, minimal risk of contamination. In particular, pressure-driven membrane processes, such as microfiltration (MF), ultrafiltration (UF), nanofiltration (NF), and reverse osmosis (RO), are already applied extensively on industrial scale in food and beverage industries for clarification, concentration, fractionation, desalting, recycling, recovery, and purification purposes [16–18]. In these processes, a mixture of different components (*feed*) is brought to the surface of a perm-selective barrier (*membrane*) under a pressure difference, referred to as *transmembrane pressure* (TMP), applied between its upstream and downstream side. As a result, the feed solution is separated into a filtrate (*permeate*), that is depleted in specific particles or molecules, and a *retentate* in which these components are concentrated.

MF membranes are used to separate particles with diameter of 0.1 to 10 mm (typically suspended solids, bacteria and colloids) from a solvent and other low molecular weight compounds. Typical operating pressures are in the range of 0.1 to 2 bar. UF membranes are typically asymmetric with pore sizes in the skin layer of 2 to 10 nm. These membranes are characterized by their molecular weight cutoff (MWCO), which is arbitrarily defined as the molecular weight at which 90% of the macromolecular solute is rejected by the membrane. The MWCO of UF membranes is between 10^3 and 10^6 Dalton making them suitable for separation and fractionation of macromolecules. Operating pressures typically used in UF are of 0.5 to 10 bar [19].

NF membranes are characterized by an intermediate separation capability between UF and RO membranes (MWCO is in the range of 200–1,000 Da). They are essentially used to separate different organic solutes with low molecular weights or to fractionate salt compounds based on cation or anion valency. The separation mechanism is based on steric, Donnan and dielectric exclusion effects. The pore size of NF membranes is in the range of 0.5 to 2 nm. Hydrostatic pressures used in the process are of the order of 5 to 40 bar [20].

RO membranes are generally used to separate low molecular weight compounds from a relatively pure solvent. The particle size for RO applications is in the range 0.1 to 1 nm and solutes with molecular weight greater than 300 Da are separated. Hydrostatic pressures to obtain significant

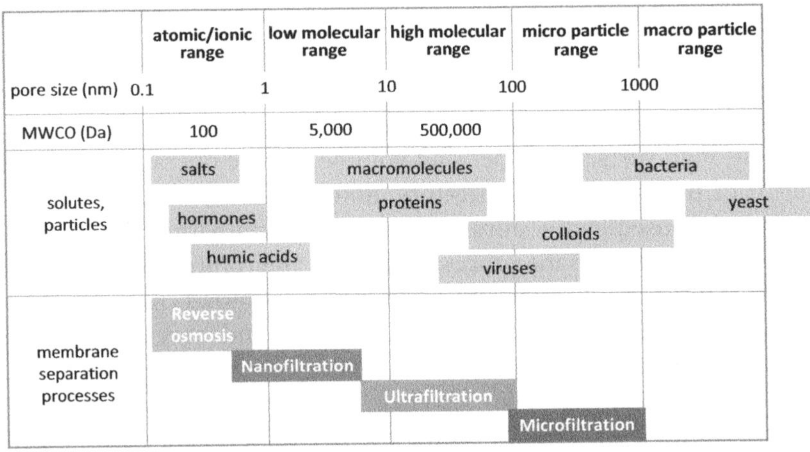

Figure 11.1 Separation characteristic of pressure-driven membrane operations.

transmembrane flux are in the range of 10 to 100 bar depending upon the osmotic pressure of the feed mixture [21]. A schematic diagram of the separation characteristics of pressure-driven membrane operations is depicted in Figure 11.1.

Most synthetic membranes are manufactured from both organic and inorganic materials. The most used organic polymers are: cellulose acetate (CA), polysulfone (PS), polyethersulfone (PES), polyamide (PA), polyvinylidene fluoride (PVDF), polytetrafluoroethylene (PTFE), polypropylene (PP) and polyethylene (PE). The base materials for the preparation of ceramic membranes are alumina ($A1_20_3$), titania (TiO_2), silica ($Si0_2$), and zirconia ($Zr0_2$).

Membranes are typically arranged in proper devices, called membrane modules, which allow to separate the feed stream into permeate and retentate streams. On large industrial scale, membrane modules are available in five basic designs: cartridge, hollow fiber, spiral-wound, tubular, plate-and-frame, and capillary. They are quite different in their design, mode of operation, production costs, and energy requirement for pumping the feed solution through the module.

The membrane performance in pressure-driven membrane systems is determined by the filtration rate and membrane separation properties are generally evaluated in terms of membrane rejection. Permeate fluxes and selectivities are both influenced by concentration polarization and membrane fouling. The former is a natural consequence of membrane selectivity. Indeed, compounds which are partially or completely rejected by the membrane tend to accumulate on the membrane surface where their concentration increases over the operation time leading to the formation of a boundary layer.

Membrane fouling refers to a long-term flux decline caused by the deposition of retained particles (colloids, macromolecules, dissolved organics, etc.) onto the membrane surface or in the membrane pores. The degree of membrane fouling determines the frequency of cleaning, lifetime of the membrane and the membrane area needed, and consequently, it has a significant effect on the cost, design, and operation of membrane plants [22]. Common compounds causing membrane fouling include particulate components, organic compunds, microorganisms (biofouling), and salts (scaling).

In addition to their applications in specific steps of the food processing industry, pressure-driven membrane operations have demonstrated to meet the requirements for the recovery, separation and concentration of bioactive molecules from agro-food by-products also through their combination in integrated systems.

This chapter aims to provide a comprehensive outlook on the use of pressure-driven membrane operations, also in integrated systems, for the separation, fractionation and concentration of bioactive compounds from agro-food by-products. Typical applications related to fruit and vegetables by-products, as well as to meat and fish by-products are analyzed and discussed.

11.2 Bioactive Compounds From Fruit and Vegetable By-Products

Today the transition to a more sustainable model of agro-food production, as well as the minimization of waste, are fundamental requirements for a modern agro-food industry.

Every year, a significant amount of fruit and vegetable liquid and solid by-products are generated by the agro-food industry that result from processing, distribution, preparation and disposal processes. Specifically, according to the Food Wastage Footprint and Climate Change Report, about 25% of all vegetables and 15% of all fruits are by-products of the agro-food production [23]. These by-products are composed of structural parts (peels, seeds, leaves, stems, pomace, extracts, and roots) such as olive pomace and grapes from oil and wine production, other fruit by-products (peaches, citrus fruits, blueberries, apples, etc.) from the jellies, juices and jams industry, as well as by-products from the processing of vegetables such as artichokes, licorice, carrots, tomatoes [24]. Generally, they are characterized by high quantity of suspended solids, high concentration of organic substances (lipids, organic acids, carbohydrates, proteins), and high biochemical and chemical oxygen demand (COD) causing serious management problems from an economic and environmental point of view [25]. Although a small fraction of these by-products are employed as livestock feed, biofuel production, and soil improvement materials, there is a great need to develop new valorization strategies to manage these wastes [26].

Fruit and vegetable by-products can be considered a precious source of a wide variety of bioactive compounds, such as polyphenols (mainly flavonoids and phenolic acids), carotenoids, vitamins, and minerals, well recognized for their antioxidant, anti-inflammatory, and anti-cancer properties, that can be easily extracted and valorized for the production of value-added products and chemicals [27, 28] (Table 11.1). For instance, they can be used in the food industry for the preparation of new functional foods, drinks, and natural colorants, as well as in the pharmaceutical and cosmetics industries as additives in many formulations, flavor fragrance, and cosmetic products offering a new approach for their recycling and providing at the same time economic resources in the context of environmental impact reduction and circular economy [44–46]. In this contest, several green procedures have been reported for the recovery of bioactive compounds from fruit and vegetable by-products, in order to fulfil the principles of green chemistry, avoiding the use of organic solvents, and using environmental friendly, sustainable, and possibly low cost techniques [47, 48]. Among them, membrane processes represent a low cost and sustainable solution due to their multiple advantages including mild operating conditions of temperature, therefore, preserving their functional activity and high separation efficiency capable to satisfying the required product quality. They do not require any additive agent, avoiding product contamination and the need for further purification. Other advantages include easy automatized control and scale-up at industrial level, low energy consumption, low operating and maintenances costs [9, 49, 50]. Particularly, pressure-driven membrane processes such as MF, UF, NF and RO, successfully satisfy the need for recovering, concentrating and purifying bioactive compounds. Nowadays, these processes, mostly in a sequential form or combined in integrated systems between them, or with other separation technologies, are employed to produce fractions enriched in bioactive compounds starting from any type of fruit and vegetable by-product [51]. MF and UF are typically used as pre-treatment and in the removal of undesired substances, while fractionation and pre-concentration are usually carried out by NF and RO processes.

Table 11.1 Bioactive compounds in fruit and vegetable by-products and their health beneficial effects.

Source	Bioactive compounds	Health beneficial effects	Reference
Apple pomace	Phenolic acids (chlorogenic acid) Flavonoids (catechin, epicatechin)	Anticancer, anti-inflammatory, anti-bacterial and anti-viral properties	[29]
Artichoke wastes	Phenolic acids (chlorogenic acid) Flavonoids (apigenin-7-O-glucoside and luteolin)	Anti-cancer, hepato-protective anti-bacterial, anti-HIV, hypocholesterolemic activity	[30]
Blue-berries pomace	Phenolic acids, flavonoids (anthocyanins and flavonols)	Antioxidant, anti-inflammatory and antimicrobial activity	[31]
Chestnut by-products	Tannins, phenolic acids (ellagic, gallic acids), flavonoids (catechin, epicatechin, apigenin)	Anti-inflammatory, anti-cancer, antioxidant, cardio-protective action	[32, 33]
Carrot pomace	β-carotene	Anti-inflammatory, anti-cancer, prevention of cardiovascular disease, immune-enhancement	[34]
Citrus peels	Flavonoids (hesperidin, narirutin, anthocyanins), carotenoids	Anti-cancer, anti-viral and anti-inflammatory activities, reduction capillary fragility, restricts human platelet aggregation	[35]

(*Continued*)

Table 11.1 Bioactive compounds in fruit and vegetable by-products and their health beneficial effects. (*Continued*)

Source	Bioactive compounds	Health beneficial effects	Reference
Grape pomace (peels, seeds, pulp, stems)	Flavonoids (anthocyanins, monomeric, catechin, epicatechin), resveratrol, tannins, gallic acid	Antioxidant, anti-cancer, anti-inflammatory and hypocholesterolemic activities	[36]
Licorice wastewaters	triterpenes (glycyrrhetic acid, glycyrrhizin), phenols (liquiritigenin, liquiritin, isoliquiritigenin, isoliquiritin)	Epatoprotective, antioxidant effects	[37]
Olive mill wastewaters	Phenolic acids (tyrosol, hydroxytyrosol, oleuropein, caffeic acid)	Antioxidant, anti-cancer, anti-inflammatory, hypolipidemic and hypoglycaemic	[38]
Olive pomace	Phenolic acids (tyrosol, hydroxytyrosol, caffeic acid, ferulic aci); flavonoids (luteolin-7-O-glucoside, apigenin-7-O-glucoside)	Antioxidant, anti-cancer, anti-inflammatory properties	[39]
Peach residues (peels, seeds, pulp)	Phenolic acids, carotenoids, vitamin C	Antioxidant, anti-hyperglycemic, anti-aging properties	[40]
Pineapple peel	Phenolic acids, Flavonoids (diosmetin, apigenin)	Antioxidant, cytotoxic, and antimicrobial activity	[41]

(*Continued*)

Table 11.1 Bioactive compounds in fruit and vegetable by-products and their health beneficial effects. (*Continued*)

Source	Bioactive compounds	Health beneficial effects	Reference
Pomegranate peels	Ellagitannin, Flavonoids (anthocyanins), Phenolic acids (gallic, ellagic, caffeic, chlorogenic, butyric, ferulic and cinnamic acids)	Anti-inflammatory, anti-cancer, anti-bacterial and cardiovascular protection.	[42]
Spent coffee grounds	Phenolic acids (chlorogenic caffeoylquinic, ellagic, trans-ferulic, feruloylquinic, gallic), Flavonoids (quercetin)	Antioxidant, anti-inflammatory, anti-microbial, cholesterol-lowering effects, prevention of degenerative diseases	[43]

The application of these processes for the recovery of bioactive compounds from agro-food wastes and by-products has been largely investigated by different authors. For example, integrated membrane sistems based on a combination of MF, UF and NF processes have been studied for a sustainable fractionation and concentration of polyphenols (including phenolic acids and anthocyanins) from grape pomace and lees, two by-products of winemaking process [52, 53].

In the work proposed by Giacobbo *et al.* [52], wine lees were pretrated by MF in order to remove suspended solids and then processed by UF and NF membranes in sequential design, in order to produce a concentrated fraction enriched of polysaccharides and polyphenols, respectively. In particular, UF membranes having MWCO of 10 and 1 kDa (Etna 10PP and 01PP, from Alfa-Laval) showed a rejection to polysaccharides higher than 77% and of about 50% to polyphenols. The selected NF membrane (NF270) exhibited rejections of 99% and 100% to polysaccharides and anthocyanins, respectively, and higher than 90% to polyphenols. On the basis of the obtained results, authors proposed an integrated membrane process for the fractionation of wine less able to produce three different valuable streams: (i) a fraction enriched in polysaccharides (UF retentate); (ii) a NF permeate reusable as process water and (iii) a concentrated fraction (NF retentate)

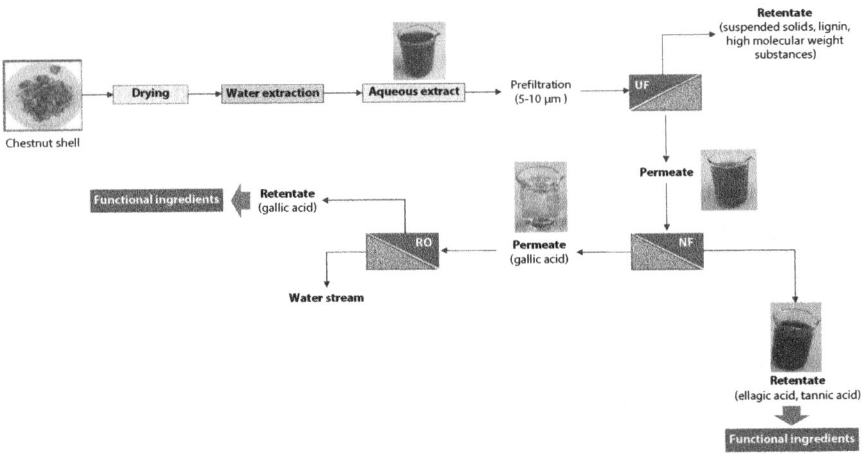

Figure 11.3 Integrated membrane process for the fractionation and recovery of bioactive compounds and water from shells of chestnut processing industry.

fiber UF membranes of 500 kDa (from Microdyn-Nadir) for removing macromolecules and suspended solids. Afterward, the UF permeate was treated with a spiral-wound NF membrane of 300 to 500 Da (Trisep XN45, from Microdyn-Nadir), and the NF permeate was finally concentrated by using a RO membrane with a $MgSO_4$ rejection of 99.4% (SC-1812 from GE Osmonics). The investigated process allowed to obtain three valuable fractions: (i) a concentrated fraction (NF retentate) containing 6.2 g/L of total flavonoids and 30.16 g/L of total polyphenols and enriched in ellagic and tannic acids (with average concentration of 3.832 and 1.49 g/L, respectively); (ii) a retentate fraction enriched in gallic acid (with concentration of 118.37 ± 4.2 mg/L); (iii) a fresh water stream (RO permeate) (Figure 11.3). The NF and RO retentate fractions enriched in bioactive compounds are suitable for the formulations of cosmeceutical, nutraceutical, and pharmaceutical products.

All the innovative integrated membrane processes discussed provide a valuable contribution to the effective valorization of fruit and vegetable by-products fulfilling the circular economy concept in terms of reduction of environmental risk disposal and resource recycling.

11.3 Bioactive Compounds From Meat By-Products

It is expected that in next three decades, world population will increase in around the 35% [56] and then, the demand of food products, including animal products, such as meat, milk, and their derivatives, will consequently rise [57–61]. Nowadays, an important part of the agro-food industry is dedicated to the production and processing of different types of animals, mainly bovine, mutton, goats, pigs, poultry, and rabbits for meat, milk, and egg production for human consumption.

The production of animal commodities includes two phases: first, the animal raising until they reach the desired weight or functionality and second, a later slaughtering in which animals are killed, cleaned, and cut into sections in accordance with commercial parameters. When the animal is alive, excreta and manure are the major waste while when the animal is dead, the residues are body tissues and parts, as well as the wastewater generated in the different operations involved [62]. It has been demonstrated that the animal excreta from farms, as well as the residues from slaughterhouses and processing industries are a relevant source of greenhouse gas (GHG) emissions [63], what is causing economic, environmental, and social concern [64, 65]. Porter and his co-authors [64] carried out an extensive work of collecting, compilation, and analyzing the average values of the emission factor for different commodities and regions (Figure 11.4). In this work it was reported that the global GHG emissions along the complete food supply for all the agro-food industry is estimated in around 591 ton of CO_2 equivalents per ton of food production, and that only the meat industry generates almost 70% of the total GHG emissions. Other studies have found that ruminants produce the highest carbon footprint values [59].

According to the Food and Agriculture Organization of the United Nations (FAO), meat and animal products represent, along the food supply

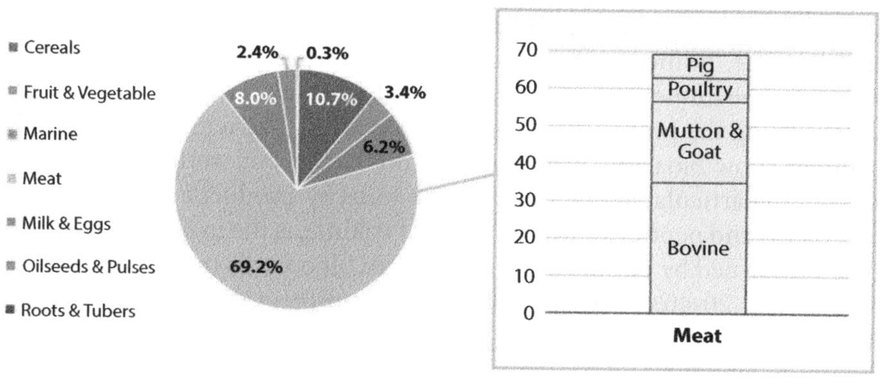

Figure 11.4 Percentage distribution of the emission factor values for different commodities in the agro-food industry (data from [64]).

chain, about the 14% of the global food waste and loss (FWL) what are detrimental to the economic benefits of industry and moreover, FWL negatively contributes to food security, safety, and world hunger issues [65]. To minimize the impact of industries involved in the animal production and processing, several regulations have been incorporated in the legislation [66], such as the Integrated Pollution Prevention and Control (IPPC) and Best Available Techniques (BAT) [67] and recently, the Sustainable Development Goals (SDGs) within the 2030 Agenda for Sustainable Development by the United Nations [68].

The industrial process to produce meat derivatives starts in abattoirs where animals (cattle, pigs, poultry, etc.) are slaughtered obtaining meat as main product. In addition, those animal fractions that are also suitable for human consumption, are processed to obtain food products (Figure 11.5a). On the other hand, around the 40-60% of the living weight of the animal is considered as residual material (blood, skin, viscera, bones, carcasses, heads, fatty tissues and feathers, depending on the animal) [69, 70]. These residuals can be grouped into co- and by-products (Figure 11.5a). Co-products are those residuals still remaining food-grade quality to be used as food or animal feed, after their proper processing [71]. The use of co-products as feed has been widely used since, at the beginning of 20th Century it was found that their addition to corn rations for pigs feeding, accelerated their growth with the consequent economic benefits [72]. On the other side, by-products are those residuals that are not suitable for human consumption for commercial and/or safety reasons. For some time now, several attempts from research and industry are focused on finding out solutions for the use of industrial by-products as raw materials [73]. The main by-products obtained during the meat production (Figure 11.5b) are blood, collagen, connective tissue, feathers, hairs, nails, and skin, which are rich in protein in a range of 15-20% (w/w), also different amino acids as isoleucine, lysine, methionine, phenylalanine, threonine, tryptophan, valine and histidine [72], as well as minerals and salts [74].

Regarding the protein content, several studies have reported that these peptides show bioactive properties as antioxidant and antimicrobial [71, 74–78]. In particular, blood is the most studied by-product for the recovery of proteins and peptides since, in terms of volume, is the major by-product. Blood is formed by plasma or serum and red blood cells in a proportion of 60-80% and 20-40%, respectively. The water/protein distribution in each fraction is different, consisting in 92% of water and 7% of protein in plasma (albumin, globulin and fibrinogen) and 70% of water and 28% of protein (mainly hemoglobin) in red blood cells [79, 80].

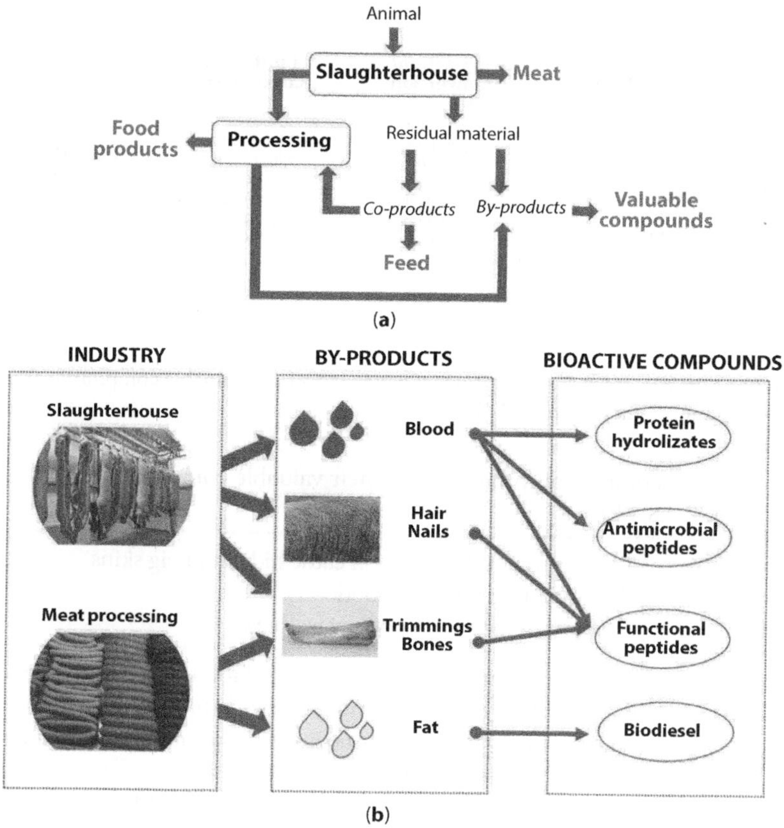

Figure 11.5 (a) Process of meat and derivatives production (adapted from Aspevik *et al.*, 2017); (b) Scheme relating the different by-products obtained from meat industry to the most valuable compounds (images under CC0 license) (adapted from [70]).

The recovery of valuable compounds from meat by-products usually follows pre-treatment, extraction/conversion and purification/concentration steps (Figure 11.6) and depending on the characteristics of the starting material and the intended applications (Table 11.2), different operations will be used.

Membrane technology is an extensive separation process that has been already studied and even industrially implemented for the recovery of bioactive compounds mainly from a great variety of agro-food industry by-products [30, 49, 54, 55, 81, 82]. Based on those successful applications, membrane technology has high potentialities and can play an important role to fractionate and concentrate proteins from meat by-products. Along

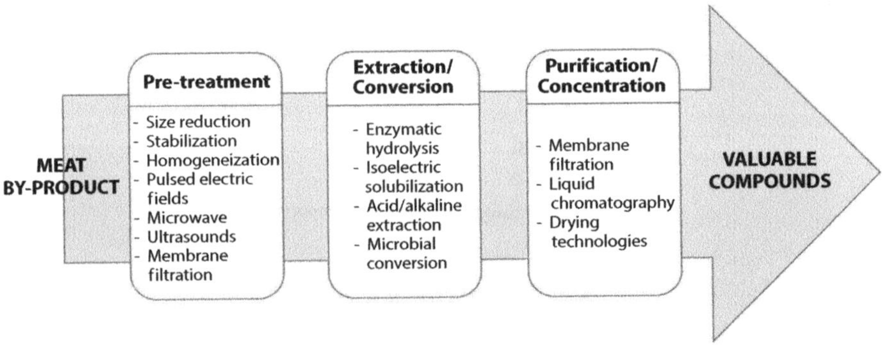

Figure 11.6 Process steps and operations for the recovery of valuable compounds from meat by-products (adapted from [77] and [78]).

Table 11.2 Uses of meat by-products and their valuable compounds in different industries [77, 78, 80].

Leather and clothing industries	Leather from salted or fresh cattle hides and pig skins. Albumen from animal blood. Wool from sheep and poultry feathers.
Chemical industry	Glue from trimmings, fleshing, fresh blood, and defatted bone. Lanolin from trimmings and fleshing; fertilizer from hydrolyzed cattle and pig hair, cattle horns, pig toenails, bonemeal and dried poultry and hatchery waste. Fire foam from extracted cattle and pig hair, cattle horns, and pig toenails. Industrial lubricant from cattle neastfood oil. Enzymes from animal glands. Biodiesel from fat and fatty acids in fatty tissues. Gelatin from cattle hides, salted or fresh pig skins, bone from cattle feet.
Pharmaceutical, medical, and cosmetic industry	Fibrin from cattle, pig and sheep blood. Tissue culture from fetal calf serum. Peptones from poultry tank waters. Enzymes from animal glands. Heart valves from pig hearts. Cholesterol from animal brains. Heparin from animal lung and animal intestines mucosa. Vitamin B12 from liver. Sutures from animal intestines. Steroids from brains and spinal cords. Biopeptide from blood, bones, collagen, liver, lung, placenta and visceral mass.

last five decades, several authors have proposed some applications and reported interesting results using membrane technology for this purpose.

For the treatment of blood plasma and red blood cells, several authors have made important advances, thus chronologically, Porter and Michaels [83], Ericksson and von Bockelmann [84] and Fernando [85], studied the separation of blood plasma and red blood cells with UF and RO membrane systems. These studies showed important findings: (i) the permeate flux is directly dependant on the temperature, but the maximum operative temperature should be fixed at 43°C because this is the temperature of blood coagulation; (ii) permeate fluxes were higher with red blood than with plasma solutions; (iii) the separation process is dramatically controlled by concentration-polarization phenomena, what leads to important permeate decline inferring that membrane fouling is an important concern for the operation and scaling-up of membrane processes. In this sense, Maartens and co-authors [86] found that the use of enzymatic solutions for the fouled membranes cleaning, allowed to restore the permeate flux. Moreover, some attempts tried to modify membranes to reduce fouling effects [78]. At the end of the 1990s, some authors used UF process as an intermediate step for the recovery and concentration of globin from blood [87, 88]. There were also some studies on the obtaining of gelatine from non-edible parts using MF and UF membranes [89].

Starting the 21st Century, studies on the treatment of by-products for the recovery of valuable compounds have been intensified. Some examples are: in 2002 Torres and co-authors studied the concentration of blood plasma from chicken [90]; it was reported the UF performance for the demineralization and concentration of proteins from plasma [91]; other extracts obtained from beef and pork lungs, were successfully purified by MF and UF processes [92]; More recently, Elmer *et al.*, 2009 [93] obtained good results in the purification of bovine hemoglobin by using an integrated membrane system consisting in a sequence of a two polysulfone membranes. MF and UF membranes were also used in the purification of extracts from deboned chicken [94] and for the recovery of hemoglobin from bovine blood [95]; Lafarga and co-authors, used UF for the purification of bioactive hydrolysates of bovine hemoglobin [96].

Some of these membrane applications for the recovery of valuable compounds from meat by-products have been industrially assayed and implemented. The company Wafilin run in 1983 an industrial multistage installation located in the Netherlands for the processing of blood plasma [80]; The Atlantic Gelatin (Kraft General), in 1984 applied the UF process at industrial scale to process gelatin solutions [89]. To extend the scaling-up of membrane technology for these purposes, there are several attempts for the obtention of low-fouling membranes that allow higher productions and improve performance of membrane separation processes in industry [72].

11.4 Bioactive Compounds From Marine By-Products

The fish processing industry is characterized by the production of huge amounts of process effluents coming from different operations including washing, chilling, thawing, rinsing, blanching, fileting and cooking [97]. These effluents are rich of bioactive compounds such as polyunsaturated fatty acids (PUFA), proteins, carotenoids, aroma and flavor compounds which can be potentially recovered and reused for the production of health sustaining pharmacological and human-food/animal-feed additives [98]. In this regard, the use of green non-thermal technologies, including membrane processes, are more and more investigated for the conversion of marine by-products into value-added compounds providing at the same time a solution for related environmental problems [8].

Afonso and Bórquez [99] evaluated the recovery of proteins from effluents of a fishmeal plant through a combination of membrane-based processes. The original effluent was pretreated by MF in order to reduce the content of suspended matter, grease and oil. Tubular UF membranes (Carbosep M2, monotubular, 15 kDa MWCO, with an active layer of ZrO_2-TiO_2 on carbon support) allowed to reduce the organic load of the MF permeate rejecting proteins in the range 49-62% depending on the operating conditions. Higher protein rejections (between 63% and 82%) were obtained when the MF permeate was treated with a ceramic multichannel NF membrane (Kerasep NanoN01A) with a MWCO of 1 kDa [100]. The economical evaluation related to a production of 544 ton/yr of fish meal, yielded a net worth of $160x10^3$ US $, an interest rate of return of 17% and a payback time of 8 years [101].

Ghalamara *et al.* [102] investigated the use of UF membranes to obtain protein peptide-enriched fractions from codfish blood and sardine cooking wastewaters. For both raw materials protein/peptides >10 kDa were rejected by using membranes with a MWCO of 10 and 50 kDa. Rejection values of 2% and 23% for peptides <1 kDa were measured with 10 kDa and 2.5 kDa membranes when treating codfish blood and sardine wastewaters, respectively. Peptide fractions from codfish blood treated with the 10 kDa membrane exhibited the highest antioxidant activity.

High concentrations of astaxanthin (10-13 μg mL^{-1}) were recovered by using 300 kDa UF membranes from shrimp cooking wastewaters due to its aggregation with high MW retained proteins [103]. Sequential UF/DF steps with membranes of 300, 100 and 30 kDa membranes resulted also an effective way for recovering proteins from these wastewaters, reaching a maximum of $27.12 \pm 0.16\, g\, L^{-1}$ protein content afterthe treatment with the 30 kDa membrane. Hydrolysates from protein-concentrated fractions showed high ACE-inhibitory and β-carotene bleaching activities compared with hydrolysates from other fish and seafood species. A similar approach was used to produce a peptide mixture with remarkable antihypertensive and antioxidant activity after enzymatic hydrolysis with alcalase of UF protein concentrates from cuttlefish industrial wastewaters [104].

UF membranes can be also used to improve the bioactivity of fish protein hydrolysates by concentrating or fractionating peptides with specific molecular weight. Chabeaud *et al.* [105] analyzed the effect of the initial peptide content, the operating pressure and volume reduction factor on the permeation flux and selectivity of a 4 kDa modified PES membrane used in the fractionation a saithe (*Pollachius virens*) protein hydrolysate. Results indicated that the selectivity is strongly affected by the pressure as well as by the VRF. Therefore, theses parameters could be properly selected to refine a fraction of interest with the aim of increasing its specific activity.

In another work Authors evaluated the impact of membrane material and MWCO on the permeation flux, retention and cleaning effectiveness in the fractionation of saithe protein hydrolysate by UF [106]. At this purpose five tubular membranes of different material (PES, modified PES and PS) ranging from 4 to 9 kDa were tested. Experimental results indicated that permeation fluxes were mainly influenced by the MWCO and in a lesser extent by the membrane material. PES membranes exhibited lower permeation fluxes and were considered less adapted for the fractionation of the hydrolysate. Authors identified the modified PES membrane with a MWCO of 4 kDa as the best suited membrane to enrich the hydrolysate in peptides lower than 2 kDa. This membrane exhibited a peptide retention of about 67% at 5 bar, producing the most enriched permeate in small peptides. The sequential combination of 4 kDa modified PES membrane and 8 kDa PS membrane was also suggested in order to recover medium-size peptides.

A 4-kDa PES membrane (MT 44-ES 404, from PCI Membranes) resulted also suitable to concentrate biologically active peptides from cuttlefish hydrolisates. Oligopeptides and free amino acids were recovered in the permeate stream which was considered suitable for animal nutrition and bioactivities [107].

According to Picot *et al.* [108], the combination of UF and NF membranes (PES, 4 kDa MWCO and PA/PES, 300 Da, respectively) allowed to obtain fractions rich in bioactive peptides from industrial fish protein hydrolysates, depending on the composition of the starting solution. In particular, an increase of the calcitonin gene-related peptide (CGRP) in the UF permeate was observed with respect to the starting fish protein hydrolysate.

Vandanjon *et al.* [109] proposed a combination of UF and NF membranes in order to concentrate and purify blue whiting peptide hydrolysates in a membrane enzymatic bioreactor. UF membranes with MWCO of approximately 20 kDa were an efficient tool to separate peptides from non-hydrolyzed proteins in the treatment of the original feed, although about 30% of peptides present in the initial mixture are retained. Peptides with high and low molecular weight were separated from the UF permeate by using UF membranes with a MWCO of 4-8 kDa. Finally, NF membranes with MWCO of 300 Da allowed to concentrate peptides larger than 1000 Da in good conditions of both fluxes and recovery rate (Figure 11.7). These results confirm

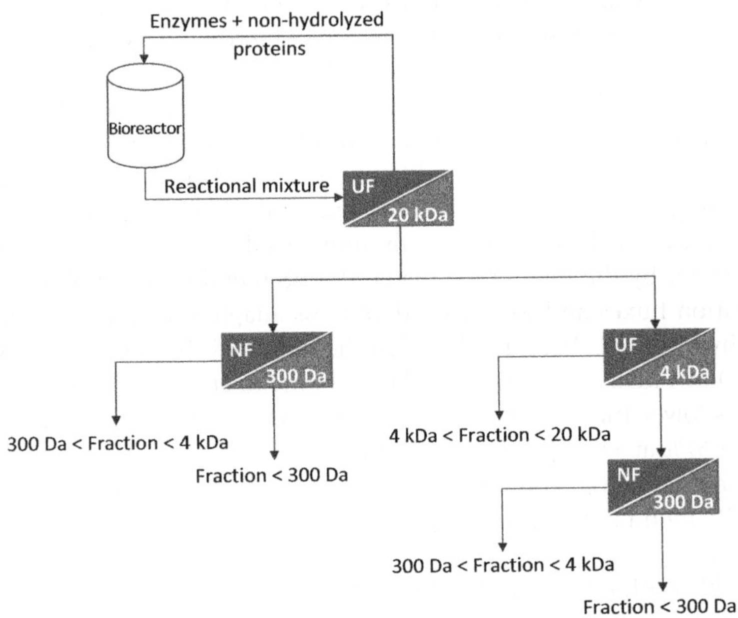

Figure 11.7 Concentration and the fractionation of marine peptides by UF and NF membranes (adapted from [109]).

the suitability of UF membranes with MWCO between 20 and 3 kDa to separate peptides from un-hydrolyzed proteins; on the other hand, membranes with MWCO between 1-3 kDa and 3.8 kDa are suitable for fractionation of bioactive peptides with desired molecular weights [110].

A similar approach was used to enrich the most bioactive peptide classes (100–500 and 1000–3500 Da) of white fish hydrolysates [111]. Specifically, 8, 4, and 300 Da flat-sheet membranes in polysulfone, polyethersulfone and polyamide, respectively, were successively used at this purpose.

Four peptide fractions (<3, 3–10, 10–30, and 30 kDa<) were obtained by Pezeshk *et al.* [112] in the UF of hydrolysates of yellowfin tuna viscera with regenerated cellulose membranes having MWCO of 3, 10 and 30 kDa, respectively. Among the fractionated enzymatic hydrolysates, the fraction with the lowest MW (< 3 kDa) had significantly the highest percentage of bacteria inhibition against Gram-positive and Gram-negative pathogenic and fish spoilage-associated microorganisms as well as the highest antioxidant activity.

A two-step membrane process for the fractionation of a protein hydrolysate produced by enzymatic hydrolysis of tuna dark muscle was investigated by Saidi *et al.* [113]. Several UF membranes with different configuration, material and MWCO were tested. Among them a three channel tubular ceramic membrane with a MWCO of 8.5 kDa (from Tami industries) was considered the best candidate for an industrial application. This membrane retained more than 80% of peptides with molecular weight greater than 4 kg mol^{-1}, without presenting a too high retention rate for the 1-4 kg mol^{-1} fractions. The UF permeate was used as feed solution for the NF step. It was performed by using a polyethersulfone membrane with a MWCO of 1 kDa in flat-sheet configuration (NP010, Microdyn Nadir). The proposed process allowed to obtain three different fractions: a fraction rich in peptides with molecular weight greater than 4 kg mol^{-1} (UF retentate); a fraction rich in peptides with molecular weight between 1 and 4 kg mol^{-1} (NF retentate), considered the most interesting for nutritional and/or pharmaceutical uses; a fraction containing free amino acids and dipeptides (NF permeate).

The combination of UF and NF membranes for the fractionation of a protein hydrolysate obtained from tuna processing by-products was also evaluated by Abejón *et al.* [114]. The maximal purity of the proposed cascade resulted of 49.3 % and the corresponding yield of the process was 62.6 %.

11.5 Conclusion and Future Trends

The recovery of high added-value compounds from agro-food by-products and wastewaters is an important challenge of the food processing industries.

This chapter attempts to provide an overview on the use of pressure-driven membrane operations in the treatment of fruit, vegetable, meat and fish by-products, aimed at recovering bioactive compounds in agreement with the principles of the circular economy. Results clearly highlight the key role of these processes, as emerging alternatives to the conventional methodologies, in enhancing the current valorization protocols within sustainable strategies for biorefinery, and the associated advantages in terms of environmental sustainability.

Innovative membrane technologies, such as forward osmosis, membrane distillation, pervaporation and electrodialysis, can provide a further contribution in the separation of electrically charged compounds, the recovery of aroma compounds and the concentration of target compounds from agro-food by-products under mild operating conditions.

References

1. Baiano, A., Recovery of bio-molecules from food wastes - A review. *Molecules*, 19, 14821, 2014.
2. Arias, A., Feijoo, G., Moreira, M.T., Exploring the potential of antioxidants from fruits and vegetables and strategies for their recovery. *Innov. Food Sci. Emerg. Technol.*, 77, 102974, 2022.
3. Ben-Othman, S., Jõudu, I., Bhat, R., Bioactives from agri-food wastes: Present insights and future challenges. *Molecules*, 25, 510, 2020.
4. Galali, Y., Omar, Z.A., Sajadi, S.M., Biologically active components in by-products of food processing. *Food Sci. Nutr.*, 8, 3004, 2020.
5. Guil-Guerrero, J.L., Ramos, L., Moreno, C., Zuniga-Paredes, J.C., Carlosama-Yepez, M., Ruales, P., Antimicrobial activity of plant-food by-products: A review focusing on the tropics. *Livest. Sci.*, 189, 32, 2016.
6. Sharma, P., Gaur, V.K., Sirohi, R., Varjani, S., Sang, H.K., Wong, J.W.C., Sustainable processing of food waste for production of bio-based products for circular bioeconomy. *Bioresour. Technol.*, 325, 124684, 2021.
7. Fierascu, R.C., Fierascu, I., Avramescu, S.M., Sieniawska, E., Recovery of natural antioxidants from agro-industrial side streams through advanced extraction techniques. *Molecules*, 24, 1, 2019.

8. Ali, A., Wei, S., Liu, Z., Fan, X., Sun, Q., Xia, Q., Liu, S., Hao, J., Deng, C., Non-thermal processing technologies for the recovery of bioactive compounds from marine by-products. *LWT-Food Sci. Technol.*, 147, 111549, 2021.
9. Castro-Muñoz, R., Cassano, A., Conidi, C., Membrane-based technologies for meeting the recovery of biologically active compounds from foods and their by-products. *Crit. Rev. Food Sci. Nutr.*, 59, 2927, 2019.
10. Castro-Muñoz, R., Díaz-Montes, E., Gontarek-Castro, E., Boczkaj, G., Galanakis, C.M., A comprehensive review on current and emerging technologies toward the valorization of bio-based wastes and by-products from foods. *Compr. Rev. Food Sci. Food Saf.*, 21, 46, 2022.
11. Ebrahimi, P. and Lante, A., Environmentally friendly techniques for the recovery of polyphenols from food by-products and their impact on polyphenol oxidase: A critical review. *Appl. Sci.*, 12, 1923, 2022.
12. Kumar, K., Srivastav, S., Sharanagat, V.S., Ultrasound assisted extraction (UAE) of bioactive compounds from fruit and vegetable processing by-products: A review. *Ultrason. Sonochem.*, 70, 105325, 2021.
13. Marić, M., Grassino, A.N., Zhu, Z., Barba, F.J., Brnčić, M., Rimac Brnčić, S., An overview of the traditional and innovative approaches for pectin extraction from plant food wastes and by-products: Ultrasound-, microwaves-, and enzyme-assisted extraction. *Trends Food Sci. Technol.*, 76, 28, 2018.
14. Pagano, I., Campone, L., Celano, R., Piccinelli, A.L., Rastrelli, L., Green non-conventional techniques for the extraction of polyphenols from agricultural food by-products: A review. *J. Chromatogr. A*, 1651, 462295, 2021.
15. Zhou, J., Gullon, B., Wang, M., Gullon, P., Lorenzo, J.M., Barba, F.J., The application of supercritical fluids technology to recover healthy valuable compounds from marine and agricultural food processing by-products: A review. *Processes*, 9, 357, 2021.
16. Cassano, A., Rastogi, N.K., Basile, A., Membrane technologies for water treatment and reuse in the food and beverage industries, in: *Advances in Membrane Technologies for Water Treatment - Materials, Processes and Applications*, A. Basile, A. Cassano, N.K. Rastogi (Eds.), pp. 551–580, Woodhead Publishing, Cambridge, 2015.
17. Lipnizki, F., Cross-flow membrane applications in the food industry, in: *Membranes for Food Applications*, K.V. Peinemann, S. Pereira, L. Giorno (Eds.), pp. 1–24, Wiley, Weinheim, 2010.
18. Pinelo, M., Jonsson, G., Meyer, A.S., Advances in the effective application of membrane technologies in the food industry, in: *Separation, Extraction and Concentration Processes in the Food, Beverage and Nutraceutical Industries*, S.S.H. Rizvi (Ed.), pp. 180–201, Woodhead Publishing Limited, Cambridge, 2010.
19. Cheryan, M., *Ultrafiltration and microfiltration handbook*, pp. 71–112, CRC Press, Boca Raton, 1998.

20. Cassano, A., Conidi, C., Ruby-Figueroa, R., Castro-Muñoz, R., Nanofiltration and tight ultrafiltration membranes for the recovery of polyphenols from agro-food by-products. *Int. J. Mol. Sci.*, 19, 351, 2018.
21. Ahsan, A. and Imteaz, M., Application and future prospects of reverse osmosis process, in: *Nanotechnology in Water and Wastewater Treatment*, A. Ahsan and M. Imteaz (Eds.), pp. 297–301, Elsevier, Amsterdam, 2019.
22. Mulder, M., *Basic principles of membrane technology*, pp. 416–464, Kluwer Academic Publishers, Dordrecht, 1996.
23. Mohd Basri, M.S., Abdul Karim Shah, N.N., Sulaiman, A., Mohamed Amin Tawakkal, I.S., Mohd Nor, M.Z., Ariffin, S.H., Abdul Ghani, N.H., Mohd Salleh, F.S., Progress in the valorization of fruit and vegetable wastes: Active packaging, biocomposites, by-products, and innovative technologies used for bioactive compound extraction. *Polymers*, 13, 3503, 2021.
24. Dilucia, F., Lacivita, V., Conte, A., Del Nobile, M.A., Sustainable use of fruit and vegetable by-products to enhance food packaging performance. *Foods*, 9, 857, 2020.
25. Capanoglu, E., Nemli, E., Tomas-Barberan, F., Novel approaches in the valorization of agricultural wastes and their applications. *J. Agric. Food Chem.*, 70, 6787, 2022.
26. Ganesh, K.S., Sridhar, A., Vishali, S., Utilization of fruit and vegetable waste to produce value-added products: Conventional utilization and emerging opportunities-a review. *Chemosphere*, 287, 132221, 2022.
27. Lau, K.Q., Sabran, M.R., Shafie, S.R., Utilization of vegetable and fruit by-products as functional ingredient and food. *Front. Nutr.*, 8, 661693, 2021.
28. Mármol, I., Quero, J., Ibarz, R., Ferreira-Santos, P., Teixeira, J.A., Rocha, C.M.R., Pérez-Fernández, M., García-Juiz, S., Osada, J., Martín-Belloso, O., Rodríguez-Yoldi, M.J., Valorization of agro-food by-products and their potential therapeutic applications. *Food Bioprod. Process.*, 128, 247, 2021.
29. Gumul, D., Ziobro, R., Korus, J., Kruczek, M., Apple pomace as a source of bioactive polyphenol compounds in gluten-free breads. *Antioxidants*, 10, 807, 2021.
30. Conidi, C., Rodriguez-Lopez, A.D., Garcia-Castello, E.M., Cassano, A., Purification of artichoke polyphenols by using membrane filtration and polymeric resins. *Sep. Purif. Technol.*, 144, 153, 2015.
31. Ross, K.A., DeLury, N., Fukumoto, L., Diarra, M.S., Dried berry pomace as a source of high value-added bioproduct: Drying kinetics and bioactive quality indices. *Int. J. Food Prop.*, 23, 2123, 2020.
32. Barreira, J., Ferreira, I., Oliveira, M., Pereira, J., Antioxidant activities of the extracts from chestnut flower, leaf, skins and fruit. *Food Chem.*, 107, 1106, 2008.
33. Cacciola, N.A., Squillaci, G., D'Apolito, M., Petillo, O., Veraldi, F., La Cara, F., Peluso, G., Margarucci, S., Morana, A., Castanea sativa Mill. shells aqueous extract exhibits anticancer properties inducing cytotoxic and pro-apoptotic effects. *Molecules*, 24, 18, 2019.

34. Sharma, K.D., Karki, S., Thakur, N.S., Attri, S., Chemical composition, functional properties and processing of carrot-A review. *J. Food Sci. Technol.*, 49, 22, 2012.
35. Rafiq, S., Kaul, R., Sofi, S.A., Bashir, N., Nazir, F., Ahmad Nayik, G., Citrus peel as a source of functional ingredient: A review. *J. Saudi Soc. Agric. Sci.*, 17, 351, 2018.
36. Teixeira, A., Baenas, N., Dominguez-Perles, R., Barros, A., Rosa, E., Moreno, D.A., Garcia-Viguera, C., Natural bioactive compounds from winery by-products as health promoters: A review. *Int. J. Mol. Sci.*, 15, 15638, 2014.
37. Yin, G., Cao, L., Xu, P., Jeney, G., Nakao, M., Lu, C., Hepatoprotective and antioxidant effects of Glycyrrhiza glabra extract against carbon tetrachloride (CCl4)-induced hepatocyte damage in common carp (Cyprinus carpio). *Fish Physiol. Biochem.*, 37, 209, 2011.
38. Tundis, R., Conidi, C., Loizzo, M.R., Sicari, V., Cassano, A., Olive mill wastewater polyphenol-enriched fractions by integrated membrane process: A promising source of antioxidant, hypolipidemic and hypoglycaemic compounds. *Antioxidants*, 9, 602, 2020.
39. Difonzo, G., Troilo, M., Squeo, G., Pasqualone, A., Caponio, F., Functional compounds from olive pomace to obtain high-added value foods–A review. *J. Sci. Food Agric.*, 101, 15, 2021.
40. Nowicka, P. and Wojdyło, A., Content of bioactive compounds in the peach kernels and their antioxidant, anti-hyperglycemic, anti-aging properties. *Eur. Food Res. Technol.*, 245, 1123, 2019.
41. Moreira, B., Pereira, E., Finimundy, T.C., Pinela, J., Calhelha, R.C., Carocho, M., Stojković, D., Sokovic, M., Ferreira, I.C.F.R., Caleja, C., Barros, L., Pineapple by-products as a source of bioactive compounds with potential for industrial food application. *Food Funct.*, 13, 9959, 2022.
42. Mo, Y., Ma, J., Gao, W., Zhang, L., Li, J., Li, J., Zang, J., Pomegranate peel as a source of bioactive compounds: A mini review on their physiological functions. *Front. Nutr.*, 9, 887113, 2022.
43. Andrade, C., Perestrelo, R., Câmara, J.S., Bioactive compounds and antioxidant activity from spent coffee grounds as a powerful approach for its valorization. *Molecules*, 27, 7504, 2022.
44. Faustino, M., Veiga, M., Sousa, P., Costa, E.M., Silva, S., Pintado, M., Agro-food by-products as a new source of natural food additives. *Molecules*, 24, 1056, 2019.
45. Flórez-López, E. and Grande-Tovar, C.D., The potential of selected agri-food loss and waste to contribute to a circular economy: Applications in the food, cosmetic and pharmaceutical industries. *Molecules*, 26, 515, 2021.
46. Kandemir, K., Piskin, E., Xiao, J., Tomas, M., Esra Capanoglu, E., Fruit juice industry wastes as a source of bioactives. *J. Agric. Food Chem.*, 70, 6805, 2022.
47. Panzella, L., Moccia, F., Nasti, R., Marzorati, S., Verotta, L., Napolitano, A., Bioactive phenolic compounds from agri-food wastes: An update on green and sustainable extraction methodologies. *Front. Nutr.*, 7, 60, 2020.

48. Carpentieri, S., Soltanipour, F., Ferrari, G., Pataro, G., Donsì, F., Emerging green techniques for the extraction of antioxidants from agri-food by-products as promising ingredients for the food industry. *Antioxidants*, 10, 1417, 2021.
49. Conidi, C., Drioli, E., Cassano, A., Membrane-based agro-food production processes for polyphenol separation, purification and concentration. *Curr. Opin. Food Sci.*, 23, 149–164, 2018.
50. Nazir, A., Khan, K., Maan, A., Zia, R., Giorno, L., Schroën, K., Membrane separation technology for the recovery of nutraceuticals from food industrial streams. *Trends Food Sci. Technol.*, 86, 426, 2019.
51. Cassano, A., Conidi, C., Drioli, E., Integrated membrane systems as an innovative approach for the recovery of high value-added compounds from agro-food by-products. *Chem. Eng. Trans.*, 87, 361, 2021.
52. Giacobbo, A., Bernardes, A.M., De Pinho, M.N., Sequential pressure-driven membrane operations to recover and fractionate polyphenols and polysaccharides from second racking wine lees. *Sep. Purif. Technol.*, 173, 49, 2017.
53. Cassano, A., Bentivenga, A., Conidi, C., Galiano, F., Saoncella, O., Figoli, A., Membrane-based clarification and fractionation of red wine lees aqueous extracts. *Polymers*, 11, 1089, 2019.
54. Conidi, C., Cassano, A., Garcia-Castello, E., Valorization of artichoke wastewaters by integrated membrane process. *Water Res.*, 48, 363, 2014.
55. Conidi, C., Donato, L., Algieri, C., Cassano, A., Valorization of chestnut processing by-products: A membrane-assisted green strategy for purifying valuable compounds from shells. *J. Cleaner Prod.*, 378, 134564, 2022.
56. Mohan, A. and Long, J.M., Valorization of wastes and by-products from the meat industry, in: *Valorization of Agri-Food Wastes and By-Products. Recent Trends, Innovations, and Sustainability Challenges*, R. Bath (Ed.), pp. 457–474, Academic Press, London, 2021.
57. Henchion, M., McCarthy, M., Resconi, V.C., Troy, D., Meat consumption: Trends and quality matters. *Meat Sci.*, 98, 561, 2014.
58. Stoll-Kleemann, S. and O'Riordan, T., The sustainability challenges of our meat and dairy diets. *Environment*, 57, 34, 2015.
59. Litskas, V.D., Platis, D.P., Anagnostopoulos, C.D., Tsaboula, A.C., Menexes, G.C., Kalburtji, K.L., Stavrinides, M.C., Mamolos, A.P., Climate change and agriculture: Carbon footprint estimation for agricultural products and labeling for emissions, in: *Sustainability of the Food System. Sovereignty, Waste, and Nutrients Bioavailability*, N. Betoret and E. Betoret (Eds.), pp. 33–49, Academic Press, London, 2020.
60. Food and Agriculture Organization of the United Nations, *Dairy market review*, 2020, https://www.fao.org/3/ca8341en/CA8341EN.pdf.
61. Leong, B.F., Chuah, W.C., Chye, F.Y., Recent advances and emerging trends in the utilization of dairy by-products/wastes, in: *Valorization of Agri-Food Wastes and By-Products. Recent Trends, Innovations, and Sustainability Challenges*, R. Bath (Ed.), pp. 371–389, Academic Press, London, 2021.

62. Zhang, Y., Zhang, J., Simpson, B.K., An introduction to agricultural and fishery wastes, in: *By-products from Agriculture and Fisheries: Adding Value for Food, Feed, Pharma, and Fuels*, B.K. Simpson, A.N. Aryee, F. Toldrá (Eds.), pp. 1–17, Wiley & Sons, Chichester, 2020.
63. Cai, Y., Tang, R., Tian, L., Chang, S.X., Environmental impacts of livestock excreta under increasing livestock production and management considerations: Implications for developing countries. *Curr. Opin. Environ. Sci.*, 24, 100300, 2021.
64. Porter, S.D., Reay, D.S., Higgins, P., Bomberg, E., A half-century of production-phase greenhouse gas emissions from food loss & waste in the global food supply chain. *Sci. Total Environ.*, 571, 721, 2016.
65. Long, J.M. and Mohan, A., Development of meat powder from beef by-product as value-added food ingredient. *LWT-Food Sci. Technol.*, 146, 111460, 2021.
66. Bello Bugallo, P.M., Cristóbal Andrade, L., Magán Iglesias, A., Torres López, R., Integrated environmental permit through Best Available Techniques: Evaluation of the fish and seafood canning industry. *J. Cleaner Prod.*, 47, 253, 2013.
67. Directive 2010/75/EU, *Industrial emissions directive*, 2010, https://ec.europa.eu/environment/industry/stationary/ied/legislation.htm.
68. United Nations, *Transforming our world: The 2030 Agenda for Sustainable Development*, 2015, https://sdgs.un.org/publications/transforming-our-world-2030-agenda-sustainable-development-17981.
69. Aspevik, T., Oterhals, Å, Rønning, S.B., Altintzoglou, T., Wubshet, S.G., Gildberg, A., Afseth, N.K., Whitaker, R.D., Lindberg, D., Valorization of proteins from co- and by-products from the fish and meat industry. *Top. Curr. Chem.*, 35, 53, 2017.
70. Alfaia, C.M., Madeira, M.S., Pestana, J., Coelho, D., Lopes, P.A., Toldrá, F., Prates, J.A.M., Pork by-products, in: *By-products from Agriculture and Fisheries: Adding Value for Food, Feed, Pharma and Fuels*, B.K. Simpson, A.N. Aryee, F. Toldrá (Eds.), pp. 21–41, John Wiley & Sons, West Sussex, 2020.
71. Toniasso, D.P.W., da Silva, C.G., de Souza Brum Junior, B., Somacal, S., Emanuelli, T., Kubota, E.H., Prestes Dornelles, R.C., Mello, R., Collagen extracted from rabbit: Meat and by-products: Isolation and physicochemical assessment. *Food Res. Int.*, 162, 111967, 2022.
72. Castro-Muñoz, R. and Ruby-Figueroa, R., Membrane technology for the recovery of high-added value compounds from meat processing coproducts, in: *Sustainable Meat Production and Processing*, C.M. Galanakis (Ed.), pp. 127–143, Academic Press, London, 2019.
73. Patrignani, F., D'Alessandro, M., Vannini, L., Lanciotti, R., Use of functional microbial starters and probiotics to improve functional compound availability in fermented dairy products and beverages, in: *Sustainability of the Food System. Sovereignty, Waste, and Nutrients Bioavailability*, N. Betoret and E. Betoret (Eds.), pp. 167–180, Academic Press, London, 2020.

74. Selmane, D., Vial, C., Djelveh, G., Production and functional properties of beef lung protein concentrates. *Meat Sci.*, 84, 315, 2010.
75. Toldrá, F., Mora, L., Reig, M., New insights into meat by-product utilization. *Meat Sci.*, 120, 54, 2016.
76. Chiralt, A., Menzel, C., Hernandez-García, E., Collazo, S., Gonzalez-Martinez, C., Use of by-products in edible coatings and biodegradable packaging materials for food preservation, in: *Sustainability of the Food System. Sovereignty, Waste, and Nutrients Bioavailability*, N. Betoret and E. Betoret (Eds.), pp. 101–127, Academic Press, London, 2020.
77. Borrajo, P., Pateiro, M., Barba, F.J., Mora, L., Franco, D., Toldrá, F., Lorenzo, J.M., Antioxidant and antimicrobial activity of peptides extracted from meat by-products: A review. *Food Anal. Methods*, 12, 2401, 2019.
78. Drummond, L., Álvarez, C., Mullen, A.M., Proteins recovery from meat processing coproducts, in: *Sustainability Meat Production and Processing*, C.M. Galanakis (Ed.), pp. 69–83, Academic Press, London, 2019.
79. Sorapukdee, S. and Narunatsopanon, S., Comparative study on compositions and functional properties of porcine, chicken and duck blood. *Korean J. Food Sci. Anim. Resour.*, 37, 228, 2017.
80. Castro-Muñoz, R., Garcia-Depraect, O., León-Becerril, E., Cassano, A., Conidi, C., Fíla, V., Recovery of protein-based compounds from meat by-products by membrane-assisted separations: A review. *J. Chem. Technol. Biotechnol.*, 96, 3025, 2021.
81. Garcia-Castello, E., Cassano, A., Criscuoli, A., Conidi, C., Drioli, E., Recovery and concentration of polyphenols from olive mill wastewaters by integrated membrane system. *Water Res.*, 44, 3883, 2010.
82. Garcia-Castello, E.M., Rodriguez-Lopez, A.D., Barredo-Damas, S., Iborra-Clar, A., Pascual-Garrido, J., Iborra-Clar, M.I., Fabrication and performance of low-fouling uf membranes for the treatment of isolated soy protein solutions. *Sustainability*, 13, 13682, 2021.
83. Porter, M.C. and Michaels, A.S., Membrane ultrafiltration III. Applications in the processing of meat by-products. *Chem. Technol.*, 1, 440, 1971.
84. Eriksson, G. and Bockelmann, I., Ultrafiltration of animal blood serum: Technology and microbiology. *Process Biochem.*, 10, 11, 1975.
85. Fernando, T., Concentration of animal blood by ultrafiltration. *Biotechnol. Bioeng.*, 23, 19, 1981.
86. Maartens, A., Swart, P., Jacobs, E.P., An enzymatic approach to the cleaning of ultrafiltration membranes fouled in abattoir effluent. *J. Membr. Sci.*, 119, 9, 1996.
87. Belhocine, D., Grib, H., Abdessmed, D., Comeau, Y., Mameri, N., Optimization of plasma proteins concentration by ultrafiltration. *J. Membr. Sci.*, 142, 159, 1998.
88. Gómez-Juárez, C., Castellanos, R., Ponce-Noyola, T., Calderón, V., Figueroa, J., Protein recovery from slaughterhouse wastes. *Bioresour. Technol.*, 70, 129, 1999.

89. Cheryan, M., *Ultrafiltration and microfiltration handbook*, pp. 421–425, CRC Press, Boca Raton, 1998.
90. Torres, M.R., Marín, F.R., Ramos, A.J., Soriano, E., Study of operating conditions in concentration of chicken blood plasma proteins by ultrafiltration. *J. Food Eng.*, 54, 215, 2002.
91. Del Hoyo, P., Moure, F., Rendueles, M., Díaz, M., Demineralization of animal blood plasma by ion exchange and ultrafiltration. *Meat Sci.*, 76, 402, 2007.
92. Selmane, D., Christophe, V., Gholamreza, D., Extraction of proteins from slaughterhouse by-products: Influence of operating conditions on functional properties. *Meat Sci.*, 79, 640, 2008.
93. Elmer, J., Harris, D.R., Sun, G., Palmer, A.F., Purification of hemoglobin by tangential flow filtration with diafiltration. *Biotechnol. Prog.*, 25, 1402, 2009.
94. Jayathilakan, K., Sultana, K., Radhakrishna, K., Bawa, A.S., Utilization of by-products and waste materials from meat, poultry and fish processing industries: A review. *J. Food Sci. Technol.*, 49, 278, 2012.
95. Stojanović, R., Ilić, V., Manojlović, V., Bugarski, D., Dević, M., Bugarski, B., Isolation of hemoglobin from bovine erythrocytes by controlled hemolysis in the membrane bioreactor. *Appl. Biochem. Biotechnol.*, 166, 1491, 2012.
96. Lafarga, T., Rai, D.K., O'connor, P., Hayes, M., Generation of bioactive hydrolysates and peptides from bovine hemoglobin with *in vitro* renin, angiotensin-i-converting enzyme and dipeptidyl peptidase-iv inhibitory activities. *J. Food Biochem.*, 40, 673, 2016.
97. Cristóvão, R.O., Botelho, C.M., Martins, R.J.E., Loureiro, J.M., Boaventura, R.A.R., Fish canning industry wastewater treatment for water reuse-a case study. *J. Cleaner Prod.*, 87, 603, 2015.
98. Venugopal, V., Valorization of seafood processing discards: Bioconversion and bio-refinery approaches. *Front. Sustain. Food Syst.*, 5, 611835, 2021.
99. Afonso, M.D. and Borquez, R., Review of the treatment of a seafood processing wastewaters and recovery of proteins therein by membrane separation processes e prospects of the ultrafiltration of wastewaters from the fish meal industry. *Desalination*, 142, 29, 2002.
100. Afonso, M.D. and Borquez, R., Nanofiltration of wastewaters from the fish meal industry. *Desalination*, 151, 131, 2003.
101. Afonso, M.D., Ferrer, J., Bórquez, R., An economic assessment of proteins recovery from fish meal effluents by ultrafiltration. *Trends Food Sci. Technol.*, 15, 506, 2004.
102. Ghalamara, S., Silva, S., Brazinha, C., Pintado, M., Valorization of fish by-products: Purification of bioactive peptides from codfish blood and sardine cooking wastewaters by membrane processing. *Membranes*, 10, 44, 2020.
103. Amado, I.R., González, M.P., Murado, M.A., Vázquez, J.A., Shrimp wastewater as a source of astaxanthin and bioactive peptides. *J. Chem. Technol. Biotechnol.*, 91, 793, 2015.

104. Amado, I.R., Vázquez, J.A., González, M.P., Murado, M.A., Production of antihypertensive and antioxidant activities by enzymatic hydrolysis of protein concentrates recovered by ultrafiltration from cuttlefish processing wastewaters. *Biochem. Eng. J.*, 76, 43, 2013.
105. Chabeaud, A., Vandanjon, L., Bourseau, P., Jaouen, P., Chaplain-Derouiniot, M., Guérard, F., Performances of ultrafiltration membranes for fractionating a fish protein hydrolysate: Application to the refining of bioactive peptidic fractions. *Sep. Purif. Technol.*, 66, 463, 2009.
106. Chabeaud, A., Vandanjon, L., Bourseau, P., Jaouen, P., Guérard, F., Fractionation by ultrafiltration of a saithe protein hydrolysate (*Pollachius virens*): Effect of material and molecular weight cut-off on the membrane performances. *J. Food Eng.*, 91, 408, 2009.
107. Soufi-Kechaou, E., Derouiniot-Chaplin, M., Ben Amar, R., Jaouen, P., Berge, J.P., Recovery of valuable marine compounds from cuttlefish by-product hydrolysates: Combination of enzyme bioreactor and membrane technologies. *C. R. Chim.*, 20, 975, 2017.
108. Picot, L., Ravallec, R., Fouchereau-Péron, M., Vandanjon, L., Jaouen, P., Chaplain-Derouiniot, M., Guérard, F., Chabeaud, A., Legal, Y., Alvarez, O.M. *et al.*, Impact of ultrafiltration and nanofiltration of an industrial fish protein hydrolysate on its bioactive properties. *J. Sci. Food Agric.*, 90, 1819, 2010.
109. Vandanjon, L., Johannsson, R., Derouiniot, M., Bourseau, P., Jaouen, P., Concentration and purification of blue whiting peptide hydrolysates by membrane processes. *J. Food Eng.*, 83, 581, 2007.
110. Hayes, M. and Gallagher, M., Processing and recovery of valuable components from pelagic blood-water waste streams: A review and recommendations. *J. Cleaner Prod.*, 215, 410, 2019.
111. Vandanjon, L., Grignon, M., Courois, E., Bourseau, P., Jaouen, P., Fractionating white fish fillet hydrolysates by ultrafiltration and nanofiltration. *J. Food Eng.*, 95, 36, 2009.
112. Pezeshk, S., Ojagh, S.M., Rezaei, M., Shabanpour, B., Fractionation of protein hydrolysates of fish waste using membrane ultrafiltration: Investigation of antibacterial and antioxidant activities. *Probiotics Antimicrob. Proteins*, 11, 1015, 2019.
113. Saidi, S., Deratani, A., Ben Amar, R., Belleville, M.P., Fractionation of a tuna dark muscle hydrolysate by a two-step membrane process. *Sep. Purif. Technol.*, 108, 28, 2013.
114. Abejón, R., Abejón, A., Belleville, M.P., Garea, A., Irabien, A., Sanchez-Marcano, J., Water recovery and reuse in the fractionation of protein hydrolysate by ultrafiltration and nanofiltration membranes. *Chem. Eng. Trans.*, 52, 283, 2016.

12

Food Industry Effluents, A Renewable Source for the Production of Porous Materials With High Added Value

Miloudi Hlaibi* and Issam Mechnou

Laboratoire Génie des Matériaux pour Environnement et Valorisation (GeMEV), Faculté des Sciences Aïn Chock BP Maârif, Hassan II University of Casablanca (UH2C), Casablanca, Marocco

Abstract

Fresh olives contain oil, vegetable water and solid matter. Processes used for the extraction of olive oils, generate significant quantities of olive mill solid waste (OMS), and olive mill wastewater (OMW), rich in valuable compounds. OMW effluents are very toxic, often rejected without treatment, causing serious problems for aquatic environments. These effluents are made up of several phyto-toxic molecules, hardly biodegradable. Current treatment methods tested are not suitable for eliminating effectively these effluents, especially OMW effluents. Often, applied methods use additives, generate polluting by-products, and require specific installations. However, these discharges represent a renewable and potential biomass resource, rich in quantities of recoverable total organic carbon. From these observations, in this work, we thought to carry out total speciation and characterization of effluents in order to valorize them by transformation according to the objectives of sustainable and green chemistry. After identification of the compositions, specific carbonization process "flash pyrolysis process" was designed and applied to transform these effluents into potentially recoverable bio coal. This compound is widely used in several promising strategic sectors, production of carbon fibers, wastewater treatment and production of energy compounds. FTIR and X-ray diffraction spectroscopic techniques were used to characterize the obtained bio coal solid phases.

**Corresponding author*: m.hlaibi58@gmail.com

Umile Gianfranco Spizzirri (ed.) Nutraceutics from Agri-Food By-Products, (391–422) © 2023 Scrivener Publishing LLC

Keywords: Olive oil extraction process, olive oil industry discharges, phytotoxic molecules, renewable biomass resources, carbonization processes, strategic sectors for bio-coal

12.1 Olive Oil Production and Its Impact on Natural Aquatic Resources

In recent years, the food industry has experienced significant growth. In particular the production of olive oils, which has experienced a significant increase over the last decade, resulting from the high consumption of this food of approximately 3,056,500 t/year [1], for its important role as preferred food in the diet of Mediterranean populations. This oil is widely used in body care, in massage, in the preparation of perfumes and in the manufacture of soaps [2]. Today, there are nearly a billion olive trees cultivated in the world, with more than 70% in the Mediterranean basin [3]. However, olives contain around 20% oil, 50% vegetable water, and 30% solids [4], regardless of the extraction process type used. The extraction of olive oils by a continuous two-phase system, generates by-products, solids (OMS), and especially liquid (OMW) very toxic for the environment.

Olive oil extraction processes are often traditional and generate large quantities of OMW effluent which can reach 1450 L for one ton of fresh olives treated (Figure 12.1) [5].

This highly toxic effluent is often released without prior treatment, causing serious environmental problems for groundwater, surface water and living species in the aquatic ecosystem, therefore intervention is needed to improve production and sustainability of this important food [6]. The

Figure 12.1 Traditional olive oil production unit and artificial lake for storing toxic effluents OMW.

toxicity of OMW effluents largely depends on their richness in organic matter, their phytotoxic nature, due to the concentration of polyphenols which can reach 12 g/l, causing destruction of the local flora and fauna, their low biodegradability, justified by a high COD/BOD5 ratio (approximately 8) due to the high concentration of long-chain lipids which can reach up to 70 g/L. The high nitrogen and phosphorus content promotes excessive growth of plants and algae, which consume large amounts of oxygen, thus causing an imbalance of the aquatic ecosystem (eutrophication) [7]. On the other hand, the elimination of OMW effluents by bio-treatment is almost impossible because their phytotoxic nature affects the used bacterial communities, and promotes the appearance of strains resistant to antibiotics, which can indirectly lead to infectious diseases, incurable to humans and animals [8].

Several studies have been carried out to treat and reduce the pollutant load of these effluents before discharging them into receiving environments. Among the adopted techniques, the treatment by electrocoagulation, this technique allows a reduction of 85% of the COD, 80% of the polyphenols, a dilution of the color of about 5, and an elimination of the suspended matter of the order 8 kg/m^3. Despite these important results, this technique requires a voltage of 20 V, approximately 14 kWh/m^3 in energy, and generates a loss in mass of the electrodes (0.7 kg/m^3) [9]. Conventional coagulation by adding aluminum sulphate compound (1.5 g/L) at neutral pH, reduces 41% of polyphenols, 40% of COD and 27% of suspended solids [10]. The use of adsorbents such as yellow clay or activated charcoal can reduce up to 100% of COD and 60% of polyphenols [11]. The treatment of OMW effluents by an advanced H2O2 (1.2 mol/L)/ Cu (II) oxidation process under microwaves makes it possible to degrade up to 62% of the phenolic compounds in a treatment time of 65 min [12].

On the other hand, these liquid effluents (OMW) represent a potential source of valorizable biomass because of their richness in organic matter and oils lost during the extraction process. Therefore, valuing this source is an opportunity for the profitability and sustainability of the olive oil industry. Several attempts have been made to enhance the value of these OMW effluents, such as their use after adequate treatment in agricultural spreading [13], the production of enzymes [14], or the methanization [15]. The proposed methods are very expensive and have low yields, and the anaerobic digestion process is unnecessary because these liquid effluents are difficult to biodegrade because of their high COD/BOD5 ratio. Therefore, adopting physico-chemical methods is not an appropriate choice, as these techniques require pre-treatments, generate by-products, require specific equipment and are economically expensive (Figure 12.2).

Figure 12.2 Very negative impact of OMW effluents on Fauna and Flora.

The vegetable waters present liquid effluents OMW rich in organic matter (around 48%), and thermal processes have become more suitable for efficiently transforming and upgrading these effluents into value-added compounds [16]. In particular, the hydro-carbonization processes of organic matter from vegetable waters in autoclaves at temperatures between 180 and 300 °C and under self-generated pressures [17]. Thus these OMW liquid effluents were filtered through a bed of olive pits; the obtained mixture was converted by hydro-thermal carbonization into bio-char. This technique exhibited a high carbon recovery of around 84.2% and an overall yield of up to 46.8%, and a reduction in volatile matter of 52.2% [18]. Hydro carbonization of vegetable waters at a temperature of 200 °C for 24 hours after drying at 120 °C, has made it possible to recover up to 57% of carbon in the form of charcoal [19]. Our direct carbonization process by flash pyrolysis of OMW liquid effluents is a new technique with several advantages including economy, cleanliness (without by-products) and speed (30 seconds). This method makes it possible to transform the vegetable waters into clean bio-coal exhibiting important properties with good carbon content and a very high porosity, as well as a carbon recovery of 70% [16]. Up to now, these OMW liquid effluents constitute a real danger for the environment and a black spot for this industry. The search for new techniques to treat, transform and enhance them is a necessity, and thermal techniques such as flash pyrolysis and hydro-carbonization have proven their effectiveness. These techniques make it possible to transform these highly toxic effluents into a strategic compound with high added value, which constitutes a real additional source of benefits for farmers and extraction units, and ultimately a clean, green and sustainable olive oil production industry.

12.2 Transformation and Valorization of Solid and Liquid Toxic Discharges From Moroccan Olive Oil Industries, in Bio Coal Phases for Several Potential Applications

12.2.1 Introduction

Today, the olive sector is of great interest to the food and cosmetics industries in the world and particularly in a Mediterranean country like Morocco. In 2018, olive trees cover almost 1 045 000 ha, about 55% of all Moroccan arboriculture (MAPMDREF/DSS). During last three years, almost 297.000 ha of olive trees have been planted, about 99.000 ha per year. Similarly, as part of the "Green Morocco" agricultural strategy, it is planned to plant one million and 220 thousand hectares, to produce around 2.5 million tons of olives each year [20].

Indeed, the increase in the production of olive oils generates and introduces problems with technical and environmental constraints that are often difficult to solve. In Morocco, the olive industry does not have appropriate techniques and olive oil extraction processes are usually discontinuous and traditional. The extraction units of olive oils do not improve their extraction processes in parallel with the significant development of olive oil production. These units produce about 200 000 tons of olive oil per year (MAPMDREF/DSS), and extraction yields do not exceed 12 % [21], while total oil content in olives is around 20% [22]. This results in 8% loss of production, which is easily evacuated in OMW and OMS discharges and accounts for between 18 and 25% of national production. This growing trend in olive oil production has led to a significant increase in quantities of different pollutant effluents resulting from this activity [23], particularly:

- Olive mill solid waste (OMS): "pulp residues, olive stones, and skins ..."
- Olive mill wastewater (OMW) "liquid organic materials, water added during the process, and washing water ..."

Current olive oil extraction processes generate about 0.9 m^3 of OMW by processing one ton of olives [24]. Therefore, for an annual production of 175 000 tons of olives, about 157500 m^3 of OMW are regenerated. This huge amount of dissolved organic compounds is spontaneously released into nature without any prior treatment. This hazardous operation causes serious and dangerous environmental problems for groundwater, surface water, fauna and flora [25, 26]. The toxicity of OMW is due to the presence of long lipid chains which can form a stagnant and impermeable film preventing the penetration of light and oxygen in aquatic environment [27]. Similarly, the high concentration of several polyphenols, stable molecules hardly biodegradable, which are considered to be phytotoxic compounds because of their inhibitory effect on bacterial growth [28], vegetative growth and plant germs. On the other hand, high level of phosphorus compounds, promotes the development of algae which cause the ***eutrophication*** phenomenon [26], characterized by increased consumption of dissolved oxygen, and serious decrease in the availability of this essential element (**O_2**) for other living organisms. This depletion of the oxygen element is the main cause of the sudden death of fish and the reduction of flora [29].

In Morocco, solid effluents are often used by some industrial units to prepare "OMS oil," mainly composed of olive oil lost during the extraction process (2 to 3%). It is extracted using chemical solvents (hexane), it has the same characteristics and the same lipid profile as oil extracted during the initial process. It would be ready for consumption and marketing, after refining and mixing with other virgin oils [30]. Due to their high calorific value, OMS effluents can also be used as fuel in industrial furnaces, boilers, and in public baths, which presents a real source of renewable energy, and additional financial source for the extraction units [31].

In addition, to limit the harmful and dangerous effects of OMW effluents on the environment, several studies have been carried out on the possibilities of their treatment and ecological valuation by many researchers. Among the best treatment techniques of OMW effluent, the process by electro coagulation, it is a process of electrolysis in the presence of iron or aluminum electrodes. This technique allows obtaining excellent results, itt can reduce 70% of COD, 70% of polyphenols, and a decrease in color [32]. In fact, for period of 15 minutes, electric current with a voltage of 30 volts, which corresponds to 2.63 kWh of energy consumed/1000 liters treated and a mass loss of 0.085 kg of Al electrode, allows deleting one kg of COD A technique similar to the previous one, coagulation-flocculation by adding coagulants, aluminum (III) or iron (II and III) was used. Studies have shown that adding $FeSO_4.7H_2O$ (6.67 g/L) and FLOCAN-23 (0.287 g/L) at neutral pH can reduce 40% of polyphenols, 72% of COD and 97% of the suspended matter [33]. In addition, the reduction of the OMW effluent polluting load by adsorption on activated clay by 71% of COD and 81% of polyphenols, compared to a 100% removal of polluting charge by adsorption on the yellow clay, or activated carbon [16]. Other studies have been carried out with the idea of valorizing OMW effluents and using them after an adequate treatment, in agricultural application [34], production of compost [34], production of enzymes [35], and adsorption of phenolic compounds [36]. In addition, transformation and valorization techniques mentioned above require unique and specific installations, with an expensive economic aspect, presenting a major drawback for these techniques.

From these observations, and to make the treatment systems more profitable, objective of our work is to develop treatment and recovery processes for OMS and OMW discharges according to the principles of ***green and sustainable chemistry***, to produce a highly adsorbent bio coal for different applications, environmental, cosmetic, pharmaceutical and industrial....

12.2.2 Materials and Methods

12.2.2.1 Samples and Characterizations

OMW and OMS effluents from olive oil extraction processes, studied and characterized, were taken from four semi-traditional extraction units, located in different regions of Morocco, rich in olive production.

These extraction units are characterized by semi-traditional processes, at the beginning the fresh olives are crushed using an electric grinder. The obtained paste is recovered and spread out in superimposed layers, separated by coconut fiber discs "scourtins," in order to carry out the extraction-filtration process using a mechanical compressor. The obtained OMS effluent is recovered directly at the end of this process, whereas the obtained oily liquid filtrate containing olive oil is placed in natural separation by difference of density in settling basins, and after this natural operation the two phases olive oil, and OMW effluent are separated [37].

Hydrogen potential: is measured using a high-performance pH meter with intuitive operation "mettler toledo" equipped with an integrated temperature sensor.

Conductivity (mS.cm^{-1}): is measured using a "HANNA" conductivity meter.

Suspended matter (g/L): 100 mL filtration of OMW effluent on Whatmann membranes (0.45 μm), and the determination of the difference in weight between new filter and after filtration, allows to calculate the rate suspended matter. Drying of the wet filter in the oven at 105 °C is necessary to remove any liquid trace. Finally this parameter is calculated according to the following relation (AFNOR T 90-105):

$$[\boldsymbol{MES}]_{g/L} = 10 \times (\boldsymbol{m\,filter}_{(after)} - \boldsymbol{m\,filter}_{(before)}) \tag{12.1}$$

BOD (mg/L): is usually the amount of oxygen required for self-oxidation of biologically degradable organic matter. The importance of determining this parameter lies in the obtained indication on the polluting organic load present. Indeed, BOD parameter is always proportional to the quantity of biodegradable organic matter present in the effluent. We calculated this parameter using an Oxitop breathing measurement system. (AFNOR, T 90-103).
COD (mg/L): we measured this parameter using the ST-COD method in closed tubes; this is an estimate of the oxygen amount required for total oxidation of the organic matter in a known volume of OMW effluent (in our case) by chemical reaction (Dichromate).
TOC (%): total organic carbon content is considered the most important of all parameters, to verify the feasibility and competitiveness of our objective. In principle, this analysis type is not specific, and does not allow determining the organic constituents present in the complex compositions of OMS and OMW effluents. The measurement of this important parameter (**TOC**) allows knowing the exact amount of organic carbon in all forms, present in OMS and OMW discharges. Indeed, this parameter was determined following the modified ANNE method, based on the oxidation of the organic matter in our samples by potassium dichromate ($K_2Cr_2O_7$) in sulfuric acid medium [38].
TOM (%): total organic matter content is always proportional to the total organic carbon content; it is generally obtained by estimating the ratio between TOM/TOC which is almost always constant equal to 1.72 [39].

12.2.2.2 Gravimetric Analyzes

Dry matter (105 °C) (%): drying operation of OMWM effluent (100 mL) in an oven at 105 °C until the stabilization of its mass, allows to determine the solids content at 105 °C, which corresponds to the whole of the content in dissolved salts, as well as the content of suspension expressed in percentage and calculated by the following relationship (AFNOR T90-029):

$$\%[\boldsymbol{DM}]_{105°C} = (\boldsymbol{m}_2/\boldsymbol{m}_1) \times 100 \tag{12.2}$$

m_1 and m_2: respective masses of the solution before and after drying.

Dry matter (180 °C) (%): to determine this parameter, we only dried in an oven at 180 °C for 24 hours the dry matter obtained during the previous experiment; this parameter is always expressed in percentage and calculated by the following relation (AFNOR T90-029):

$$\%[\boldsymbol{DM}]_{180°C} = (\boldsymbol{m}_3/\boldsymbol{m}_1) \times 100 \qquad (12.3)$$

$\boldsymbol{m}_1$ and $\boldsymbol{m}_3$: initial and final masses of the solution.

Volatile matter (550 °C) (%): determined by the calcinations at 550 °C for 4 hours the dry matter obtained by drying OMWW effluent at 105 °C, in a programmable oven and preheated, expressed as a percentage relative to dry matter at 105 °C and calculated by the following relation (NF EN 12880):

$$\%[\boldsymbol{VM}]_{550°C} = [(\boldsymbol{m}_2 - \boldsymbol{m}_4)/\boldsymbol{m}_2] \times 100 \qquad (12.4)$$

$\boldsymbol{m}_4$: the obtained mass after calcinations at 550 °C.

Volatile matter (900 °C) (%): determined by the calcinations for 2 hours at 900 °C of the solid matter obtained by calcinations of OMW effluent at 550 °C, in a programmable oven and preheated, expressed as a percentage relative to dry matter at 105 °C and calculated by the following relation (ISO 562: 2010):

$$\%[\boldsymbol{VM}]_{900°C} = [(\boldsymbol{m}_2 - \boldsymbol{m}_5)/\boldsymbol{m}_2] \times 100 \qquad (12.5)$$

$\boldsymbol{m}_5$: the obtained mass after calcination at 900 °C.

12.2.2.3 Preparation of OMW Solutions

Due to the large amount of water contained in the OMW effluents, it is very useful to dry them for some time before the pyrolysis step. This step is fundamental to remove a large part of the water in order to obtain thick and sticky biomass arrangements suitable for carbonization preparation.

Kinetics and modeling: A graduated test tube of 100 mL filled with a OMW emanating arrangement was put beneath the impact of the warm in a programmable broiler for diverse temperatures progressively 80 °C, 95 °C, 100 °C and 105 °C by considering the advancement of the arrangement volume as a work of time (hours) for each of considered temperatures. From the gotten comes about, the dynamic modeling of invented

Table 12.1 Empirical and semi-empirical models chosen for the kinetic modeling of the drying process of OMW effluent solutions.

Model name	Model function	Reference
Lewis (Newton)	$X^{*}= e^{-kt}$	[40]
Henderson and Pabis	$X^{*}=\mathbf{A} \times e^{-kt}$	[40]
Modified Page	$X^{*}= e^{-(kt)^{n}}$	[41]
Wang and Singh	$X^{*}=1+at+bt^{2}$	[42]
Two-term exponential	$X^{*}=a.\exp(-kt)+(1-a).\exp(-k.at)$	[42]
Diffusion approximation	$X^{*}=a.\exp(-kt)+(1-a).\exp(-k.bt)$	[43]
Verna *et al.*	$X^{*}=a.\exp(-kt)+(1-a).\exp(-k't)$	[44]
Two-term	$X^{*}=a.\exp(-kt)+b.\exp(-k't)$	[45]

X*: reduced volume at time ***t*** relative to the initial volume (V_t/V_0).

drying of OMW profluent arrangements was inspected and carried out agreeing to a few primary scientific models depicted within the writing (Table 12.1).

Verification of models: to verify a model, we chose the following parameters, coefficient of determination R^2, p-value, Khi square X^2 and the square root of the mean squared error (RMSE) as criteria for evaluating and comparing the examined models.

The model with the highest determination coefficients, and p-value and the lowest Khi-square and RMSE is the most verified model. In short, the determination coefficient and the p-value are obtained automatically during the regression and the adjustment of the models by the software "statistica" [46], whereas the Khi square X^2 and the RMSE are calculated according to the following relationships.

$$\textbf{(Khi square)}X^2 = [\Sigma_{i=1}^{N}(X_{ei}^{*} - X_{pi}^{*})^2]/(N-n) \tag{12.6}$$

$$RMSE = \sqrt{[\Sigma_{i=1}^{N}(X_{ei}^{*} - X_{pi}^{*})^2/N]} \tag{12.7}$$

TOC stableness: a 100 ml graduated test tube filled with OMW was thermally impacted in a programmable grid at different temperatures, progressively 80 °C, 95 °C, 100 °C, and 105 °C, to examine the change in overall natural carbonaceous substance in the OMW effluent as a function of the volume of water lost, in accordance with the "ANNE adjusted" strategy described above.
Evolution of the density and its influence: For the pyrolysis process of the OMW solution, the density of the liquid sample plays an important role on its stability during the high-temperature carbonization process and indirectly on the transformation yield. This is because the pyrolysis procedure of a low density liquid sample results in its spillage inside the furnace. In order to examine this influence of the liquid density, we determine the mass of the solid sample at each 5% of the volume vaporized of the studied OMW effluent solution.

12.2.2.4 Carbonization Processes

The technique used to carry out charcoal is pyrolysis, which is in principle the transformation of the organic substance under the thermal effect at high temperature and in the absence of oxygen [47]. Indeed, it allows obtaining products such as coal, pyrolysis oils, as well as gaseous products (CH_4. CO_2...). In the publications, three types of pyrolysis processes are often cited [48].

Slow pyrolysis: it is principally the breakdown of organic matter at a relatively high temperature (350-500 °C) without the presence of oxygen.
Fast pyrolysis: In contrast to the previous type, this fast operation is carried out at a higher temperature (about 500 °C), its transformation time is very brief (a few seconds), and it occurs under an intense and swift outer heat transfer [49].
Flash pyrolysis: breakdown of organic matter in no oxygen at very high heat (between 800 and 1000 °C). This kind of process is carried out with a very short transformation time and a high production of condense water [50].
Carbonization of OMS: Because of the high temperature stability of the OMS effluent and to operate at very short transformation processes, in order to get a significant char yield, we have chosen to use a flash burn at 950 °C [51] in a laboratory furnace, previously warmed up for the transformation of the OMS effluent into char. The conversion and maximization of the amount of char resulting from the adopted process is done according to the present report: We placed our sample of OMS effluent in a single pot of

metal, with a diameter of 55 mm, a base size of 23.75 cm^2, and a volume of 50 mL. Mass weighing of the produced char is essential to study the effect of surface finish and conversion time in the kiln, on both the firewood oxidization process and the char mass yield obtained from the char.

Carbonization of OMW effluents: After drying to a volume reduction of 1/2, the OMW solutions have been successfully burned, in order to study the impact of the filling factor of the pot as well as the interaction of the pyrolysis temperature on the carbonization efficiency of these wet samples. Moreover, it should be noted that the following steps are performed in the flash pyrolysis process of a solution of OMW effluents:

1st step: dehydration of the sample justified by the sudden release of water vapor outside the furnace OMW effluent sample:

LIQUID (1) + water vapor.

2nd step: thermolysis of the sample justified by a white smoke release outside the furnace due to the incomplete combustion of organic matter:

LIQUID (1) => **LIQUID (2) + GAS.**

3rd step: transformation of the resulting liquid into coal, justified by the release of brownish smoke outside the furnace, due to the incomplete combustion of the liquid:

LIQUID (2) => **Bio coal + GAS**

In addition, the processing time in the furnace of treated OMW effluent solutions, always depends on the cessation of the smoke released during the carrying out of the process.

12.2.3 Speciation Results and Discussions

Viscosity: Reportedly, after decantation, the OMW stream has a mixed macroscopic composition, a light stagnant part that holds the oils lost during the stripping action, a Heavy fraction that contains mainly solids in suspension, and an abundant third portion of water, disseminated solids, and biological species that have nearly the same consistency as water. Therefore, by virtue of their high degree of moisture and the neutralizing influence of bulk density of suspended material and light oils, OMW fractions have a density close to that of the water [52].

Hydrogen potential: The lower average pH value (Table 12.2) observed is primarily due to the availability of active organic substrates of an acidic nature, namely: essential fatty acids [43] and those phenolic acids produced by the transformation of phenolic alcohols in storage [53].

Conductance and saltiness: The conductance in the main cases is proportionate to the level of dissolved salts, for the OMW effluents studied the

Table 12.2 Physical and chemical characteristics of OMW effluents according to the different sampling regions.

Region	Density	pH	Conductivity	Chloride (g/L)	NaCl (g/L)
SEBT GZOULA	1.1	4.8	38.00	10.7	17.7
ATAAOUIYA	1.0	4.8	26.00	8.6	14.2
OUAZZANE	1.0	4.9	22.00	5.3	8.8
AOUTAT EL HAJ	1.1	4.7	28.00	9.6	16.0

conductances obtained are high (Table II-12.2), in fact this is certainly due to the large quantities of salts added by the farmers during the storage of fresh olives [54].

The arrangement of estimates made for thickness and pH of the proflu-ent OMW arrangements are characterized by a small standard deviation of 0.044 and 0.059 separately. These values demonstrate the non-dispersion of the results obtained in agreement with the examination districts. In expansion the pH and thickness values of the considered arrangements of these distinct olive oil generation places in Morocco are compared for the analyzed distinct emanating discharges of OMW. With a normal thickness close to 1 [55, 56] and a normal acuity pH = 4.77, notwithstanding the test location [56]. On the other hand, the arrangement of estimates made for electrical conductivity is characterized by a moderately high standard deviation. Thus, it appears that the estimates for testing locations are highly dispersed, with values varying from one location to another (Table 12.2). Thus, this heterogeneity in conductivity for these harmful OMW effluents in Morocco is fundamentally due to the method used.

12.2.3.1 *Chemical and Biochemical Oxygen Demand*

Because of the high biomass content of OMW wastewater streams, the biological and chemical oxygen demands of the investigated and analyzed OMW effluent stream solutions are vastly higher than (Table 12.3) suggest that the COD/BOD5 ratio is more than 3 in all areas of sampling. Clearly, as a result, the organic matter in OMW effluents is difficult to achieve biodegradability [57], and this presents a dual major challenge for the management of these harmful effluents. This is because current methods (lagooning, bio-disc, trickling bed, etc.) are unable to properly treat these waste streams. Conversely, OMW effluent solutions are very stable when

Table 12.3 Chemical parameters of OMS and OMW effluents generated by olive oil industries and collected in the prospected producing regions.

Region	Safi		Marrakech		Tanger-Tétouan-Al Hoceïma	
Waste effluents	**OMW**	**OMS**	**OMW**	**OMS**	**OMW**	**OMS**
COD g(O_2)/L	116.25	X	97	X	92.65	X
BOD5 g(O_2)/L	14.5	X	7	X	7.5	X
COD/BOD5	8.2	X	13.8	X	12.2	X
TOC (%)	24.37	48.48	14.96	51.45	14.05	53.56
TOM (%)	42.02	83.58	25.79	88.71	24.23	92.34
OM (g//L)	48.41	X	37	X	35.88	X

they are stored in tanks for an easier treatment of these toxic effluents, devastating for the wildlife and the plants, and a greater natural vaporization of the aqueous phase.

These OMW effluents have a headstrong COD (Table II-12.3), legitimized by the nearness of overwhelming natural particles (tannins nearly 12g/L, phenolic esters, sugars, etc.), macromolecules such as lignin, fragrant compounds (caffeic corrosive, corrosive 4-hydroxyphenylalcohol, 3.4-dihydroxyphenylethanol) and oleuropein, which has more than 2% of the weight of new olives [58].

12.2.3.2 *Total Matter and Organic Carbon*

The TOC content of OMW effluent solutions is high because they are made of a lot of biomass (38% on average). Likewise, the TOC level in OMW effluent discharges from the olive oil industry is high because it is very dry and there is a lot of vegetation in the area.

12.2.3.3 *Preparation of OMW Effluent Solutions for Treatment*

12.2.3.3.1 Kinetic Study of Drying Process

The graph in Figure 12.3 shows the evolution of the liquid volume content of OMW effluent solutions, under the action of convective drying as a function of time (days) at different temperatures. It appears that the drying speed increases with the temperature factor. For about two days in an oven

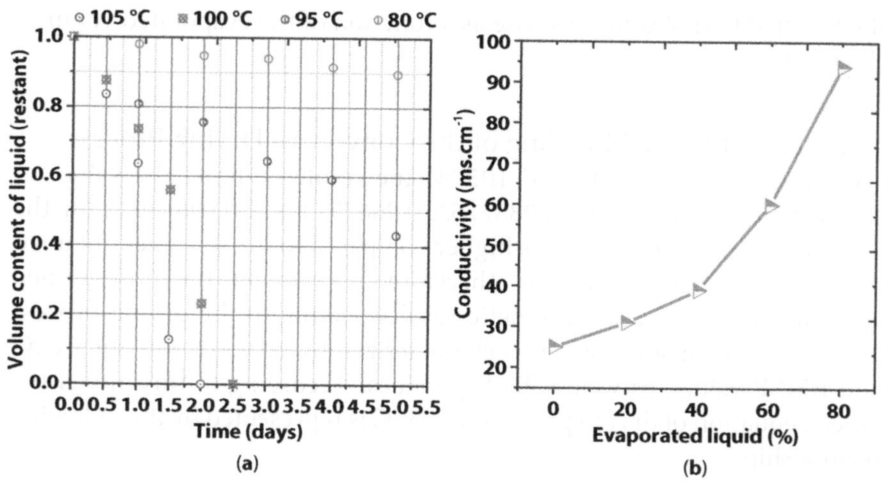

Figure 12.3 (a) evolution of liquid volume content for OMW effluent, (b) salinity evolution of treated OMW effluent during the drying process operation at temperatures: 80°C., 95°C., 100°C., and 105°C.

at 105°C, a sample of 100 mL/7cm^2 is completely evaporated and there remains only a negligible volume of a very viscous solution physically similar to petroleum. Therefore, a temperature of 100°C is sufficient to evaporate the entire liquid volume of an OMW effluent solution in two days and a few hours.

Further study is required for the permanent drying of OMW effluents solutions. For temperatures lower than 90°C, the boiling of the OMW slurry solution is very poor, needing more than five days. However, for temperatures above 100°C., the boil off is rapid (Figure 12.3a), although the freed vapors are saturated with organic matter. Such water discharged from the fumes represents air pollution and a significant loss of organic matter for the respective treated OMW process solutions. On the other hand, a prominent result shows that close to 95°C, a decrease of about fifty percent of the water volume is witnessed, with a clean water steam containing very little urban organic matter. The failure to vaporize these liquid effluents at low temperatures is surely due to the thin skin of oil on the surface of the OMW solutions, which precludes the visual evaporation of the water. On the other hand, this is achieved rapidly but with little pollution at any point above 100°C, with the attendant loss of a large amount of organic matter. Therefore, the preferred technology for the drying of OMW that has been adopted for all subsequent studies is 95°C with a reduction rate of half in 3 days. All these studies were supported by the fact that the salinities of the handled OMW solutions rose as a function of the evaporated volumes (Figure 12.3).

12.2.3.3.2 Kinetic Modeling of the Convective Drying Process

For temperatures 80, 100 and 105 °C the results indicate the values of $\mathbf{R^2}$ coefficient successively: 0.9887763. 0.9667774 and 0.99648022 for the Wang and Singh model, always higher compared to the R^2 values obtained for the other examined models. Therefore, this model is the most verified and suitable for processing kinetic results relating to the drying process of OMW effluent solutions. It is characterized by very low Khi-square $\mathbf{X^2}$ and **RMSE** values, as well as high **p**-values. According to this model the kinetic equation of drying process at 105 °C is represented by the following relationship:

$$\boldsymbol{X^*}= -0.0854193\boldsymbol{t^2} - 0.358168\boldsymbol{t} + 1 \quad \text{at } 105°\text{C} \tag{12.8}$$

For drying process the results obtained at 80.,95 and 100 °C temperatures are well verified by this same kinetic model, and allow obtaining the best modeling. So for studied process, Wang and Singh model adapts well to the experimental results compared to the other examined models. It allows establishing the following equations for all study temperatures

$$\boldsymbol{X^*}= -0.09914783\boldsymbol{t^2} - 0.178122\boldsymbol{t} + 1 \quad \text{at } 100°\text{C} \tag{12.9}$$

$$\boldsymbol{X^*}= 0.00590062\boldsymbol{t^2} - 0.13723\boldsymbol{t} + 1 \quad \text{at } 95°\text{C} \tag{12.10}$$

$$\boldsymbol{X^*}= 0.000571428\boldsymbol{t^2} - 0.0234286\boldsymbol{t} + 1 \quad \text{at } 80°\text{C} \tag{12.11}$$

In addition, we got a good correlation between experimental results and theoretical values predicted by Wang and Singh model. Resolution of previous equations shows that the convictive continuous drying process of OMW effluent solutions at 95 °C requires up to 18 days evaporating 50% of the total volume. While for lower temperatures, drying operation of these solutions is practically impossible. Therefore, based on these results, we can confirm that evaporation or natural solar drying of these solutions is impossible. In fact, this thermal stability increases the pollution risks for groundwater and rivers by infiltration of polluting compounds contained in these OMW effluents, often discharged into nature without prior treatment.

12.2.3.3.3 Evolution of TOC Content and Density in Dried OMW Effluent Solutions

These curves in Figure 12.4 portray the % TOC leakage across the OMW system throughout the cone drying induced process. The trend in these graphical contours reveals that this curing process leads to a limited loss of total organic coal for every testing temperature. This TOC is mostly due to the existence of considerable amounts of organic compounds that boil close to the analyzed temperatures. Examples include poultry fatty acids, spirits produced by acidogenic bacteria, and acetous acid produced by acetoic bacteria in the context of anestatic fermentor digestion [57].

Our analysis of the results in (Table 12.4) illustrates that the extent of TOC loss is proven to be proportional to the rise in temperature achieved in the drying process. With temperatures close to a maximum of 105 °C, when 50% of the liquid component of the OMW is boiled off, 35% of the TOC is looted. Therefore, it is advisable to dry the OMW evaporated solutions at 95 °C, in the sense to avoid air pollution and to keep a proper amount of organic matter for the pyrolysis process.

The effluent substances of the OMW are a residual liquid; experience suggests that such liquid compounds are volatile during the pyrolysis cycle, mainly during the deswatering stage. The reason for this property is the large amount of water occurring in these effluent solutions, having a water-usage density. Consequently, the evolution of this aspect during the

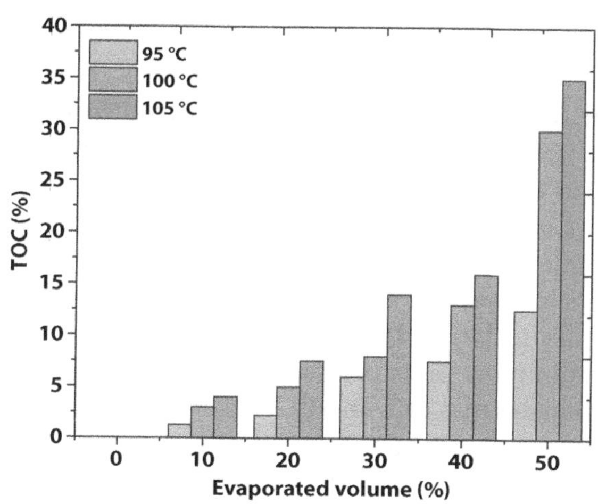

Figure 12.4 Evolution of TOC loss rate in dried solutions versus T°C and volume of evaporation during the drying process.

Table 12.4 Results of gravimetric analyzes for the determination of contents in MS at 105 and 180 °C, and in MVS at 550 and 900 °C according to the sample regions.

Temperature °C				
	MS (%)		MVS (%)	(MS105 °C)
	105 °C	180 °C	550 °C	900 °C
SEBT GZOULA	12.30	8.5	88.29	96.29
ATAAOUIYA	8.01	6.51	89.35	96.97
OUAZZANE	8.25	6.48	90.50	95.80
WTAT el HAJ	7.52	4.85	94.53	97.19
Average	**9.03**	**6.58**	**90.67**	**96.57**

dry-down section has been monitored. The density of the OMW sewage solutions is almost 1.0 kg/L, while after finishing the drying process and reaching 80% evaporation, this density becomes equal to 1.1 kg/L (Figure 12.4). Therefore, the effect of evaporation on the density parameter of these solutions is minimal [59, 60].

12.2.3.3.4 Gravimetric Analyses

In this paragraph, we have studied the complete combustion of dried OMW effluent solutions in a previously unheated furnace to avoid any thermal shock. These experiments allow determining respectively the dry matter content at 105 °C and 180 °C, and the volatile dry matter content at 550°C and 900°C. Finally, the process of progressive pyrolysis up to 550 °C of the dried matter at 105°C causes the decomposition of organic matter, part of the salts and part of the mineral matter, corresponding to about 90.67% of the dry matter at 105 °C. In addition, the pyrolysis process up to 900°C allows in most cases the decomposition of organometallic complexes and carbonates into CO_2 and CO. All of these elements decomposed at high temperature present 6% of the dry matter obtained at 105°C [61, 62] (Table 12.4).

12.2.4 Flash Pyrolysis Process (Thermal Shock) and Interpretation of Results

12.2.4.1 Pyrolysis of OMS Effluents

To achieve the best efficiency in converting TOC to coal, optimization of the parameters that affect the pyrolysis of OMS exhaust flash process is necessary, complying with the different environmental, cost and other quality aspects [63]. As shown in the graph in Figure 12.5, the reported results of the flash pyrolysis process of OMS effluents reveal that the efficiency of the biochar is growing in direct proportion to the incremental amount of surface area of the processed OMS effluents. Certainly, this gain is due to the diminution of the amount of molecular area compared to the mass of the treated sample in the kiln. This is why the application of the flash pyrolysis process becomes the best method for the preparation of OMS when the ratio of the available oxygen to the amount of material introduced in the oven is very little. In these situations, a clean, high quality and ashless biochar can be obtained from the transformation of these solid effluents [64].

Insufficient conversion per treatment area would cause a considerable loss of organic matter, yielding very little coal of lower material quality. This is for example the case of transformation time corresponding to 25 s as shown in the graph of Figure 12.5.

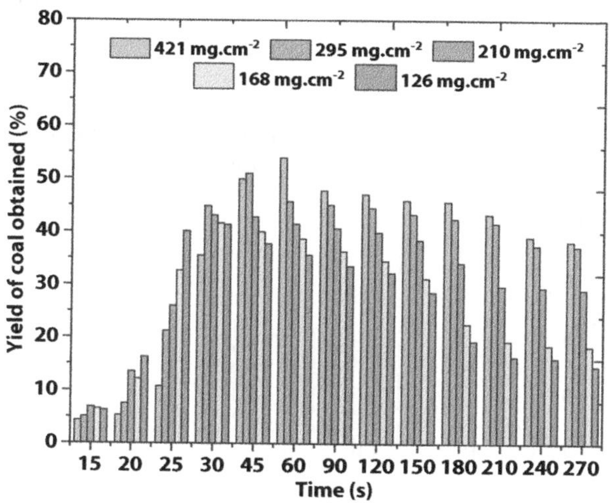

Figure 12.5 Evolution of bio coal yield obtained according to the introduced quantity for treatment and the transformation time in the furnace.

Effectively, these effluent discharges of OMW are transformed into biochar in proportion to the surface mass of solid effluent that needs to be treated according to this specific process of pyrolysis by thermal shock. However this is only valid for surface matters higher than five grams/twenty-three hundred and seventy-five square centimeters. On the other hand, for smaller surface quantities, the ideal transformation time is constant around and around 30 (s). In fact, this result can be supported by the inhibiting force exerted by the superficial layers of these treated effluent waste. Thus, if the surface mass of the treated OMW effluents is important, the heat propagation rate through these solid phases is slow.

12.2.4.2 Transformation of Dried OMW Effluent Solutions by Flash Pyrolysis Process

For this flash pyrolysis transformation of WIP solutions, if the feed quantity into the cell is not completely controlled, the solutions spill over outside the cell. This will provoke a complete loss of organic content in the kiln, with a poor and slow efficiency of the process and a consequent poor quality of the biochar obtained. The graphic in Figure 12.6 describe the impact of the filling rate on the efficiency of the biochar produced by the processing of the OMW streams at different capacities. Indeed, this particular pyrolysis system for the thermic shock treatment of OMW sludge is found to have an admissible yield of 9.2% for a rate of filling of the cells

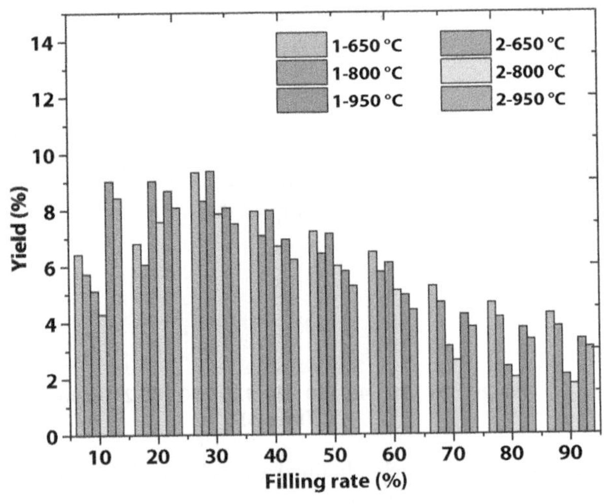

Figure 12.6 Evolution of bio coal yield obtained by single and double flash pyrolysis process according to the filling rate and at studied temperatures.

of about 30% and with temperatures ranging between 600 and 800 °C. However, the optimal filling rate for the heat-shock process of wastewater is not attained. On the other hand, the maximum rate of filling evolves disproportionately with the increase of the temperature, and at a very high temperature the flash pyrolysis is instantaneous and powerful, the spillover phenomenon is avoided, and the thus obtained biochar is clean and of very high quality. In reality, the yield of 9.2% is calculated with regards to the TOM% in the cleaned solutions, whereas the real yield should be about 18.4% with regard to the TOM% in the OMW exhaust solutions before the drying (Table 12.3).

12.2.4.3 *Form of Bio Coal Obtained by Flash Pyrolysis of OMW Solutions*

By transformation of OMW wastewater solutions by the flash process, the bio-coal appears as black, extremely slender and humid grains, easily agglomerated and at high temperatures, the result is a high content of compounds that are volatile. These solid phases clump together quite easily and at high temperatures, the resulting bio-char indicates a high proportion of volatile substances. In order to provide a crushing dry biochar, we made a second process of pyrolysis of these fibers to produce a compact mixture. After the compact mass was ground, a very dry and partially lustrous black coal powder was obtained, representing a natural coal.

12.2.5 Characterization of Bio Coal Obtained by Single and Double Flash Pyrolysis Processes

12.2.5.1 *Characterization by FTIR Infrared Spectroscopy*

The graphic in Figure 12.7 depicts the FTIR spectrum relating to the biochar particle phase assay performed by the single and double flash pyrolysis processes of OMW effluent and dried OMW effluent solutions.

12.2.5.1.1 Bio Coal Obtained by Transformation of OMS Effluents

Bio-coal obtained by transformation of OMW is characterized by: the presence of OH bonds related to the presence of moisture, at the wave number of 3730 cm^{-1}. The presence of the bond C=C at the wave number 1600 cm^{-1}, the presence of the bond C-NO_2 at the wave number 1400 cm^{-1}, the presence of the bond C-O at the wave number 1000 cm^{-1}. The results show that there is a strong possibility that the biochar made from

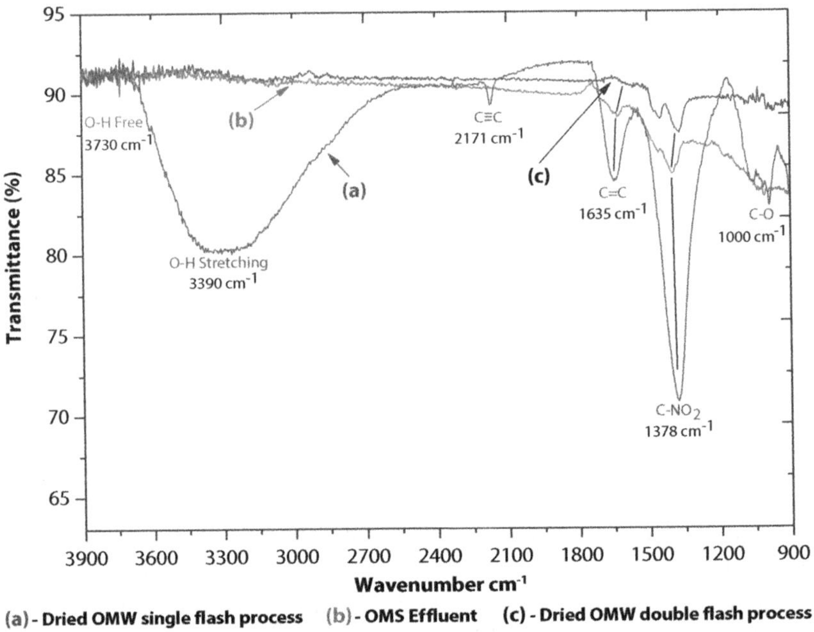

Figure 12.7 Assignment of FTIR infrared spectra for the bio coal solid phases obtained by single and double flash pyrolysis processes.

OMW solid fractions is carbon with double and singly bonded. It has small amount of hydrogen carbonate ions (HCO_3^-) with large amount of associated NO_2 groups due to the abundance of nitrogen-containing elements in the treated effluent.

12.2.5.1.2 Bio Coal Obtained by Single Flash Pyrolysis of Dried OMW Solutions

Bio-coal obtained by transformation of OMW is characterized by: the presence of OH bonds at the wave number of 3730 cm^{-1}, the presence of C≡C bonds at the wave number of 2771 cm^{-1}, the presence of C=C bonds at the wave number of 1600 cm^{-1}, the presence of C-NO_2 connections at the wave number of 1400 cm^{-1}, the presence of C-O bonds connections at the wave number of 1000 cm^{-1}.

The results indicate that the solid phases of the biochar resulting from a one-shot pyrolysis process of dried OMW are moist and have a few built-in impurities. Indeed, the oven is closed during the transformation, so the water vapor and gases emerged during the process are captured and remain in contact with the formed biochar. In view of confirming the presence of wetness and impurities and achieving a high quality biochar, we subjected the solid phases obtained by simple flash to high temperature to get rid of the wetness and degrade the maximum of impurities. Previous studies have shown almost the same result for an activated carbon obtained by processing bitter rapeseed (Brassica napus) stems [65–67].

12.2.5.1.3 Bio Coal Obtained by Double Flash Pyrolysis Process of Dried OMW Solutions

Bio coal obtained by double flash pyrolysis is characterized by: the presence of Csp3-H bonds at the wave number of 1450 cm^{-1}, the presence of C-NO_2 connections at the wave number of 1400 cm^{-1}, the presence of C-O bonds connections at the wave number of 1000 cm^{-1}. Coal solid phase, generated by the whole transforming of dried OMW effluent by a double flash pyrolysis, is mainly constituted of carbon atoms, a little impurity (HCO_3^-) and associated NO_2 groups. The outcomes are showing that this recovery product is a very good performing biochar, which is suitable for a variety of applications.

12.2.6 Characterization by X-Ray Diffraction Spectroscopy

XRD patterns of the resulting biochar phases from OMW effluent and OMW solutions cured by both simple and twin flash pyrolysis are plotted in Figure 12.8.

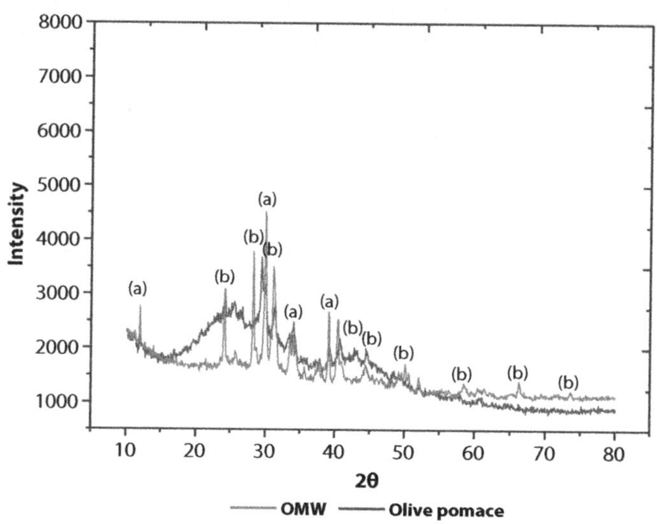

Figure 12.8 Diffractograms of bio coal solid phases from single and double flash pyrolysis of OMS and OMW effluent discharges.

These spectra reveal that the analyzed biochar contains carbonate phases noted (a). The diffractogram of the biochar phase from processing the dried OMW samples, in contrast, presents intense fine peaks marked (b) with diffracting angles approaching those of the pure NaCl and KCl segments. Therefore, the flash pyrolysis heat shock process for the dried OMW wastes leads to a biochar having an amorphous structure rich in carbonates and salts, which confirms the above results.

12.2.7 Conclusions

All results obtained and described in the present work are very interesting and show that original technique of *flash pyrolysis process* designed and adopted, could be a suitable and reliable solution for treating OMS effluents and especially OMW solutions from olive oil industries, very polluting for aquatic environment. Indeed, the treatment and valorization of effluent discharges from olive oil industries as the main objective of this work has been achieved. For the treatment and transformation of these effluents into bio coal with specific properties for different applications, the studies were carried out in two major steps:

- Speciation and characterization of OMS and OMW effluents discharges in order to determine their richness in organic matter and their polluting loads.
- Design and feasibility studies for the technique of thermal shock flash pyrolysis process. Adoption and optimization of this special process for treating analyzed and prepared effluents. Determination and optimization of transformation yields and finally characterization of obtained bio coal phases.

During the first step, several parameters characterizing effluent discharges were determined, especially for OMW effluent solutions which are very polluting and difficult to eliminate or recover. Among the important parameters that were specified, total organic carbon content (% [TOC]), and total organic matter content (% [MOT]). After this speciation phase, detailed study on the drying processes of OMW effluent solutions was carried out in order to reduce the humidity. In fact, for an effective treatment and valorization of liquid effluents according to *flash pyrolysis process by thermal shock*, the density of liquid effluent solutions to be treated is an important parameter. Detailed kinetic studies with the adoption of several models allowed determining optimum transformation time and temperature in order to obtain denser solutions by reducing humidity rate about 50%. Without modifying the quantity of the recoverable organic matter (elimination of clean water vapors).

After this speciation step and preparation of solid and liquid samples, we have started the most important step of treatment and valorization of analyzed and prepared samples at high temperature and in absence of oxygen according to the flash pyrolysis process designed in the laboratory. Influence studies of several factors on the treatment and recovery of these effluents were carried out in order to elucidate the optimal operating conditions. In particular, transformation temperature, transformation time in the furnace, and filling rate of the treatment cell. Obtained results show that for temperatures between 600 and 800 °C, a treatment time of approximately 50 seconds and a filling rate of 30 to 40 %, the treatment process by total elimination and valorization of studied effluent discharges are effective and profitable. Indeed, under these optimal conditions, total yield of produced bio coal approximately 10 % has been achieved, relative to total mass of treated effluent sample. Knowing that for dried OMW effluent solutions the TOC rate is around 20%.

In addition, analysis studies relating to compositions of produced bio coal solid phases by FTIR and X-ray diffraction spectroscopes were carried out. Results show that the bio coal solid phase obtained by treatment of OMS effluents is amorphous with reduced humidity. On the other hand, the bio coal solid phase obtained by single flash pyrolysis treatment of OMW solutions is too wet. While the treatment by double flash pyrolysis process of dried effluent solutions allows to obtain a bio coal, amorphous, dry and homogeneous, consisting of fine grains with very high quality after grinding operation.

After these detailed studies and promising obtained results, we can affirm that it is possible to find solutions to the environmental problems caused by toxic effluent discharges, according to the principles of *green and sustainable chemistry*, whatever their origins their compositions and their natures. Indeed, these toxic OMW effluents which presented a great ecological risk because of their evacuation in surface and underground waters without prior treatment [68]. In reality they constitute a renewable source for the production of high quality bio coal, if the treatment process is well designed and adapted. Knowing that, this produced precious amorphous compound is widely adopted for several applications such as, water treatment, pharmaceutical industry, cosmetic, storage of toxic compounds, and carbon fibers.

References

1. Kalogeropoulos, N. and Tsimidou, M., Antioxidants in Greek virgin olive oils. *Antioxidants*, 3, 387–413, 2014, https://doi.org/10.3390/antiox3020387.
2. Torbati, M., Arzanlou, M., Azadmard-damirchi, S., Babai-ahari, A., Alijani, S., Effect of fungal species involved in the olive fruit rot on the qualitative properties of olive oil. *Arch. Phytopathol. Plant Prot.*, 47, 292–297, 2014, https://doi.org/10.1080/03235408.2013.809183.
3. Arampatzis, G., Hatzigiannakis, E., Pisinaras, V., Kourgialas, N., Psarras, G., Kinigopoulou, V., Panagopoulos, A., Koubouris, G., Soil water content and olive tree yield responses to soil management, irrigation, and precipitation in a hilly Mediterranean area. *J. Water Clim. Change*, 9, 672–678, 2018, https://doi.org/10.2166/wcc.2018.224.
4. Nergiz, C. and Engez, Y., Compositional variation of olive fruit during ripening. *Food Chem.*, 69, 55–59, 2000, https://doi.org/10.1016/S0308-8146(99)00238-1.
5. Elabdouni, A., Haboubi, K., Merimi, I., El Youbi, M.S., Olive mill wastewater (OMW) production in the province of Al-Hoceima (Morocco) and their physico-chemical characterization by mill types. *Mater. Today Proc.*, 27, 3145–3150, 2020, https://doi.org/10.1016/j.matpr.2020.03.806.
6. Enaime, G., Baçaoui, A., Yaacoubi, A., Belaqziz, M., Wichern, M., Lübken, M., Phytotoxicity assessment of olive mill wastewater treated by different technologies: Effect on seed germination of maize and tomato. *Environ. Sci. Pollut. Res.*, 27, 8034–8045, 2020, https://doi.org/10.1007/s11356-019-06672-z.
7. Hodaifa, G., Malvis, A., Maaitah, M., Sánchez, S., Combination of physicochemical operations and algal culture as a new bioprocess for olive mill wastewater treatment. *Biomass Bioenergy*, 138, 105603, 2020, https://doi.org/10.1016/j.biombioe.2020.105603.
8. Silvan, J.M., Pinto-Bustillos, M.A., Vásquez-Ponce, P., Prodanov, M., Martinez-Rodriguez, A.J., Olive mill wastewater as a potential source of antibacterial and anti-inflammatory compounds against the food-borne pathogen Campylobacter. *Innov. Food Sci. Emerg. Technol.*, 51, 177–185, 2019, https://doi.org/10.1016/j.ifset.2018.05.013.
9. Hanafi, F., Sadif, N., Assobhei, O., Mountadar, M., Traitement des margines par électrocoagulation avec des électrodes plates en aluminium. *Rev. Des. Sci. L'Eau*, 22, 473–485, 2009, https://doi.org/10.7202/038326ar.
10. Achak, M., Ouazzani, N., Yaacoubi, A., Mandi, L., Caractérisation des margines issues d'une huilerie moderne et essais de leur traitement par coagulation-floculation par la chaux et le sulfate d'aluminium. *Rev. Des. Sci. L'Eau*, 21, 53–67, 2008, https://doi.org/10.7202/017930ar.
11. Zerrouq, F., Kherbeche, A., Chaqroune, A., L'effluent des huileries d'olive, traitement par adsorption et par oxydation catalytique. *Cah. l'Association Sci. Eur. Pour l'Eau La Santé*, 7, 53–60, 2002, https://doi.org/10.1051/asees/20020701053.

12. Esteves, B.M., Rodrigues, C.S.D., Maldonado-Hódar, F.J., Madeira, L.M., Treatment of high-strength olive mill wastewater by combined Fenton-like oxidation and coagulation/flocculation. *J. Environ. Chem. Eng.*, 7, 4, 103252, 2019, https://doi.org/10.1016/j.jece.2019.103252.
13. Rinaldi, M., Rana, G., Introna, M., Olive-mill wastewater spreading in southern Italy: Effects on a durum wheat crop. *Field Crops Res.*, 84, 319–326, 2003, https://doi.org/10.1016/S0378-4290(03)00097-2.
14. Ertuğrul, S., Dönmez, G., Takaç, S., Isolation of lipase producing Bacillus sp. from olive mill wastewater and improving its enzyme activity. *J. Hazard. Mater.*, 149, 720–724, 2007, https://doi.org/10.1016/j.jhazmat.2007.04.034.
15. Pontoni, L., D'Antonio, G., Esposito, G., Fabbricino, M., Frunzo, L., Pirozzi, F., Thermal pretreatment of olive mill wastewater for efficient methane production: Control of aromatic substances degradation by monitoring cyclohexane carboxylic acid. *Environ. Technol.*, 36, 1785–1794, 2015, https://doi.org/10.1080/09593330.2015.1012179.
16. Mechnou, I., Mourtah, I., Raji, Y., Chérif, A., Lebrun, L., Hlaibi, M., Effective treatment and the valorization of solid and liquid toxic discharges from olive oil industries, for sustainable and clean production of bio-coal. *J. Cleaner Prod.*, 288, 125649, 2020, https://doi.org/10.1016/j.jclepro.2020.125649.
17. Wang, J., Sun, C., Lin, B.-C., Huang, Q.-X., Ma, Z.-Y., Chi, Y., Yan, J.-H., Micro- and mesoporous-enriched carbon materials prepared from a mixture of petroleum-derived oily sludge and biomass. *Fuel Process. Technol.*, 171, 140–147, 2018, https://doi.org/10.1016/j.fuproc.2017.11.013.
18. Enaime, G., Baçaoui, A., Yaacoubi, A., Wichern, M., Lübken, M., Hydrothermal carbonization of the filter bed remained after filtration of olive mill wastewater on olive stones for biofuel application. *Biomass Convers. Biorefin.*, 12, 1237–1247, 2020, https://doi.org/10.1007/s13399-020-00743-9.
19. Azzaz, A.A., Jeguirim, M., Kinigopoulou, V., Doulgeris, C., Goddard, M.L., Jellali, S., Matei Ghimbeu, C., Olive mill wastewater: From a pollutant to green fuels, agricultural and water source and bio-fertilizer – hydrothermal carbonization. *Sci. Total Environ.*, 733, 2020, 139314, 2020, https://doi.org/10.1016/j.scitotenv.2020.139314.
20. El Mouhtadi, I., Agouzzal, M., Guy, F., La filière Oléacgineuse en Afrique, Oilseeds fats. *Crops Lipids*, 21, 21–23, 2014.
21. Smail, A., *Introduction de l'olivier (Olea europaea L.) à Oued Souf: Situation actuelle et perspectives de développement, cas de l'exploitation Daouia*, Université Kasdi Merbah, Ouargla, 2018.
22. Moreno-Castilla, C., Carrasco-Marín, F., López-Ramón, M.V., Alvarez-Merino, M.A., Chemical and physical activation of olive-mill wastewater to produce activated carbons. *Carbon N. Y.*, 39, 1415–1420, 2001, https://doi.org/10.1016/S0008-6223(00)00268-2.
23. Dermeche, S., Nadour, M., Larroche, C., Moulti-Mati, F., Michaud, P., Olive mill wastes: Biochemical characterizations and valorization

strategies. *Process Biochem.*, 48, 1532–1552, 2013, https://doi.org/10.1016/j.procbio.2013.07.010.
24. Vlyssides, A.G., Loizides, M., Karlis, P.K., Integrated strategic approach for reusing olive oil extraction by-products. *J. Cleaner Prod.*, 12, 603–611, 2004, https://doi.org/10.1016/S0959-6526(03)00078-7.
25. Khdair, A.I., Abu-Rumman, G., Khdair, S.I., Pollution estimation from olive mills wastewater in Jordan. *Heliyon*, 5, e02386, 2019, https://doi.org/10.1016/j.heliyon.2019.e02386.
26. Babić, S., Malev, O., Pflieger, M., Lebedev, A.T., Mazur, D.M., Kužić, A., Čož-Rakovac, R., Trebše, P., Toxicity evaluation of olive oil mill wastewater and its polar fraction using multiple whole-organism bioassays. *Sci. Total Environ.*, 686, 903–914, 2019, https://doi.org/10.1016/j.scitotenv.2019.06.046.
27. Lacomelli, A., Olive mill wastewater & SAP adopted by BACRON contracting parties. Work Shop Agro-Industry Dev. Coast. Areas, with Spec. Focus Olive Oil Ind., pp. 4–6, 2000.
28. Colarieti, M.L., Toscano, G., Greco, G., Toxicity attenuation of olive mill wastewater in soil slurries. *Environ. Chem. Lett.*, 4, 115–118, 2006, https://doi.org/10.1007/s10311-006-0050-5.
29. Noe, V., Restauration des milieux lagunaires vis à vis de l'eutrophisation en Languedoc-Roussillon, 2007.
30. Zamora, R., Navarro, J.L., Hidalgo, F.J., Identification and classification of olive oils by high-resolution 13C nuclear magnetic resonance. *J. Am. Oil Chem. Soc.*, 71, 361–364, 1994, https://doi.org/10.1007/BF02540514.
31. Mechnou, I., Meskini, S., El Ayar, D., Lebrun, L., Hlaibi, M., Olive mill wastewater from a liquid biological waste to a carbon/oxocalcium composite for selective and efficient removal of methylene blue and paracetamol from aqueous solution. *Bioresour. Technol.*, 365, 128162, 2022, https://doi.org/10.1016/j.biortech.2022.128162.
32. Hanafi, F., Assobhei, O., Mountadar, M., Detoxification and discoloration of Moroccan olive mill wastewater by electrocoagulation. *J. Hazard. Mater.*, 174, 807–812, 2010, https://doi.org/10.1016/j.jhazmat.2009.09.124.
33. Papaphilippou, P.C., Yiannapas, C., Politi, M., Daskalaki, V.M., Michael, C., Kalogerakis, N., Mantzavinos, D., Fatta-Kassinos, D., Sequential coagulation-flocculation, solvent extraction and photo-Fenton oxidation for the valorization and treatment of olive mill effluent. *Chem. Eng. J.*, 224, 82–88, 2013, https://doi.org/10.1016/j.cej.2012.11.047.
34. Galliou, F., Markakis, N., Fountoulakis, M.S., Nikolaidis, N., Manios, T., Production of organic fertilizer from olive mill wastewater by combining solar greenhouse drying and composting. *Waste Manage.*, 75, 305–311, 2018, https://doi.org/10.1016/j.wasman.2018.01.020.
35. Salgado, V., Fonseca, C., Lopes da Silva, T., Roseiro, J.C., Eusébio, A., Isolation and identification of Magnusiomyces capitatus as a lipase-producing yeast from olive mill wastewater. *Waste Biomass Valorization.*, 11, 3207–3221, 2019, https://doi.org/10.1007/s12649-019-00725-7.

36. Frascari, D., Bacca, A.E.M., Zama, F., Bertin, L., Fava, F., Pinelli, D., Olive mill wastewater valorisation through phenolic compounds adsorption in a continuous flow column. *Chem. Eng. J.*, 283, 293–303, 2016, https://doi.org/10.1016/j.cej.2015.07.048.
37. Leone, A., Tamborrino, A., Zagaria, R., Sabella, E., Romaniello, R., Plant innovation in the olive oil extraction process: A comparison of efficiency and energy consumption between microwave treatment and traditional malaxation of olive pastes. *J. Food Eng.*, 146, 44–52, 2015, https://doi.org/10.1016/j.jfoodeng.2014.08.017.
38. Aubert, G., *Méthodes d'analyse des sols*, CRDP, Marseille, France, 1978.
39. Luo, Z., Wang, E., Sun, O.J., Can no-tillage stimulate carbon sequestration in agricultural soils? A meta-analysis of paired experiments. *Agric. Ecosyst. Environ.*, 139, 224–231, 2010, https://doi.org/10.1016/j.agee.2010.08.006.
40. Lewis, W.K., The rate of drying of solid materials. *Ind. Eng. Chem.*, 13, 427–432, 1921, https://doi.org/10.1021/ie50137a021.
41. Overhults, D.G., White, G.M., Hamilton, H.E., Ross, I.J., Drying soybeans with heated air. *Trans. Am. Soc. Agric. Eng.*, 16, 112–113, 1973, https://doi.org/10.13031/2013.37459.
42. Sharaf-Eldeen, Y.I., Blaisdell, J.L., Hamdy, M.Y., Model for ear corn drying. *Trans. Am. Soc. Agric. Eng.*, 23, 1261–1265, 1980, https://doi.org/10.13031/2013.34757.
43. Kassem, A.S., Comparative studies on thin layer drying models for wheat. *13th Int. Congr. Agric. Eng.*, p. 2, 1998.
44. Verma, L.R., Bucklin, R.A., Endan, J.B., Wratten, F.T., Effects of drying air parameters on rice drying models. *Trans. Am. Soc. Agric. Eng.*, 28, 296–301, 1985, https://doi.org/10.13031/2013.32245.
45. Henderson, S.M., Progress in developing the thin layer drying equation. *Trans. Am. Soc. Agric. Eng.*, 17, 1167–1168, 1974, https://doi.org/10.13031/2013.37052.
46. STATsoft, STATsoft electronic statistic textbook, Statistica - Southern African Analytics, 2011, http://www.statsoft.com/textbook/stathome.html.
47. Faubert, P., Barnabé, S., Bouchard, S., Côté, R., Villeneuve, C., Pulp and paper mill sludge management practices: What are the challenges to assess the impacts on greenhouse gas emissions? *Resour. Conserv. Recycl.*, 108, 107–133, 2016, https://doi.org/10.1016/j.resconrec.2016.01.007.
48. Foong, S.Y., Liew, R.K., Yang, Y., Cheng, Y.W., Yek, P.N.Y., Wan Mahari, W.A., Lee, X.Y., Han, C.S., Vo, D.V.N., Van Le, Q., Aghbashlo, M., Tabatabaei, M., Sonne, C., Peng, W., Lam, S.S., Valorization of biomass waste to engineered activated biochar by microwave pyrolysis: Progress, challenges, and future directions. *Chem. Eng. J.*, 389, 124401, 2020, https://doi.org/10.1016/j.cej.2020.124401.
49. Bridgwater, A.V., Czernik, S., Piskorz, J., An overview of fast pyrolysis, in: *Prog. Thermochem. Biomass Convers*, vol. 30, pp. 977–997, 2008, https://doi.org/10.1002/9780470694954.ch80.

50. Win, M.M., Asari, M., Hayakawa, R., Hosoda, H., Yano, J., Sakai, S.I., Gas and tar generation behavior during flash pyrolysis of wood pellet and plastic. *J. Mater. Cycles Waste Manage.*, 22, 547–555, 2020, https://doi.org/10.1007/s10163-019-00949-8.
51. Alvarez, J., Amutio, M., Lopez, G., Barbarias, I., Bilbao, J., Olazar, M., Sewage sludge valorization by flash pyrolysis in a conical spouted bed reactor. *Chem. Eng. J.*, 273, 173–183, 2015, https://doi.org/10.1016/j.cej.2015.03.047.
52. Boubaker, F. and Ridha, B.C., Modelling of the mesophilic anaerobic co-digestion of olive mill wastewater with olive mill solid waste using anaerobic digestion model No. 1 (ADM1). *Bioresour. Technol.*, 99, 6565–6577, 2008, https://doi.org/10.1016/j.biortech.2007.11.035.
53. Leouifoudi, I., Zyad, A., Amechrouq, A., Oukerrou, M.A., Mouse, H.A., Mbarki, M., Identification and characterisation of phenolic compounds extracted from moroccan olive mill wastewater. *Food Sci. Technol.*, 34, 249–257, 2014, https://doi.org/10.1590/fst.2014.0051.
54. Issaoui, M., Dabbou, S., Mechri, B., Nakbi, A., Chehab, H., Hammami, M., Fatty acid profile, sugar composition, and antioxidant compounds of table olives as affected by different treatments. *Eur. Food Res. Technol.*, 232, 867–876, 2011, https://doi.org/10.1007/s00217-011-1455-3.
55. Amaral, C., Lucas, M.S., Coutinho, J., Crespí, A.L., do Rosário Anjos, M., Pais, C., Microbiological and physicochemical characterization of olive mill wastewaters from a continuous olive mill in Northeastern Portugal. *Bioresour. Technol.*, 99, 7215–7223, 2008, https://doi.org/10.1016/j.biortech.2007.12.058.
56. Paredes, C., Cegarra, J., Roig, A., Sánchez-Monedero, M.A., Bernal, M.P., Characterization of olive mill wastewater (alpechin) and its sludge for agricultural purposes. *Bioresour. Technol.*, 67, 111–115, 1999, https://doi.org/10.1016/S0960-8524(98)00106-0.
57. Koutsos, T.M., Chatzistathis, T., Balampekou, E.I., A new framework proposal, towards a common EU agricultural policy, with the best sustainable practices for the re-use of olive mill wastewater. *Sci. Total Environ.*, 622–623, 942–953, 2018, https://doi.org/10.1016/j.scitotenv.2017.12.073.
58. Hamdi, M., Nouvelle conception d'un procédé de dépollution biologique des margines, effluents liquides de l'extraction de l'huile d'olive, Thèse Dr. En *Biol. Cell. Microbiol.*, 168, 1991, http://horizon.documentation.ird.fr/exl-doc/pleins_textes/doc34-01/34545.pdf.
59. Zagrouba, F., *Séchage mixte par convection et un apport rayonnant micro-ondes des milieux déformables. Modélisation des phénomènes de transferts de chaleur et de matière*, Univ-Lorraine, France, 1993.
60. Kechaou, N., *Séchage des gels fortement déformables: Études de la diffusion interne de l'eau et modélisation*, Vandoeuvre-les-Nancy, INPL, France, 1989.
61. Frost, R.L., Hales, M.C., Martens, W.N., Thermogravimetric analysis of selected group (II) carbonate minerals - implication for the geosequestration

of greenhouse gases. *J. Therm. Anal. Calorim.*, 95, 999–1005, 2009, https://doi.org/10.1007/s10973-008-9196-7.

62. Carraher, C.E., Thermal characterizations of inorganic and organometallic polymers. *J. Macromol. Sci. Part A - Chem.*, 17, 1293–1356, 1982, https://doi.org/10.1080/00222338208074401.
63. Elkhalifa, S., Al-Ansari, T., Mackey, H.R., McKay, G., Food waste to biochars through pyrolysis: A review. *Resour. Conserv. Recycl.*, 144, 310–320, 2019, https://doi.org/10.1016/j.resconrec.2019.01.024.
64. Zhao, B., O'Connor, D., Zhang, J., Peng, T., Shen, Z., Tsang, D.C.W., Hou, D., Effect of pyrolysis temperature, heating rate, and residence time on rapeseed stem derived biochar. *J. Cleaner Prod.*, 174, 977–987, 2018, https://doi.org/10.1016/j.jclepro.2017.11.013.
65. Kadri, Z., Mechnou, I., Zyade, S., Migration of bisphenol A from epoxy coating to foodstuffs. *Mater. Today Proc.*, 45, 7584–7587, 2021, https://doi.org/10.1016/j.matpr.2021.02.581.
66. Raji, Y., Mechnou, I., Yassine, W., Kadri, Z., Oumghar, K., Cherkaoui, O., Zyade, S., Extraction of the natural indigo carmine pigment from the Isatis plant, characterization and dyeing of wool. *IOP Conf. Ser. Mater. Sci. Eng.*, 948, 012017, 2020, https://doi.org/10.1088/1757-899X/948/1/012017.
67. Kadri, Z., Zyade, S., Mechnou, I., Study of the behavior of an Epoxy coating food contact. *Mor. J. Chem.*, 9, 681–698, 2021, https://doi.org/10.48317/IMIST.PRSM/morjchem-v9i3.22253.
68. Elhag, M., Bahrawi, J.A., Galal, H.K., Aldhebiani, A., Al-Ghamdi, A.A.M., Stream network pollution by olive oil wastewater risk assessment in Crete, Greece. *Environ. Earth Sci.*, 76, 1–12, 2017, https://doi.org/10.1007/s12665-017-6592-y.

Index

Printed and bound by CPI Group (UK) Ltd, Croydon, CR0 4YY
19/09/2026
14965102-0001